WX 85

D1349592

Long-Term Care in Europe

The project 'Health systems and long-term care for older people in Europe. Modelling the INTERfaces and LINKS between prevention and rehabilitation, quality and informal care' was funded by the European Commission under the Seventh Framework Programme (Grant agreement no. 223037)

Long-Term Care in Europe
Improving Policy and Practice

Edited by

Kai Leichsenring
European Centre for Social Welfare Policy and Research, Austria

Jenny Billings
University of Kent, UK

and

Henk Nies
Vilans, The Netherlands

First published 2013 by
PALGRAVE MACMILLAN

Palgrave Macmillan in the UK is an imprint of Macmillan Publishers Limited, registered in England, company number 785998, of Houndmills, Basingstoke, Hampshire RG21 6XS.

Palgrave Macmillan in the US is a division of St Martin's Press LLC, 175 Fifth Avenue, New York, NY 10010.

Palgrave Macmillan is the global academic imprint of the above companies and has companies and representatives throughout the world.

Palgrave® and Macmillan® are registered trademarks in the United States, the United Kingdom, Europe and other countries.

ISBN 978–1–137–03233–1

This book is printed on paper suitable for recycling and made from fully managed and sustained forest sources. Logging, pulping and manufacturing processes are expected to conform to the environmental regulations of the country of origin.

A catalogue record for this book is available from the British Library.

A catalog record for this book is available from the Library of Congress.

10   9   8   7   6   5   4   3   2   1
22 21 20 19 18 17 16 15 14 13

Printed and bound in Great Britain by
CPI Antony Rowe, Chippenham and Eastbourne

Contents

List of Figures, Tables and Boxes

Figures

Tables

Boxes

Preface

The impetus for this book stemmed from a three-year project titled 'INTERLINKS – Health systems and long-term care for older people. Modelling INTERfaces and LINKS between prevention and rehabilitation, quality assurance and informal care'. This project was funded under the European Commission's Seventh Framework Programme. It was carried out by 15 research organisations from 13 European countries:

- European Centre for Social Welfare Policy and Research, Vienna (Austria)
- Ecole d'études sociales et pédagogiques, Lausanne (Switzerland)
- University of Southern Denmark (Denmark)
- Institut de Recherche et Documentation en Economie de la Santé – IRDES, Paris (France)
- National Institute for Health and Welfare – THL, Helsinki (Finland)
- Institut für Soziale Infrastruktur – ISIS, Frankfurt a.M. (Germany)
- Wissenschaftszentrum Berlin für Sozialforschung – WZB, Berlin (Germany)
- CMT Prooptiki ltd., Athens (Greece)
- University of Valencia – ERI Polibienestar (Spain)
- Studio Come S.r.l., Rome (Italy)
- Vilans – Centre of Expertise for Long-term Care, Utrecht (The Netherlands)
- Institute for Labour and Family Research, Bratislava (Slovak Republic)
- Forum for Knowledge and Common Development, Stockholm County Council (Sweden)
- University of Kent – Centre for Health Services Studies, Canterbury (United Kingdom)
- University of Birmingham – HSMC (United Kingdom)

The challenges of long-term care in ageing societies are well documented. The purpose of this book is to draw together significant themes from the INTERLINKS project and couple them with related evidence and thinking. This overview should indicate ways forward to construct new approaches and to improve policy and practice in long-term care.

Notes on Contributors

Kerry Allen (UK) is a Research Fellow at the Health Services Management Centre, University of Birmingham. With a background in the sociology of health and illness her research focuses on policy and governance issues in chronic illness and prevention. Her publications include 'The Billion Dollar Question: Embedding Prevention in Older People's Services – Ten High-Impact Changes' (with J. Glasby).

Rastislav Bednárik (Slovakia) is Senior Researcher at the Institute for Labour and Family Research, Bratislava. He has coordinated national projects and participated in international research projects on social protection, the labour market and social care. He is a national correspondent of the European Commission's MISSOC initiative and has recently published 'The state of social protection in Slovakia'.

Jenny Billings (UK) is a Reader in Applied Health Research in the Centre for Health Service Studies at the University of Kent. Her strength lies in methodological design and she has an interest in health and social care improvement of older people. She has been involved in several EU projects resulting in publications such as 'Integrating Health and Social Care Services for Older People' (ed. with K. Leichsenring).

Stephanie Carretero (Spain) worked as a Researcher at the Polibienestar Research Institute at the University of Valencia and is now a Scientific Officer at JRC-IPTS in the European Commission in Seville. She has been involved in European research on social innovation for long-term care since 2001 and has published in a number of international journals for geriatric psychiatry and gerontology.

Patrizia Di Santo (Italy) is a sociologist and Managing Director of Studio Come Srl. in Rome, where she is also Professor of Equal Opportunities at Lumsa University (Master in Social Services) and La Sapienza University (Master in Communication and Management of Human Resources). She has conducted numerous projects and published extensively on innovation in local institutions, equal opportunities and organisational development, and integrated service networks.

Anja Dieterich (Germany) is a Physician with a Masters degree in Public Health. She is a Health Policy Advisor at the Diakonie Bundesverband, the social welfare organisation of the Protestant Church in Germany, and previously worked as a researcher at the Social Science Research Centre in Berlin.

She has conducted several cross-national research projects with a focus on integrated care for older people, resulting in a number of publications.

Thomas Emilsson (Sweden) works as a Research Assistant at the Forum for Knowledge and Common Development in the Stockholm County Council. He is particularly knowledgeable in quality of care issues regarding vulnerable older people and their informal carers, testing and evaluating new instruments with qualitative methodologies.

Jorge Garcés (Spain) is Professor at the University of Valencia, University of Innsbruck and Erasmus University of Rotterdam. He is an expert in sustainability and transformations in European public policies and Director of the Polibienestar Research Institute.

Jon Glasby (UK) is the Director of the Health Services Management Centre and Professor of Health and Social Care at the University of Birmingham. Editor of the 'Better Partnership Working series' (Policy Press), his research interests include joint working between health and social care, community care and personalisation.

Pierre Gobet (Switzerland) is a sociologist and a registered nurse. He specialises in health policy and health economics and has conducted research on managed care organisations, case management and on epistemological issues in nursing care and integrated long-term care. He is a Professor and Researcher at the Lausanne School of Social Work.

Teija Hammar (Finland) has a D.Sc. in Health Economics and works as a Senior Researcher at the National Institute for Health and Welfare, Helsinki. She has been involved in many research and development projects in the area of health and social care, especially home care, resulting in publications such as 'The cost-effectiveness of integrated home care and discharge practice for home care clients' (with P. Rissanen and M.-L. Perälä).

Elisabeth Hirsch Durrett (Switzerland) was trained in Occupational Therapy and Sociology (London School of Economics). She specialised in Sociology of Health (Boston University) and in Gerontological Policy (Brandeis University). Active in health policy and gerontology, she is a Professor and researcher at the Lausanne School of Social Work.

Laura Holdsworth (UK) is a Research Associate at the University of Kent. She has worked on a variety of projects within long-term care and palliative care, producing publications such as 'A retrospective analysis of preferred and actual place of death for hospice patients' (with S. Fisher).

Georgios Kagialaris (Greece) is a Public Health Nurse working at the Hellenic navy as an officer. He collaborates with CMT Prooptiki and has performed a number of research projects and clinical trials, focusing on older people.

Susanne Kümpers (Germany) is Professor for Qualitative Health Research, Social Health Inequalities and Public Health Strategies at the University of Applied Sciences in Fulda. Before this she worked as a Senior Researcher at the Social Science Research Centre in Berlin and coordinated research projects on old age, integrated care, inequality and health.

Kai Leichsenring (Austria) is Associate Senior Researcher at the European Centre for Social Welfare Policy and Research, Vienna. He has coordinated a range of EU research and development projects in the area of health and social care, resulting in publications such as 'Integrating Health and Social Care Services for Older People' (ed. with J. Billings).

Sabina Mak (The Netherlands) is a Project Manager for the programme 'Quality and Innovations in Elderly Care' at Vilans, Centre of Expertise for Long-Term-Care, Utrecht. Her main focus is on optimising care for older people with dementia and on quality measurement in nursing homes, resulting in publications such as 'Measuring Progress: Indicators for care homes'.

Tasos Mastroyiannakis (Greece) is a member of CMT Prooptiki, Athens. He is a Health Economist and has participated in a range of EU research and development projects in the health sector and on social issues.

Satu Meriläinen-Porras (Finland) works as a Researcher at the National Institute for Health and Welfare, Helsinki, Finland. She has participated in national and European research and development projects in the area of health and social care, in particular focusing on care for older people.

Michel Naiditch (France) is a statistician, mathematician, medical doctor and public health expert and works as an Associate Researcher at IRDES (Research and Documentation Institute in Health Economy) in Paris. He has conducted research linked to home care, the assessment and implementation of health and social networks for chronic diseases, and the conditions for facilitating coordination and cooperation, resulting in a large range of publications.

Henk Nies (The Netherlands) is CEO of Vilans, the Netherlands Centre of Expertise for Long-Term Care. He is Professor of Organisation and Policy Development in Long-term Care at the VU University Amsterdam. Throughout his career he has worked between the boundaries of policy, practice and research, publishing books, articles and blogs.

Kvetoslava Repková (Slovakia) is Director of the Institute for Labour and Family Research in Bratislava. She has coordinated and collaborated in a number of national and international research projects focusing on disability and long-term care resulting in monographs such as 'Long-term care in the context of integrated social work'.

Georg Ruppe (Austria) is a medical doctor and holds a Masters degree in Medical Anthropology (University of Amsterdam). He worked as a Researcher at the European Centre for Social Welfare Policy and Research from 2008 to 2012 and is now Managing Director of the Austrian Interdisciplinary Platform on Ageing (ÖPIA). His research interests include interdependencies between the socio-cultural context and the organisation of health care services, and between medical intervention and quality of life in old age.

Francisco Ródenas (Spain) is lecturer at the University of Valencia and Senior Researcher at the Polibienestar Research Institute. His main research topic is the sustainability, efficiency and quality of long-term care systems, resulting in publications such as 'Readings of the social sustainability theory' (ed. with J. Garcés and S. Carretero).

Ricardo Rodrigues (Austria) is a Research Fellow at the European Centre for Social Welfare Policy and Research, Vienna, where he has been conducting comparative research in the area of long-term care for older people. He has participated in a number of European funded projects covering issues such as equity, expenditure on long-term care, care markets and user choice policies.

Karin Stiehr (Germany) is an Associate Partner and Managing Director of ISIS, Institut für Soziale Infrastruktur, Frankfurt am Main. As a researcher, her work focuses on issues of care and volunteering. In these fields, she has coordinated numerous projects at European and national levels and is a regular consultant to the Governments of Hessen and North-Rhine-Westphalia.

Judy Triantafillou (Greece) is a medical practitioner and member of the Administrative Council of the NGO 50plus Hellas, which collaborates with CMT Prooptiki. She has participated in a number of EU and national projects on health, services and informal care of older people, with relevant publications in these areas.

Roelf van der Veen (The Netherlands) is Senior Programme Officer at Vilans, Centre of Expertise for Long-Term Care in Utrecht. He has managed a range of projects in the area of care for older people. He has most recently been involved in integrated dementia care as well as in innovation and improvement projects in the context of the Dutch national quality improvement programme 'Zorg voor Beter'.

Lis Wagner (Denmark) is Professor and research leader at the Research Unit of Nursing, University of Southern Denmark and former director of WHO EURO nursing and midwifery office. She has been involved in a wide international research network within gerontology resulting in publications describing long-term care in the Danish health care system since the 1980s.

Barbara Weigl (Germany) is a gerontologist and worked as a researcher at the Social Science Research Centre Berlin until January 2012. She is now Lecturer for Nursing Sciences, Care Management and Gerontology at the Protestant University of Applied Sciences in Berlin.

Acknowledgements

This book has uncounted mothers and fathers who lent their minds and bodies to conceive and accompany its progress and growth. Apart from the individual authors, we have to thank a large number of technical assistants, colleagues and other staff working in the 15 organisations that participated in the INTERLINKS project and contributed with their ideas and support to its success.

We cannot thank native speakers in the INTERLINKS consortium enough for their linguistic editing skills and additional efforts provided to ensure that high standards of mutual understanding were achieved throughout; in particular, this refers to Jenny Billings and Laura Holdsworth (University of Kent), Kerry Allen (University of Birmingham), Judy Triantafillou (CMT Prooptiki), Elisabeth Hirsch Durrett (University of Applied Sciences Western Switzerland) and Lorna Campbell (University of Southern Denmark).

In addition, we would like to thank all members of the 13 National Expert Panels who provided their knowledge, time and expertise. Their contributions as representatives of national stakeholder organisations, policy and research to identify practice examples and in validating the INTERLINKS Framework for long-term care were a most valuable resource.

We are also indebted to the representatives of European stakeholder organisations who provided valuable feedback and support as a sounding board, namely AGE Platform, Alzheimer Europe, Centraal Planbureau (representing ANCIEN), EAHSA – European Association of Homes and Services for the Ageing, E.D.E – European Association for Directors of Residential Care Homes for the Elderly, the European Federation of Nurses Associations (EFN), the European Foundation for the Improvement of Living and Working Conditions, Eurocarers, the European Social Network (ESN), Eucomed, EURAG – The European Federation of Older Persons (also Member of the European Social Platform), Homecare Europe (represented by the Caritas of the Archdiocese of Vienna and Familiehulp v.z.w.), the Federal Planning Bureau (representing ANCIEN), the International Association of Geriatrics and Gerontology (IAGG – European Region), the Ministry of Health and Social Affairs, Division Social Services (during the Swedish EU Presidency), OECD, WHO Regional Office for Europe, EASPD, Health Forum Gastein, DG Employment, DG Research and the VU University Medical Center (representing SHELTER).

Finally, we would like to thank the European Commission, Directorate General Research and its staff for facilitating INTERLINKS as a project, and therefore also this book, through generous funding and continuous support during the project period.

We hope that our results will be able to support the engagement of all these national and international stakeholders in improving long-term care all over Europe – and beyond.

Part I
Introduction

1
Addressing Long-Term Care as a System – Objectives and Methods of Study

Jenny Billings, Kai Leichsenring and Lis Wagner

A systemic perspective of long-term care: objectives and definitions

One of the major achievements of humanity over the past 50 years has been the steadily rising life expectancy to which health systems all over the world have significantly contributed. Popular media coverage of this advancement however has tended to neglect the benefits of scientific progress that improves human longevity and instead portray catastrophic scenarios such as 'demographic nightmares' or the 'ageing tsunami'. It is undisputed, however, that societies are struggling heavily with the consequences of demographic changes, particularly with respect to health and social care needs of a rising proportion of older people across Europe. The urgent need for policy and practice to respond to this demand has been the main trigger for the authors gathered in this book to elaborate on issues to improve and further develop long-term care (LTC) for older people in a systemic perspective. LTC in general remains a fragmented area; there is a lack of shared definitions within and between European countries, and in many countries it is only just beginning to emerge as an idea at the interfaces between informal and formal care, and between health and social care services. The economic crisis enveloping Europe with its consequent restrictions on public expenditure may also stifle momentum towards improvement and much needed change in LTC services. A number of other key challenges facing policy and practice in this area have been well documented, focusing for example on the lack of clear governance and finance mechanisms, increasing reliance on untrained migrant care workers, poor coordination between services, the burden on informal carers, and the lack of user involvement in care (European Union, 2012; Colombo *et al.*, 2011; Bettio *et al.*, 2006; Costa-i-Font and Courbage, 2011; Lamura *et al.*, 2008; Bewley *et al.*, 2011; Hofmarcher *et al.*, 2007).

However, the aim of this book is to 'invert' the negativity inherent within the well-articulated challenges by presenting the most relevant themes and key issues within LTC in Europe and focusing on progression and improvement for policy and practice. These cases build on the state of the art and harness the most current evidence in Europe to describe how and in which direction the construction of LTC systems can be taken forward. The cases include examples of the contextual conditions under which good practice can be transferred between countries and sustained, in order to ensure that those working in this field will benefit from the necessary knowledge and techniques discussed to influence and improve care of older people. In addition, methods of how LTC projects can be evaluated were included, in order to arm practitioners with the ability to provide evidence of effectiveness of their own initiatives.

Underpinning this book is research undertaken within a Seventh Framework EU-funded project 'Health systems and long-term care for older people in Europe – Modelling the INTERfaces and LINKS between prevention, rehabilitation, quality of services and informal care' (INTERLINKS). INTERLINKS has assembled a range of themes, sub-themes and 135 key issues into a web-based Framework for LTC that is illustrated by over a hundred examples of validated practice in LTC for older people. This knowledge base will allow those working in this field to assess and develop their own practice using this evidence. By clustering the examples, INTERLINKS has brought into sharp focus key innovations and forward movement in how services for older people are developing and being provided across the EU, and this analysis served as a building block for the cases and issues discussed within this book.

The book is thus gathering the benefits of the advancement in knowledge and potential practice improvements drawn from the following subject matter:

- a debate concerning what is currently understood by the 'identity' of LTC and its relationship to progressing policy and practice in the care of older people in the coming years;
- a focus on descriptions and critical analysis of new ways of overcoming the divisions between health and social care services and formal and informal care. This includes improvements in user-centred care, support for informal carers, different types of joint working, care coordination and new approaches to integrated care provision;
- a description and critical review of innovative cases in the construction of LTC systems, introducing fresh topics such as prevention and rehabilitation, volunteering, and advancements in quality development and information technology, but also a review of funding and governance mechanisms;
- a debate regarding critical issues such as migrant care workers in LTC and respective consequences for sending and receiving countries, and ethical as well as practical questions regarding palliative care at the end of life.

Perspectives on LTC must always consider the health and social service aspects in equal parts and in their entirety, and this has often not been addressed sufficiently well, yet interdisciplinary working is the cornerstone of progress in this area. Thus the contributions in this book shed light on how these perspectives merge within the descriptive cases. Central to this book is the visibility of the user and the informal carer, where carers in particular are acknowledged both as co-providers of care and as clients with their own needs for support. This is a necessary step to gain insight into 'whole systems' construction with respect to LTC for older people. Geographically, empirical evidence with respect to descriptive examples is drawn from 14 countries representing the different 'cultures of care' in Europe, but also literature from a wider international perspective has been included.

The purpose of this introductory chapter therefore is to outline the background of the publication, the aims of INTERLINKS and the justification for addressing LTC focusing on links and interfaces between social/health and formal/informal care divisions. It will also include a section on terminology and the methodological process behind the INTERLINKS Framework for LTC, presenting a rationale for the basis of the individual cases described throughout this book.

Addressing the INTERfaces and LINKS between long-term care and the health system

The rising demand for LTC calls for policy approaches allowing for holistic and inclusive views that integrate the role of different public programmes, sectors of society, and private initiatives. Moreover, there is growing evidence of the discrimination against older people in need of care concerning their access to mainstream health care and to prevention and rehabilitation. The current state in all European countries with fragmented policies and organisations providing varying degrees of support to mainly informal care shows that LTC is still under construction and calls for a systematic developmental approach that needs to be addressed by both health *and* LTC reforms.

INTERLINKS as an EU-funded project was therefore designed to elaborate on the INTERfaces and LINKS between prevention and rehabilitation, quality, informal care and governance, and to guarantee the confluence of the single elements in a general framework for describing and analysing LTC. The main objectives of this project, accomplished between November 2008 and December 2011 were:

- To develop a concept and methodology to describe and analyse long-term care and its links with the health system. This methodology was to facilitate cross-national comparisons between Member States, enabling them to assess their developmental status and to identify future areas for national development;
- To identify a set of practical tools to describe the evidence base of practice examples;

- To identify acknowledged and established good policy and practice that was transferable across Member States, particularly with respect to assessing and monitoring quality of care, promoting prevention and rehabilitation and supporting informal carers as well as addressing respective governance and financing issues.

Although the individual aspects of health and social care services for people who depend on continuous support are now an area of extensive research in many countries, the concepts, indicators and models for international comparisons and for the identification of good practice across countries are still very much in their infancy. This is particularly the case for existing evidence and model ways of working towards prevention and rehabilitation in LTC, the quality of services (such as organisational development towards more coordinated and integrated working), monitoring governance and financing, and the specific role of informal care provided by family members, friends, neighbours and volunteers. Even at a national level, methodology and measurement are often deficient to address these aspects in a coherent way. In addition, there still a large gap in LTC research and practice between Southern and Northern Europe, particularly that which addresses the formal and informal care interfaces and inherent cultural differences that impact on how care is provided.

The focus of this project was therefore to draw the existing elements together in a 'state of the art' European framework for analysing LTC provision. Given the huge variety of health and social care systems in Europe, such a framework had to be constructed by considering pathways of reform policies, economic and other incentives and thresholds for improvement at any stage of a national system's development.

Figure 1.1 locates the virtual current and potential future position of an integrated LTC system with its links and interfaces to social and health care systems.

This figure is based on the observation that a differentiation between health, social care and long-term care of older people has taken place that leads to bottlenecks and difficulties identified and experienced by users and other stakeholders involved. In Fuchs' (2006, p. 38) theoretical essay about the health system, he observes that 'it remains to be elucidated, how "care" fits into the system, whether it should be framed as a subsystem of the health system or as a functional system of its own'. As a result, theories and methods to improve coordination, linkage and networking have been proposed and implemented – but remained fragmented (Johri *et al.*, 2003; Nies, 2004b; 2007; Lüdecke, 2009; Leutz, 1999; Kodner, 2009; Dickinson and Glasby, 2010; Minkman, 2011) or restricted to pilot projects even if these were able to show clear evidence for improvements.

The continuing 'emergence' of functional differences when it comes to LTC has been identified repeatedly in theory and practice over the past

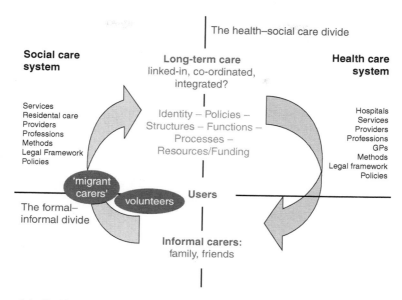

The health–social care divide

Social care system

Services
Residential care
Providers
Professions
Methods
Legal Framework
Policies

Long-term care
linked-in, co-ordinated, integrated?

Identity – Policies –
Structures – Functions –
Processes –
Resources/Funding

Health care system

Hospitals
Services
Providers
Professions
GPs
Methods
Legal framework
Policies

'migrant carers'   volunteers   **Users**

The formal–
informal divide

Informal carers:
family, friends

Figure 1.1   Positioning integrated long-term care between health and social care systems
Source: http://interlinks.euro.centre.org

decade (Leichsenring *et al.*, 2005; Fuchs, 2006; European Centre, 2010). This concerned for instance new approaches to dementia care, fresh funding opportunities, but also paradigmatic changes to improve quality of life and improve the focus on person-centredness, including support for informal carers. INTERLINKS has therefore striven to systematically frame and analyse those elements and key issues that are critical for this functional restructuring process at the systems and the organisational level. Indeed, in terms of Luhmann's theory of social systems (Luhmann, 1995), we may observe the construction of a (sub-)system, in this case LTC, if we are able to distinguish between the system and its environment. This implies the development of 'a sense of identity' as an important precondition (see Chapter 2). The respective indications, symbols and signs for such distinctions that can be observed in the context of LTC form the basis of the purpose of this book, to show how LTC systems are emerging in Europe.

However, compared to economic, political or educational systems, health and social care systems remain theoretically 'underdetermined' (Fuchs, 2006). Medical care has developed a wide range of specialisations over several centuries and, in particular, a relatively clearly defined binary code that allows for a distinct construction of reality in terms of 'ill vs healthy' (Luhmann 1990, p. 184) or, as Pelikan (2007, p. 296) proposed 'ill vs not-ill'. With this inclusion or exclusion criterion in mind, the system of medical care has achieved tremendous success in 'cure', but it has particular difficulties in dealing with end-of-life care, disabilities or long-term care (Fuchs, 2006). Following the above-mentioned paradigm shifts in LTC, in these cases, 'patients' are no longer 'patients' but must be defined as 'clients', 'users', 'consumers', 'citizens' or just 'people with LTC needs'. They cannot be 'cured' but need care, assistance, social support and rehabilitation to accomplish instrumental activities of daily life even with impairments, so they require functions for which health care organisations are poorly prepared. Moreover, LTC also has to address functions that are crucial to the meaningfulness of life and dignity. These functions might be more familiar to social care, but also in this context, people with LTC needs often do not find an appropriate solution. This may be due to the fact that social care is still frequently connected to rationales of combatting poverty, means-testing, rather than needs-testing, and welfare rather than social inclusion.

The increasing distinction between health, social and long-term care may also be viewed in terms of a transition period during which it becomes necessary to create loosely coupled subsystems for which the coexistence of closeness and openness as well as of distinctiveness and responsiveness seem to be important. Notwithstanding the process of differentiation, the distinct systems and organisations, in this case LTC, should be able to secure autonomy and develop a specific identity even though they need to interact with medical care and related realms (Orton and Weick, 1990, p. 205). From an older person's point of view, cure is relevant within or during an LTC process, such as in the presence of an acute health problem within a chronic disease. From an organisational or system point of view, knowledge exchange or organisational development can benefit from the adjacent system. This increases the opportunity to develop proper identities for structurally differentiated LTC systems and organisations, with their own visions, and specific values, such as autonomy and dignity. Also it permits more effective and meaningful interaction with respect to the needs of people with care needs who are, in health care settings, primarily perceived as 'problem cases' or 'bed-blockers' in hospitals.

This so-called transition period can also be characterised by failure, inappropriate solutions and developments that challenge the social tissue in general. In the context of emerging LTC systems, the phenomenon of live-in migrant care workers who provide personal assistance to older people living at home (see Chapter 10) may serve as an example for such a 'dysfunctional solution'. Due to the lack of available and affordable services, families in Austria, Germany, Greece, Italy, Spain and other countries used the opportunity to find carers from low-wage countries who have been ready to provide care around the clock at affordable prices. These arrangements, coupled with different forms of volunteering and care by families, friends and neighbours, show the idiosyncrasies and contradictions by which LTC remains branded. These concern in particular the blurred boundaries between paid and unpaid, trained and untrained, formal and informal, and even between legal and 'illegal' care.

The INTERLINKS contribution in this respect has been to start to cement the stylised 'ideal' position of LTC in Figure 1.1 through research on innovations and analyses of practice examples. This includes, for instance, policies and legal regulations, innovative ways of working, innovative modes of governance and financing or support measures for informal carers and older people in need of care that shed light on how the health and social care divide as well as the formal–informal care divide can be addressed. Such contribution was achieved through the integration of the professional and non-professional domain as well as the perceptions, interests and perspectives of a wide range of stakeholders. This included political and administrative decision-makers at different levels, and professional federations as well as organisations representing providers and carers.

The INTERLINKS methodology and terminology

INTERLINKS was therefore conceived as an interactive study of applied social research which was realised by a number of milestone events during which an informed public (national experts from research and practice, high-level policymakers, EU institutions and European level non-governmental organisations and providers) were involved in the validation of findings and in the elaboration of individual themes and key issues of the INTERLINKS Framework for LTC. It has been a special feature of the project to identify and to involve 'change agents' in policy and practice (people who are open for change and able to implement learning and evidence) as much as possible – in the gathering of data and in the validation of findings.

INTERLINKS was carried out by a consortium that originally consisted of 16 partners from universities, national and international research institutes with international and interdisciplinary expertise, also in cross-national research. The consortium represented 13 EU Member States (Austria, Denmark, Finland, France, Germany, Greece, Italy, the Netherlands, the Slovak Republic, Slovenia, Spain, Sweden and the United Kingdom with a focus on England) as well as Switzerland, covering different welfare regimes and geographical domains to allow for the regional and developmental, path-dependent differences to be addressed.

The project was coordinated by the European Centre for Social Welfare Policy and Research, a UN-affiliated intergovernmental research organisation concerned with all aspects of social welfare policy and research (www.euro.centre.org).

In order to coordinate scientific work with more than 40 scholars over 3 project phases, the general approach within the spiral configuration of the project design (see Figure 1.2) and methodological guidelines were specified during a first meeting of a Scientific Management Team (SMT)

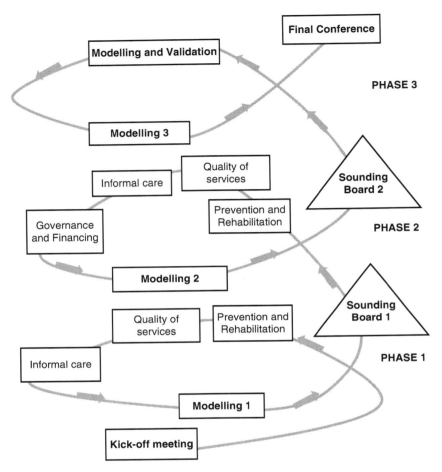

Figure 1.2   The spiral configuration of the INTERLINKS project design
Source: http://interlinks.euro.centre.org

convened from work package leaders. In order to guarantee involvement by all beneficiaries in a participative approach, more than a dozen working and discussion papers as well as templates and other methodological tools were developed and elaborated on by mutual exchange.

The results and products that emanated from INTERLINKS were based on a sound methodological framework and guidance over the entire project period. During the very first phase of the project, this consisted of focusing on the development of a shared understanding of LTC, its elements and pathways, necessary to address LTC needs from a user perspective.

Based on the 'vignettes' method, two vignettes and respective posters describing different pathways of Mrs. L.T. Care and Mr. L.T.C. Dementia (see for examples from various countries: http://interlinks.euro.centre. org/countries) were produced in the participating countries to develop a common understanding.

Furthermore, templates for national background reports were provided in the initial phase to guide research in respective teams focusing on prevention and rehabilitation (Kümpers *et al.*, 2010), quality management and quality assurance (Nies *et al.*, 2010), informal care (Triantafillou *et al.*, 2010) as well as on governance and financing of LTC (Allen *et al.*, 2011).

The modelling process of envisaging how a methodologically sound and universally understood framework could be developed to describe, analyse and improve the complexities of LTC across Europe commenced with a series of discussion papers. Starting with a systems approach, the purpose of these papers was to unravel thought processes and gain consensus on fundamental principles in order to establish the building blocks of framework conceptualisation.

Furthermore, *National Expert Panels* (NEP) were established in all participating countries. These involved researchers, policymakers, and representatives of provider and user organisations who were invited to provide support in identifying model ways of working, elements of good practice and other relevant information and expertise. During the second phase of the project, their task was to validate the INTERLINKS Framework and practice examples. NEPs were invited to two to three meetings over the project period; in some countries contact was maintained by phone and email exchange. On a European level, a *Sounding Board* was established by inviting approximately 20 experts and representatives from policy and practice. During two Sounding Board Conferences they were asked to provide feedback on presented findings and input for further improvement of the Framework. In addition, in their capacity as 'change agents' in European policymaking, they were invited to disseminate INTERLINKS findings.

During the second phase, the first draft of the INTERLINKS Framework for LTC (see below) was constructed with contributions from all partners on the basis of findings from the first phase. This included the definition of a common template to be used for describing and analysing practice examples, and a first test period of a related peer-review process. The results of this modelling process were presented for validation to NEPs and the Sounding Board so that their feedback could be integrated during the third phase of the project.

During the third phase, the INTERLINKS Framework for LTC was finalised, translated into an interactive website and underpinned with almost 100 practice examples – these had undergone an internal peer-review process for their evidence and consistency. Given this knowledge base, dozens of presentations and dissemination activities were organised, including publications and articles in the popular press, culminating in a public Final Conference with about 90 participants, including representatives of the NEPs and the Sounding Board.

The INTERLINKS Framework for long-term care

The INTERLINKS Framework has been developed to illustrate which elements have to be in place to address the links and interfaces between social and health care systems and to construct an integrated long-term care system. This framework is structured by four levels, namely themes, sub-themes, key issues and practice examples. Each theme contains an introductory explanation of its purpose in an ideal-type LTC system, and is further defined by a number of sub-themes that, again, are specified by relevant key issues. Each of the six themed sections contain a rationale of how the construction of sub-themes and key issues was informed by initial work undertaken within the first two project phases. Key issues also reflect the primary foci of INTERLINKS, namely prevention and rehabilitation, informal care, quality development and governance and finance. The selected practice examples have been described and analysed with a view to these key issues that, in general, serve to identify the links and interfaces or respective gaps addressed by the practice example. The latter have been described and analysed by means of a template designed by partners and validated by National Expert Panels and the Sounding Board. By their very nature, practice examples address several themes and sub-themes and can hence be described from different angles and/or illustrate several (sub)themes. Indeed, they were selected, described and analysed with a focus on their most relevant and innovative feature in addressing one or several key issues within specific sub-themes (see Box 1.1 as an example for the theme 'Identity'; see Appendix A for full INTERLINKS Framework).

Box 1.1 The INTERLINKS Framework for long-term care (excerpt)

1 Identity of Long-Term Care

1.1 Values

  a) how key principles that characterise LTC are expressed; what values dominate for which stakeholder perspectives (e.g. economic and quality perspectives, citizenship);
  b) whether there are sets of values that are shaping political, organisational and individual choices in LTC (e.g. through surveys of older people's specific needs for care and support);
  c) how informal care and/or family ethics are addressed in legislative frameworks;
  d) how issues of dignity, quality of life and empowerment are described in policy papers;
  e) how values relating to prevention and rehabilitation are considered;

f) how values embrace the diversity of users and carers (according to gender, culture and social inequalities) and support the specific needs of hard-to-reach groups.

1.2 Mission statements

a) organisations that explicitly address problems at the interfaces between, for example, formal/informal and health and social care, prevention and rehabilitation or the use of migrant care workers;

b) a description of how the problems are addressed and what particular factors distinguish an organisation as an 'LTC organisation', rather than a health or social care organisation;

1.3 Organisational definitions

a) how LTC is defined within or between organisations;

b) how health care providers define their purpose in relation to LTC (cure vs care; needs vs supply; expectations vs preferences);

c) how clients are defined and positioned.

Source: http://interlinks.euro.centre.org

This Framework is targeted at policymakers, managers and practitioners as well as researchers and developers working with or in LTC and can be used in various ways to describe, analyse and improve long-term care systems, in particular by means of the interactive website (http://interlinks.euro.centre.org).

The INTERLINKS Framework is the first European-wide approach to systematically describe and analyse the elements of LTC and serves as a knowledge base to facilitate the integration and improvement of LTC as a system of its own – with a specific focus on the links and interfaces between health and social care as well as between formal and informal care.

The issues addressed in this book

While the INTERLINKS Framework aims to facilitate the description, analysis and improvement of LTC at the individual, organisational and systems level, the following chapters will underpin and illustrate its most salient themes and key issues to provide a better understanding of their rationales. One of the most original features of the INTERLINKS Framework consists of highlighting the need for a proper identity for LTC.

In Part I, complementing the introduction to this chapter, *Henk Nies, Kai Leichsenring* and *Sabina Mak* will explore the justification, empirical developments and opportunities linked to the emergence of features that may lead to such an identity in Europe. This is based on the assumption that, at the boundaries between health and social care systems and between formal and informal care, a specific approach beyond 'medicalised' and 'purely social' concepts of care, activation, involvement and dignity are needed to face the growing demand for LTC. Indeed, first indications for such approaches can be retrieved from the INTERLINKS practice examples empirically by policies that elaborate on users' rights and participation, by organisational approaches to boost community-based care, by new types of job profiles and by new forms of funding. Preconditions and relevant framework conditions, as well as the pros and cons of a distinct identity for the largest growing area of societal care will be addressed. It will be argued that the national, regional and local idiosyncrasies are also a prerequisite: blueprints do not exist that currently fit LTC.

Part II will address issues in policy and practice that overcome both the health–social care divide and that of formal and informal care. *Michel Naiditch, Judy Triantafillou, Patrizia Di Santo, Stephanie Carretero* and *Elisabeth Hirsch-Durrett* view the LTC system through the eyes of consumers of services – these are not only older people with LTC needs but also those individuals, relatives and friends who provide informal care but may also be clients of formal services with their own needs for support. Based on a litera-ture review and an analysis of practice examples, they critically describe the extent to which service provision inspired by the 'active consumer' paradigm has had an impact on empowerment of users, and the degree to which it is connected to the implementation and conceptualisation of supportive measures for informal carers. Taking the example of cash benefits, the authors also consider whether cash benefits, as an 'emblem' of consumer choice, have fulfilled their promise of offering greater choice and enhanced participation for older users and their carers.

Kerry Allen, Jon Glasby and *Ricardo Rodrigues* then focus on key themes, barriers and enablers when countries have sought to promote more effective partnership working between health and social care. Drawing on concepts of linkage, coordination and integration, they use practice examples and key findings from National Reports on specific issues addressed during the INTERLINKS project. After a brief introduction to challenges and key contex-tual factors shaping policy and practice in selected countries, three key sections explore examples that offer solutions for developing partnership at an individual level, local organisational level, and at a structural/social policy level.

Various methods to better link social and health care systems and informal care are developed and implemented all over Europe – with surprisingly converging tendencies even between Nordic and Mediterranean approaches, as *Jorge Garcés, Francisco Ródenas* and *Teija Hammar* show in their chapter. As similar methods and approaches can be identified in apparently different contexts such as the Nordic and Mediterranean LTC systems, two examples will be showcased in a comparative perspective. The Finnish practice example, the PALKO model, promotes different aspects of integrated care that, in particular, helped to clearly define the place and tasks of all stakeholders across the care pathway. The professional care/case manager, the service organisation and the multidisciplinary team are able to integrate various services effectively so that service users perceive their care as 'seamless'. Also the 'Sustainable Social and Health Model' in Valencia applies the case management methodology to design personal pathways responding to the specific needs of older and chronically ill people, using a network of social and health services. Also in this context case management has been identified as an effective strategy to facilitate care of older people in their neighbourhood, to arrange hospital discharge, admission to services and further processes through a common network of health and social services.

Issues of service integration are at the very heart of developing LTC systems. For *Pierre Gobet* and *Thomas Emilsson* it is therefore necessary to understand the empirical indications of a lack of integration, the kinds of problems it may generate and how stakeholders involved try to overcome bottlenecks and shortcomings. This chapter therefore focuses on 'what happens' and what is done by the stakeholders at the interfaces and boundaries between health and social care and formal and informal care in terms of removing, crossing, shifting boundaries or even in relation to raising new boundaries. These practices and their underpinning concepts are illustrated by practice examples from 13 European countries. The chapter concludes with suggestions for further research that focuses on successful solutions at these redefined boundaries.

Part III will then focus on crucial new developments in the construction of LTC in Europe. Without doubt, prevention and rehabilitation for older people within long-term care have an important, though only slowly emerging potential to innovate LTC, as *Susanne Kümpers, Anja Dieterich, Georg Ruppe* and *Lis Wagner* reveal in their chapter. They address the common issues at stake when it comes to prevention and rehabilitation within LTC in Europe, in particular explicit interventions versus systemic strategies, inequalities in access, cultural issues and governance mechanisms as promoting or inhibiting factors for preventive and rehabilitative orientations, and self-determination and self-care as normative orientations. The chapter specifically highlights solutions identified by practice examples to illustrate the important potentials of strategies to maintain social inclusion of frail older people in the community.

Kai Leichsenring, Henk Nies and *Roelf van der Veen* describe and analyse the quest for quality in emerging and changing LTC systems in Europe. Common trends and differences between quality management and quality assurance in health and LTC are presented together with an overview of the main methods and idiosyncrasies in ten European countries, based on INTERLINKS and other European comparative research projects. Given the current developments and restraints, the chapter ends with potential future developments of quality management in the care of older people in Europe.

The increased demand for LTC services that is likely to arise from an ageing population will enhance the importance of ensuring that measures taken are cost-effective, and that outcomes are maximised in relation to the resources employed. *Ricardo Rodrigues* and *Henk Nies* explain what cost-effectiveness in LTC could mean by detailing some of the inherent challenges, namely that the person receiving care is often also the 'co-producer of care', and that the quantification of resources employed in LTC often remains inaccurate. Building on the INTERLINKS Framework for LTC and relevant practice, existing evidence (or lack of) for cost-effectiveness is highlighted. This analysis draws on the potential value for money of some of the examples, but also the methods employed and difficulties encountered in assessing their cost-effectiveness. The chapter's concluding recommendations outline possible indicators and approaches to improve the assessment of cost-effectiveness in LTC.

A wide range of deficits such as lack of services, changing family structures, family ethics and inequalities are characterising the realm of LTC in Europe. These have resulted in extraordinary and sometimes desperate societal solutions by which care activities that had formerly been provided within the family have been shifted into care arrangements with migrant carers. These arrangements may be formally supported by specific legal regulations, but in most cases they remain undocumented. The consequences of these practices are far reaching from the perspective of both sending and receiving countries. *Rastislav Bednárik, Patrizia Di Santo* and *Kai Leichsenring* shed light on the role of migrant carers in patching up the 'care gap' in a chapter that deals with this specific characteristic of emerging LTC systems at the interface between formal and informal care. Apart from the description of these practices and their impact in sending countries (e.g. Slovak Republic), and receiving countries (e.g. Italy, Germany, Austria), the chapter showcases several practice examples that strive to improve this situation for users, family carers and migrant carers.

Another characteristic of LTC is the role of volunteering and the mismatch between the needs of voluntary organisations and the aspirations of new generations of volunteers, calling for social innovation in terms of professionalising the management of volunteers. In this chapter *Kvetoslava Repková, Karin Stiehr* and *Barbara Weigl* describe volunteering as a part of LTC for older people, emphasising the importance of raising the profile of volunteers and improving supporting framework conditions. The description is based on various sources illustrating the practice of volunteering in LTC to understand different approaches and how selected countries deal with the new challenges in this field. The analysis highlights policy aspects as well as features at the organisational level and issues at the individual level. Furthermore, practice examples that address mechanisms for public support of voluntary activities in LTC, strategies for the professionalisation of volunteering management, but also the risk of using voluntary work as a tool to reduce paid staff in times of economic crisis are discussed.

A wide range of optimistic expectations are linked to the expansion of information technology and innovative applications in LTC for older people. *Jenny Billings, Stephanie Carretero, Georgios Kagialaris, Tasos Mastroyiannakis* and *Satu Meriläinen-Porras* assess the role of information technology in this area, and its potential to permit a more person-centred approach, assist carers in helping to reduce the burden, and to support older people with a wide variety of long-term physical and cognitive conditions such as diabetes and dementia. Apart from a critical literature review this chapter provides an analysis of practice examples in relation to their approach, impact, sustainability and transferability, concerning their potential to overcome bottlenecks at the interfaces between social and health care as well as to improve links between formal and informal care.

LTC for older people also means providing services to and *acting on* people at the end of their life. This entails that LTC in all its dimensions is regularly confronted with sensitive and ethically precarious situations. Decisions on the nature and dimension of medical interventions, on care procedures under different individual circumstances as well as on appropriate circumstances at the very last moments of a person's life have to be taken. A balanced mix and proper coordination of social, medical, formal and informal services is needed in these situations. However, even in professional geriatric care, end-of-life scenarios as well as dying are still widely seen as taboo. *Laura Holdsworth* and *Georg Ruppe* explore pathways between cure, care and palliation and review the extent to which LTC structures and services across Europe are prepared to deal with related challenges. The chapter outlines how end-of-life and palliative care practices are generally recognised and organised in the health and social care policies of selected countries. It then presents a selection of initiatives and practice examples from different parts of Europe highlighting their specific strengths, weaknesses, opportunities and threats. With a focus on care needs, conclusions consider common barriers as well as promising future strategies to improve the integration of health and social care services with users' and their families' expectations at the end of life.

In the fourth and final part, *Jenny Billings* focuses on the association between evidence of good practice and LTC. One characteristic of emerging practices in LTC is the lack of evidence concerning specific interventions that have been applied due in part to the tensions that exist in how evidence is understood and used for both health and social care practice and policy. The chapter outlines the challenges and opportunities of more evidence-based approaches in health and social care evaluation and provides an account of and rationale for a more pluralist approach to gathering evidence in LTC, as adopted within the INTERLINKS project. The chapter also provides a critical transversal analysis of those practice examples that reported project effects, giving an account of how they were evaluated, what sources of evidence were used, the project effects on users, carers and organisations, followed by an interpretation and critique of the extent to which INTERLINKS examples have contributed towards the evidence in LTC debate.

Finally, *Kai Leichsenring, Jenny Billings* and *Henk Nies* reflect upon the main messages and draw some conclusions in relation to future research outlining the challenges and opportunities of approaches in LTC, summarising the findings and putting forward recommendations for policy and practice.

2
The Emerging Identity of Long-Term Care Systems in Europe

Henk Nies, Kai Leichsenring and Sabina Mak

Introduction

Long-term care is a concept everybody is familiar with because we read about it in the newspapers or because we know someone who receives services or works in long-term care (LTC). Nowadays it affects the lives of many people. But what is LTC exactly? Which organisations and people provide this care? Many people still associate LTC primarily with care in residential homes, but it is much more than that.

More users of LTC receive care at home than in institutions (Colombo *et al.*, 2011). Informal carers are the backbone of the LTC system (Triantafillou *et al.*, 2010). Without this important resource, health and social care systems in Europe would not be able to function as they do now. As a consequence, the connections between these formal systems and the resources of informal care have not yet developed into an integrated LTC system. Working towards such a system, however, will be necessary to deal with the challenges of an ageing society, the burden on caregivers and high care costs.

The theme 'identity of LTC' has therefore been defined as one of the cornerstones of the INTERLINKS Framework for LTC. In this chapter, the rationale of the various sub-themes and key issues of the theme 'identity' in the INTERLINKS Framework for LTC will be explained and discussed – this includes the different facets and some inspiring examples of integration between social care and health care, and between formal and informal care. This integration is important and must succeed if frail older people in need of LTC are to continue having a fulfilling life. Furthermore, the 'identity of LTC' depends on the contribution and roles of professionals and informal carers in LTC as a blend of services and stakeholders who provide care. This implies reflecting upon the following questions: What are organisational definitions, values and missions specific to LTC? What are the motives and drivers for people to work in LTC as well as informal carers and volunteers to carry out their tasks? Do these drivers make LTC feasible and therefore sustainable? These questions will be discussed and analysed on the basis

of illustrative examples from the INTERLINKS Framework for LTC (http://
interlinks.euro.centre.org) and other sources that show methods and instruments, tangible progress and innovative ways to develop the identity of LTC
further as a differentiated functional structure at the systems, organisational
and individual levels.

By focusing on key issues within the INTERLINKS Framework for LTC,
the chapter concludes with recommendations for developing LTC as an
integrated system, in particular with a view to the opportunities of an innovative LTC identity in terms of activation, social solidarity and inclusion.

What is long-term care?

The identity of LTC systems for older people is only beginning to emerge.
It is placed at the junction between health and social care. In some countries it
is taking shape as an incipient system, sometimes with a growing formal
identity (http://interlinks.euro.centre.org).

LTC includes activities undertaken for people who require complex care
for extended periods of time. Such aspirations are well formulated in the
definition of the World Health Organization (WHO) and the Milbank
Memorial Fund (WHO, 2000, p. 6): 'The system of activities undertaken by
informal caregivers (family, friends and/or neighbours) and/or professionals
(health and social services) to ensure that a person who is not fully capable
of self-care can maintain the highest possible quality of life, according to his
or her individual preferences, with the greatest possible degree of independence, autonomy, participation, personal fulfilment and human dignity'.

This description emphasises that LTC is to a large extent provided by
informal caregivers (family, friends and neighbours), but also by formal
caregivers including professionals and auxiliaries (health, social, and other
workers), and by volunteers. LTC can be provided at home, in institutional
or in semi-institutional settings, such as day care or intermediate services.
This type of care can be provided by public, not-for-profit and for-profit
providers (see also Chapter 9 in this book), with services ranging from alarm
systems to 24/7 personal care (Colombo *et al.*, 2011).

Often it needs to be combined, preceded or followed up by other services,
such as prevention and early diagnostics or case finding, acute physical
or mental health care and rehabilitation. Outside the medical realm, LTC
is often linked with financial, social and legal support, and with housing
measures, for instance adaptations and technology such as smart housing
and ambient technology (Nies *et al.*, 2009). Informal caregivers also require
access to supportive services, including information on and assistance in
securing help, training and respite care (WHO, 2000).

The above definition also illustrates that LTC has normative connotations:
independence, autonomy, participation, personal fulfilment and human
dignity are values that are culturally bound. They are linked to social,
moral and ethical norms, government policy and other country-specific

circumstances (Ngai and Pissarides, 2009). These values are reflected in people's attitudes towards care and in the policy infrastructures in various countries and local governments.

Lamura *et al.* (2007) suggested a typology of European welfare regimes based on three key factors: demand for care (e.g. percentage of population above 80 years of age, economic wealth), provision of informal care (e.g. diffusion of extended households and informal care), and provision of formal care (e.g. public expenditure on LTC for older people, proportion of older people receiving formal home care or residential care). Table 2.1 gives an overview of the five types of LTC regimes that can be distinguished in Europe.

This typology clarifies at least some of the national, regional and local governments' priorities. It also demonstrates to what extent governments actually influence (or do not influence) day-to-day practice. For instance, in countries with high demand and low or medium provision of formal care, alternatives have developed beyond the realm of public government and existing social policy. In countries such as Greece, Italy, Spain, Portugal, Austria and Germany, large numbers of informal care workers from either recently accessed EU Member States or non-European countries (North Africa, South America) have entered the market, in many cases without legal regulation, as substitutes for family and/or professional carers. A substitution effect also appears to take place in the newly accessed Member States in Eastern Europe, where 'black markets' emerge consisting of migrant care workers from countries such as Romania, Ukraine and Russia (Di Santo and Ceruzzi, 2010; Triantafillou *et al.*, 2010). While these trends develop in society beyond the sphere of governments, occasionally governments decide to extend their control. For instance, a great number of migrant workers have been legalised in Italy since 2002 and have virtually become a part of the public system involved in organising LTC. So, due to a policy intervention

Table 2.1 A typology of European long-term care regimes

	Demand for care	Provision of informal care	Provision of formal care	Countries
Standard-care mix	medium/high	medium/low	medium	Germany, Austria, France, Italy, UK
Public-Nordic	medium	low	high	Sweden, Denmark, Netherlands
Family based	high	medium	low	Spain, Portugal, Ireland, Greece
Transition	low	high	medium/low	Hungary, Poland, Czech Republic, Slovakia

Source: adapted from Lamura *et al.* (2007).

these informal care workers (a preferable term could be 'non-formal care workers') have become part of the formal care provision system.

The typology also illustrates that, for instance, Public-Nordic countries emphasise professional solutions to satisfy LTC demands, even for supporting informal carers. In other words, the focus and pace of developing sustainable and well-integrated LTC systems differ considerably across EU Member States according to their points of departure. There is also a wide variety of attitudes towards optimum or avoidable solutions to LTC needs of older people. In countries like Denmark, France, the Netherlands and Sweden, less than 40 per cent of the population consider family care as the best solution for an older parent with LTC needs who lives alone. While they prefer professional care, in Eastern (e.g. Bulgaria, Slovakia, Poland, Hungary, Czech Republic) and Southern European countries (Portugal, Spain, Malta, Turkey) the opposite is the case. This also reflects different filial obligations towards older relatives. In countries like Poland and Italy, the reasons for stopping informal care for a relative after one year were due to the fact that, in more than 45 per cent of surveyed cases, the care was transferred to other relatives, while in countries like Sweden and Germany the primary reason, in more than two-thirds of the cases, was the transfer to a nursing home (Huber *et al.*, 2009). It must therefore be noted that there are different values that are guiding, influencing and governing LTC across Europe, but also within each country across social classes, which influence expectations and individual decisions on who should and will take on care responsibilities and what form the care should take. In addition, these values are not static but develop over time and contribute to changes in underlying paradigms that are at the origins of LTC.

Changing paradigms and values

The need for developing LTC as a social system is closely related to the ageing of the population and the increase of chronic conditions, in particular in the older population. This increase is, in turn, a consequence of improved medical care which leads to much better survival rates for previously lethal diseases (Huber *et al.*, 2009; Colombo *et al.*, 2011). Therefore, a rising number of older people live for an extended period of their lives in frailty and with reduced physical, mental, and/or cognitive functional capacities – a situation which is often aggravated by weak social and economic conditions. Although many capacities could and can be regained by adequate rehabilitation (Jané-Llopis and Anderson, 2005; Kümpers *et al.*, 2010), one of the most important challenges is the increase in mental health problems, in particular in relation to the rising prevalence of dementias due to the increasing number of people reaching advanced ages (Wittchen *et al.*, 2011; www.alzheimer-europe.org). In general, however, the type of care needed and its duration are often difficult to predict and call for a multitude of flexible and differentiated support mechanisms (for an illustration of this problem and the challenge of providing personalised care see the flash movie at http://interlinks.euro.centre.org/project).

Historically the traditional societal response to these needs came from various directions: long-term wards in hospitals or nursing homes that were built according to a hospital model; organisations of volunteers and charities that met needs of loneliness and poverty; housing facilities, such as residential homes, to combine efficient housing with care and assistance; and an expanding private nursing home and lodging sector for people who were unable to take care of themselves (Kendall, 2000; Johnson *et al.*, 2010). Over the past two decades, however, there has been growing awareness that, through these models of care, old age and ageing are, at the same time, both being medicalised and defined as a socially constructed problem (Powell and Hendricks, 2009; Kelley-Moore, 2010; Walker and Walker, 1998).

For instance, it has been argued that the interaction of social and biological factors 'create' disability: '(...) the category "age" has been transformed within a short period during which the medical successes in the treatment and management of lifespan limiting diseases was celebrated, to something that identifies age as a risk factor (for example in the context of different forms of dementia)' (Naue and Kroll, 2010, p. 2). Furthermore, it is increasingly recognised that it is important to consider what *happens* to the 'group' of frail older people in the course of social interaction. In this perspective, dementia or disability are identified as social constructions, including terms such as 'dependence', 'loss of autonomy' and 'control' (Naue and Kroll, 2010). This implies that concerns about frail older people with multiple needs are increasingly focusing on the way these needs are to be conceived and how they can be met in a flexible and differentiated way.

As outlined above, the aforementioned definition of WHO and the Milbank Memorial Fund (WHO, 2000, p. 6) reflects the normative aspects of the social construction of needs. It is also voiced by users' organisations and other stakeholders (see Box 2.1 and Box 2.2).

Box 2.1   LOC voice in healthcare

This Dutch organisation, consisting of clients' councils of LTC organisations, expressed a perspective on LTC that reflects values of the twenty-first century: individualisation and prosperity, leading to independence and tailor-made care. It states that the individual is an integral whole, and the LTC system needs to serve physical, psychological and social needs. It contributes to the person's quality of life, thereby also serving the well-being of society as a whole of which the service user is an organic and systemic member.

LOC adheres to the WHO definition of healthy ageing, which states: 'Mental health is (...) a state of well-being in which every individual realises his or her own potential, can cope with the normal stresses of life, can work productively and fruitfully, and is able to make a contribution to her or his community' (LOC, 2009a, p. 14). LOC focuses on health and capacity, rather than illness and impairments. Therefore, the organisation focuses on 'valuable healthcare' which is about the value of people and about their values. Valuable care should focus on values of people, of what is important to them. It gives a sense of purpose to their lives and helps them feel their life is meaningful. Thus care can make people feel valuable to society and responsible for their own life, despite vulnerability.

Source: LOC (2009a; 2009b).

Box 2.2   My Home Life (United Kingdom)

This British initiative of scholars, Age UK and the Joseph Rowntree Foundation, aims to reflect – among others – the interests of care home providers, commissioners, regulators, care home residents and relatives by emphasising quality of life as a key concept in residential care. This concept depends a great deal on how individuals view what makes life enjoyable, meaningful and worth living for. It implies that staff listen to what people say about their preferences and disregard their assumptions and stereotypes. It emphasises the identity of residents and care should aim to assist residents to maintain their sense of identity. It involves making sure that people can achieve what they want in their lives to make their lives worth living, including making relationships and being part of a community.

Source: My Home Life (2007).

Contrary to the traditional concept of health care by which the patient is conceived as a passive recipient of care (being *patient*), current views on LTC – for instance chronic care management (Wagner *et al.*, 2001) and Guided Care (Boyd *et al.*, 2010; Boult *et al.*, 2011) – emphasise the person as being an active and empowered service user. This implies shared decision-making between professionals, older people and their informal carers about the care and services to be delivered. In other words, professional authority is ceasing to lead in the provision of care and services (Nies, 2007). As a consequence, quality of care is no longer the only leading paradigm when it comes to defining the identity of LTC, but quality of life, self-care and dignity. Human rights are increasingly taken as a point of reference to define the objectives for care for older people (Townsend, 2006). An example of this at the national level is the German Charter of Rights for People in Need of Long-Term Care and Assistance (Tesch-Römer, 2007; German Federal Ministry of Family Affairs *et al.*, 2007;

www.pflege-charta.de/die-pflege-charta.html) that also served as a basis for a respective European Charter (see Box 2.3). A wide range of implementation activities such as conferences, ambassadors, a service desk, pilots and publications have enabled further dissemination of these Charters. Person-centred care is thus moving from widely acknowledged principles to daily practice.

Box 2.3   The European Charter of the Rights and Responsibilities of Older People in need of LTC and Assistance (2010)

This Charter was developed by ten national and two international non-governmental organisations (NGOs) representing the interests of older people and residential homes (AGE, EDE). It is now a partnership of 18 partners from 12 countries. It formulates ten principles to ensure that older people in need of LTC can lead a life with dignity and independence and participate in social and cultural life. It emphasises that 'advancing age does not involve any reduction of a person's rights, duties and responsibilities'. In ten articles it formulates rights and obligations of services users of LTC. As you grow older and may come to depend on others for support and care, you (continue to) have the right to:

- respect for your human dignity, physical and mental well-being, freedom and security;
- make your own life choices and have respect for your free will. This right extends to an appropriate third-party of your choice;
- respect for – and protection of – your privacy and intimacy;
- high quality, health promoting care, support and treatment tailored to your personal needs and wishes;
- seek and receive personalised information and advice about all of the options available to you for care, support and treatment in order to be able to make informed decisions;
- interact with others, and to participate in civic life, lifelong learning and cultural activity;
- live according to your convictions, beliefs and values;
- die with dignity, in circumstances that accord with your wishes and within the limits of the national legislation of your country of residence;
- redress in case of mistreatment, abuse or neglect;

As you grow older and may come to depend on others for support and care, you should:

- respect the rights and needs of other people living and working within your environment;
- respect the general interests of the community in which you live; your rights and freedoms should be only limited by the need to respect similar rights of other members of the community;

- respect the rights of carers and staff to be treated with civility and work in an environment free from harassment and abuse;
- plan for your future and take responsibility for the impact of your actions and lack of action on the part of your carers and relatives;
- inform the relevant authorities and those around you about a situation of abuse, mistreatment or neglect that you experience or witness.

Source: Luherne (2011); see also www.age-platform.eu

The identity of long-term care within an integrated system

These visions and values reflect the changing paradigms that increasingly underpin the development of LTC as opposed to the traditional paradigms of medical care and social care. To achieve quality of life according to older people's preferences has become the overarching ambition of LTC. This implies the construction of LTC as an integrative field of social welfare systems through linking, coordinating and integrating services (Leutz, 1999). The big challenge of LTC is to overcome divisions in:

- The health, social care and LTC sectors (whenever the latter is organised separately), as well as social support, housing, work, social participation;
- Informal and formal care;
- Care at home and care in (semi-)institutional settings;
- Public and private (for-profit and not-for-profit) provision of professional care.

Traditionally, these sectors are organised in silos and fragmented social systems (Allen *et al.*, 2011; Hudson *et al.*, 1997; Alaszewski *et al.*, 2004; Nies, 2004a; 2006). So older people, when in need of complex packages of care and support, are confronted with a diverse array of organisational structures, statuses, procedures, funding systems and professions. This fragmentation appears at the level of the individual, the organisation and the system (Dickinson and Glasby, 2009; see also Chapter 4 in this book). The challenge consists therefore of constructing societal and organisational changes to overcome these barriers and to integrate services at all levels (see Chapter 1).

Following the above-mentioned paradigm shift in terms of values and aims, the basic principles of the identity of LTC that apply to older people and their carers can be formulated as follows (http://interlinks.euro.centre.org):

- Self-direction is important for quality of life;
- The needs of frail older people should be addressed in their social, physical and societal context, and;
- Care provision should empower clients, strengthen their self-direction and support them to live a valuable and dignified life.

In order to translate these principles into practice at the individual level, they have to be recognised and supported at both system and organisational levels. These two levels will be elaborated upon in the following to (a) showcase some appropriate instruments to promote and underpin this specific identity of LTC such as policy papers, legislation, mission statements, civil society initiatives and education, and (b) discuss the roles of the various stakeholders in making this new identity of LTC a reality in daily life.

Instruments and methods to promote the new identity of LTC

The above-mentioned paradigm shifts and their translation into a 'new identity' of LTC have only recently started to show their impact in the reality of LTC delivery. However, several instruments and methods can be identified in most European countries that already promote the values and approaches outlined above in their daily practice. Some of those initiatives from various countries were also described, analysed and validated as good practice to illustrate the INTERLINKS Framework for LTC. In the following, some of these examples will be outlined to show how policymakers and organisations may learn from methods and innovative approaches to promote a more visible identity of LTC.

Involving stakeholders in public debates and legislation

To improve LTC in Europe, it is important to develop systematic approaches that publicly acknowledge and define LTC as a specific area of social risk. Although – and because – the issues around LTC governance and financing are a complex amalgamation of ideological, professional and structural elements, the involvement of all relevant stakeholders is an important aspect to create public awareness and democratic legitimacy for LTC. Such involvement may take place on the local level (Weigl, 2011; Repetti, 2011) or even on a national level (Box 2.4).

Box 2.4   The 'Big Care Debate' – public engagement in LTC funding (England)

The 'Big Care Debate' was a national initiative to raise awareness of the current LTC system, to gather views on the current provision and to involve service users, key stakeholders and the wider public in debates about future funding options for LTC. The initiative was launched by the previous Labour government (1997–2010) to consult the entire population. Over 28,000 individuals and organisations responded to the debate. While this is clearly a small number overall, this is likely to be a much higher response rate than previous consultations on such issues. Although detailed policy proposals were shelved when a new government was elected in 2010, the funding of LTC is a complex issue and such a national debate was arguably needed to help raise awareness, stimulate debate, explore potential trade-offs and seek to build consensus about next steps. The debate had the added benefit of raising awareness of the current system among all stakeholders.

Source: Glasby (2011c).

Implementing mission statements that focus on the characteristics of long-term care

Being situated between health and social care systems, LTC organisations are facing difficulties in defining themselves as a specific entity with a mission that is different from, for instance, hospitals or social work for specific target groups. The idiosyncratic features of LTC organisations that have to coordinate, link and network between different types of organisations and professions are therefore often underestimated and not sufficiently addressed. This is particularly true for many 'traditional' providers, who have gone through rapid changes over the past 15 years. For instance, traditional old-age homes that used to host relatively autonomous older people saw themselves being turned into nursing homes with a high proportion of residents suffering from dementia. Another example is traditional home care services which are increasingly confronted with having to organise complex cases that call for new types of interventions beyond 'normal' working hours. Some initiatives have already pioneered ways of addressing these challenges (Box 2.5).

Box 2.5   Bridging the gap between nursing home and community care (Denmark)

This project in the former Skævinge Municipality (Denmark) was one of the first initiatives to develop an integrated system of long-term care in a local context. Between 1984 and 1987 a pilot project was launched by local politicians and care professionals with the specified mission to create a preventive health care approach and better environments for older people in the municipality. By means of a bottom-up approach involving all staff groups and users, new creative ideas were implemented in combining health and social care as well as residential care. The change process was based on the self-care approach by which older people are supported to learn to do as much as they can by themselves, but with staff who provide support only when needed. The involvement of all relevant stakeholders generated ownership and helped shape new local organisational structures in a cost-effective manner.

Source: Wagner (2011a).

Another group of examples that illustrate the growing awareness for the need to develop and implement new types of mission statements in LTC is focusing on specific target groups, such as by adding new resources and skills that explicitly address the challenge to overcome bottlenecks between the health and social care systems (Box 2.6).

Box 2.6 Case managers for people with dementia and their informal caregivers (The Netherlands)

The main challenges for people with dementia and their caregivers are how to deal with the disease and where to find adequate help. Informal caregivers need additional information and advice, such as how to cope with their relatives' behavioural problems and about the development of the illness trajectory. This specific challenge was addressed in 2004 by an experiment initiated by the Netherlands' National Dementia Programme (LDP) with 16 frontrunner regions. The aim was to put co-operation in community dementia care into practice through the introduction of case managers as professional figures to coordinate care for dementia clients. It was left to regions to involve the relevant stakeholders, for example home care organisations, general practitioners, geriatricians, psychologists, etc., in a multidisciplinary team. By mid-2010, the experiment had been rolled out to 57 regions, based on experiences from the frontrunner regions, with the implementation of case managers as the most important consequence of this experiment. The case manager sets up a diagnosis trajectory, coordinates the care provided, and provides information and advice to the client and the caregiver. In this way, the case manager meets their specific needs (see also: Minkman *et al.*, 2009).

In practice, the mission of case managers is to organise care around the assessed needs of the individual user, to consider the interests of both the care recipient and the informal carer (family), and to consider all types of local resources to organise care in the community. They have a central position in the collaboration with the various health and social care providers and facilitate seamless care to ensure continuity and coordination. The positive effect of this kind of support on the clients' well-being, the reduction of the burden on informal caregivers and delayed admission to a nursing home has been evidenced both in the Dutch (Groenewoud *et al.*, 2008) and in other contexts (Callahan *et al.*, 2006).

Source: Mak (2011b).

In some countries, new organisations and initiatives with specific mission statements have been established as a response to completely new phenomena such as the necessity to integrate private care assistance provided by migrant care workers into the traditional array of services, i.e. at the interface between formal and informal LTC. This is a new approach to provide user-centred solutions and dignity for all stakeholders involved in a context of scarce formal home care services, where matching supply and demand gets a completely new meaning, such as through providing lists of appropriate domestic assistants, information and consultation points, and training programmes for migrant care workers (Box 2.7).

Box 2.7   ROSA – A network for employment and care services to promote the regulation of undeclared work and to improve the quality of care work (Italy)

This project has been promoted by the Equal Opportunities and Rights Department of the Italian Government and is hosted by the Puglia Region. In response to the large number of home care requests in Puglia, ROSA aims to increase domestic assistance services by integrating migrant care workers into local home care service networks and by ensuring adequate and appropriate professional solutions. The project's mission consists of, first, to build a public network of services able to promote welfare and social inclusion in the regional catchment area and, second, to encourage the regulation of undeclared work in home care, principally provided by migrant women, and third, to promote the reconciliation between the demands of employment and care which is usually still mainly provided by women. The creation of a public coordination mechanism that aims to match supply and demand in home care work (i.e. for 'regularised' migrant care workers) was complemented by measures to ensure quality of care, mainly through courses to develop migrant care workers' skills in line with labour market requirements (lifelong learning). The new network of services allows women to better balance work and family life. The complexity of different aims and interests called for a strong involvement of all local stakeholders: public bodies, trade unions, older people and family associations. In every province a forum was held to negotiate specific contents and targets. To further encourage families to establish regular contracts with migrant care workers, a care voucher scheme, supported by an information campaign aimed at families and migrant care workers, has been introduced during the project.

Source: Di Santo (2011b).

Value-driven tools and organisations to strengthen the identity of LTC

The development of specific methods and instruments to address challenges at the interface between social and health care has gained momentum over the past two decades through initiatives that address the actual working environment of professionals in LTC, for example with respect to person-centred dementia care (Kitwood, 1997), dementia care mapping (Kuhn *et al.*, 2000) or psycho-biographic working (Böhm, 1999; see also Feil, 2002). Furthermore, defining, assessing and improving the quality of LTC has become an issue of specific attention, in particular with respect to the adaptation of classical quality management tools to the characteristics of LTC. While this has triggered a range of models that are applied by individual provider organisations, in particular care homes, quality assurance across the range of LTC providers has hitherto been scarce (Nies *et al.*, 2010), although first steps towards such a development, driven by generic values attributed to LTC, can be identified (Box 2.8).

Box 2.8   The E-Qalin quality management system (Austria, Germany, Italy, Luxembourg, Slovenia)

This model had been initially developed as a result of a European Commission-funded Leonardo da Vinci project (2004–2007) for care homes, but special versions for home care facilities and services for people with disabilities are also available in several languages. E-Qalin was developed with pilot organisations in Austria, Germany, Italy, Luxembourg and Slovenia as a reaction to generic quality management systems that do not address the characteristics of LTC facilities. E-Qalin is based on the fundamental principles of human rights and pledges itself to the 'European Charter on rights and freedom of older people in care homes' (E.D.E., Maastricht 1993). Ethical stances and values such as dignity, honesty, tolerance, a readiness to enter into dialogue and to confront conflict situations, empathy, freedom and self-determination, and personal integrity are the agreed values for E-Qalin to ensure humane living and working in care homes and social services. Professionals are qualified to apply E-Qalin through special training that enables E-Qalin process managers to carry out a self-assessment process in the organisation. By involving all stakeholders in the self-assessment and the continuous improvement of quality, E-Qalin strives to strengthen the individual responsibility of staff and their ability to cooperate across professional and hierarchical boundaries. The principle of ongoing participative development is implemented through the basic value of 'involvement', that is by making all relevant stakeholders aware of their current approach and methods in order to move towards further improvement. Working with this model helps the organisation and relevant stakeholders to understand how far it has progressed, what works well within the organisation and in which areas there is a need for action or improvement. There is a growing number of organisations, in particular in Austria, that provide both residential and home care services and have started to apply E-Qalin across this traditional divide.

Source: Leichsenring (2011a).

The traditional organisation of home care, and LTC in general, has been developed by adapting a 'Taylorist' division of work with distinct tasks for home helps, geriatric aides, registered nurses, therapists and a range of other specialised job profiles. This often results in dissatisfaction of users and carers, but also of professionals themselves, in particular those driven by values on which holistic and person-centred care concepts are based. It is interesting to note that, in the Netherlands, these values triggered a group of community nurses to found an organisation that established an innovative approach to deliver user-centred home care in the neighbourhood (Box 2.9).

Box 2.9   The values of community nursing (The Netherlands)

Buurtzorg ('care in the neighbourhood') originated in 2006 from staff's dissatisfaction with the traditional organisation of home care that was perceived as too bureaucratic, forcing community nurses to work in isolation from other care providers and, above all, neglecting their professional competencies. The 'Buurtzorg' approach has triggered a successful change process within a short period of time. Today, the organisation employs more than 4,000 professionals and is proud of high satisfaction rates of both users and staff. The organisational model of Buurtzorg delivers care through small self-managing teams consisting of a maximum of 12 professional carers, and to keep organisational costs as low as possible, partially by using ICT for the organisation and registration of care. From the LTC process point of view, Buurtzorg introduces a systematic attempt to contact and integrate with other local carers and with informal caregivers. Also, Buurtzorg aims to deliver care to a client for as short a period as possible, by involving and reinforcing the client's resources.

Buurtzorg combines values such as user-centred care, local networking and financial sustainability in a successful model that responds to the client's care needs, and tries to find solutions together with the clients and their informal carers as well as with other formal carers involved. Furthermore, Buurtzorg has underpinned its strive for quality with evidence-based quality indicators and regular monitoring. The satisfaction of users and staff is extremely high, yet costs have been reduced to almost 50 per cent compared to usual home care organisations.

Source: Huijbers (2011).

The identity of users, carers and staff

To make progress in furthering the development of a proper 'LTC identity' it is important to recognise the specific interests and idiosyncrasies of users, carers and professionals. While in reality these are far from homogeneous, this section will highlight some basic characteristics that have to be considered by any initiative at the systems or organisational level.

Users of long-term care

LTC is mostly concerned with older people who need assistance with activities of daily living (ADL), such as bathing, dressing, and getting in and out of bed. In 1950, less than one per cent of the global population was aged over 80 years. By 2050, the proportion of people aged 80 years and over is expected to increase to nearly 10 per cent. Most users of LTC are women aged over 80 years and a substantial proportion of LTC recipients suffer from dementia-related problems (Colombo *et al.*, 2011).

Chronological age is often linked with frailty and multi-morbidity. However, there are wide variations between individuals of the same age. There are, in fact, many older people living in the community leading a full and active life. Despite a dramatic increase in the use of the term 'frailty' among health professionals, there is still a lack of consensus in the literature on its meaning, and there are no widely accepted conceptual guidelines for identifying older adults as frail. However, instead of being an unavoidable consequence of advancing chronological age, frailty has been well recognised as an independent geriatric syndrome (Morley *et al.*, 2006) and is described here as a progressive physiological decline in multiple body systems. Frailty is marked by loss of function, loss of physiological reserve, an increased vulnerability to disease and death, and an increase in susceptibility to acute illness, falls, disability, institutionalisation, and death (Fried *et al.*, 2001).

Although frailty is more prevalent in older people and in those with multiple medical conditions (Fried *et al.*, 2001; Newman *et al.*, 2001), it can exist independently of age, disability or disease, and may be an independent physiological process involving multiple systems (Fried *et al.*, 2001; Hamerman, 1999). Caring for frail older people is challenging because they suffer from an increased burden of symptoms, that are medically, psychologically and socially complex (Fried *et al.*, 2001; Newman *et al.*, 2001).

Informal carers, volunteers and migrant workers

Family carers are the backbone of any LTC system. It is estimated that 80 per cent or more of LTC is provided by family and other 'informal' carers: spouses, children, neighbours, volunteers, the vast majority of them being women (Huber *et al.*, 2009; Colombo *et al.*, 2011; Triantafillou *et al.*, 2010). Most care is provided by family and friends out of love or duty not just temporarily, but often for many years, and intensively (Beneken, named Kolmer *et al.*, 2008). Across the Member States of Organisation for Economic Co-operation and Development (OECD), more than one in ten adults is involved in informal, typically unpaid, caregiving, defined as providing help with personal care or basic ADL to people with functional limitations. Southern European countries have the highest percentage of informal caregiving (Colombo *et al.*, 2011).

Policy initiatives to support LTC and service providers often focus exclusively on the person in need of care and neglect the dynamics of the 'triangle of care' between the formal carer, the informal carer and the older person in need of care (Triantafillou *et al.*, 2010). This can imply, on the one hand, that the informal carer might exclude care recipients from access to support schemes and services to which they are entitled. On the other hand, if care is shared between the formal and informal care sectors, there are often problems in the way professional and informal carers work together and share responsibility for implementing and supervising the care process. Such conflicts may impede care coordination, with the family carer often obliged to act as an implicit but not recognised case manager. So there is clear evidence that professionals need better preparation in how to assume responsibility for managing the care process in collaboration with the family system. There is also strong evidence for frequently underestimated conflicts between the older person's and the informal carer's needs and expectations, for example the older person's choices may not always be in line with the needs or expectations of their informal carers and vice versa. Such conflicts may lead to both elder and carer abuse, but also to a suboptimal use of respite care and other support services. This raises the issue of whose needs are being addressed when policies to support informal care are being formulated and how to optimally link measures targeting informal carers with those focusing on older people within overall LTC policies (see Chapter 3).

Volunteers are another important resource for LTC (see Chapter 11). When they start their caring tasks, volunteers in LTC often do not know the person they care for. Their main motives to engage in care are described in the literature as important for understanding volunteer's behaviour and include mainly self-development, affiliation and altruism (Bonjean *et al.*, 1994; Nelson *et al.*, 2004). In particular younger volunteers express motives of self-development (skills and capabilities) as they might be trying to launch careers. Motives of affiliation reflect the volunteers' need for desirable social contact, and motives of altruism reflect their value of helping others. Organisational commitment reflects the volunteers' loyalty to the agency's core values and goals by performing on behalf of the organisation.

Due to the shortcomings in service provision and changing family patterns, another phenomenon has grown in importance to complement and/or replace the lack of formal and informal resources. Over the past 15 years, in particular in Central and Southern Europe where neighbouring countries with a high wage differential are geographically close, migrant care workers have played a progressively important role as 'functional equivalents' to family carers (see Chapter 10). In some countries, like Italy, it is estimated that every third household with a person in need of care resorts to support from (temporary) migrant carers who are mostly middle-aged women from Central or East European countries, including New EU Member States (Di Santo and Ceruzzi, 2010; Fujisawa and Colombo, 2009).

However, in some countries where migrant workers come from African and Asian countries with a perspective to stay in the receiving country, such as in the United Kingdom, recent studies show that migrant care workers are aged between 20 and 35 years (Cangiano *et al.*, 2009).

This phenomenon is further blurring the boundaries between formal and informal care as migrant carers are often undocumented, paid but not formally employed and in most cases not at all linked to formal care providers or quality control mechanisms. There is no doubt that the vast majority of migrant carers are nevertheless providing decent care and assistance, but under general conditions that run counter to the development of an adequate identity of LTC based on dignity, mutual respect and social inclusion. In the long run the question will be whether these arrangements can be sustainable for both families and migrant carers, but due to diminishing wage differentials the extension of the phenomenon might to a certain degree only be temporary (NNA, 2009).

Professional workers in LTC

The LTC workforce is expected to increase very rapidly if it follows the extrapolations of an increasing demand. However, many countries already expect or experience recruitment and retention problems in the LTC sector. The challenge will therefore be to develop a sustainable and highly qualified workforce that can meet growing demand (Colombo *et al.*, 2011). LTC workers will be partly responsible for ensuring quality care and quality of life for many of these older adults but it remains to be seen whether health or social care workers will define themselves as 'long-term care' workers in the future.

Despite this the LTC sector is a major source for female employment in many OECD countries (Fujisawa and Colombo, 2009). In the Netherlands, every seventh woman in employment is working in the care and welfare sector (van der Windt *et al.*, 2009). Many LTC workers work part-time, especially when they work in home care settings. Although most service users live in the community, most LTC workers are employed in institutional settings. LTC workers, on average, have lower qualifications than health workers; in many countries those working in institutional care have higher qualifications than those working in home care. In general, wages in LTC are low, compared to the health care sector. In most countries LTC workers earn somewhat more than the average for low-skilled workers (Fujisawa and Colombo, 2009; Colombo *et al.*, 2011).

Despite often poor working conditions, LTC workers in many countries consider their work meaningful and rewarding and an option for growth (Barnett, 2004, p. 27; see also Kushner *et al.*, 2008):

- 'I stay because I have grown to love the work, the long-term relationships, and the satisfaction I gain from the positive impact I can make upon residents' lives.'

- 'I have been satisfied with my work in long-term care because I feel I make a difference.'
- 'I found out that if you want to make a difference in someone's life and form relationships with your patients and their families, long-term care nursing is the way to go.'

They appreciate their caring responsibilities and teamwork in institutions, giving cared-for people dignity and respect as well as a sense that they are not alone. There is also the family's satisfaction with the job done, and learning from residents' life experiences (Cangiano *et al.*, 2009). They also appreciate the flexibility of the work, but younger workers experience a lack of prospects (Hotta, 2010).

First steps towards the above-mentioned shifts in paradigms can also be retrieved by searching for (or looking at) new approaches adopted by individual professionals who try to establish a dialogue with the client and the informal carer to negotiate about what kind of care they prefer. The previously mentioned German Charter of Rights for People in Need of Long-Term Care and Assistance, and the tools developed and disseminated by My Home Life are examples of such new visions. For instance, *Buurtzorg Nederland* (see above) is a care provider organisation offering high-level care and giving as much responsibility as possible to care workers who involve relevant stakeholders in planning and implementing user-centred care. The self-managing teams are engaged to not only provide quality care, but also to promote communication across the care chain and the development of shared values within the community (Huijbers, 2011).

Discussion and conclusions

As we speak of 'emerging' LTC systems, it should be emphasised that the process of finding this system's identity and the specific construction of the meaning of 'sense' is also still under way. This can be defined as what can and cannot be expected when and by whom, and to be able to select those tasks that are part of the system and those that are not (Kneer and Nassehi, 1993). From the perspective of systems theory, identity and sense (German: *Sinn*) are central elements of what makes a social system distinct from other systems, and to understand a system, knowledge of its guiding values and sense is a precondition (Faß, 2009, 50).

As we have seen above, the values of LTC are becoming clearer, representing increasingly user-centred perspectives, aiming at aspects such as autonomy, independence, self-direction, quality of life, meaningfulness, participation and empowerment and human dignity. Moreover, the role and needs of informal carers are increasingly recognised. However, many values are still 'indeterminate', implicit and poorly implemented. This theoretical and empirical 'indetermination' (Luhmann, 1995, 131) also holds true for other stakeholders, such as government agencies dealing with LTC, commissioners, inspection agencies and umbrella organisations. The translation of these values to legislation, funding mechanisms and modes of operation of care-providing organisations and professionals is only at the beginning. On the one hand current systems and professional attitudes come from a supply-driven tradition; moreover, existing barriers need to be removed in order to implement new systems and modes of operation. On the other hand, awareness is increasing, but the – demand driven – alternatives that are in line with the new values and sense, are often not clear-cut. They need to be experimented with and new evidence needs to be collected. The INTERLINKS Framework and this chapter provide a number of new examples, but often with scarce evidence.

Incompatibilities and shortcomings in relation to the new LTC paradigms

Professional level

In the case of LTC systems, a wide range of incompatibilities and shortcomings exist, as stakeholders are innovating, seeking and negotiating appropriate types and ways of communicating and working in partnership. Common values and missions that organisations and professionals adhere to in principle can be either missing or not fully embedded in practice, for instance if care homes are not appropriately ensuring dignity in end-of-life-care (Pleschberger, 2007) or if the organisation of home care is restricted to meeting physical needs rather than enhancing self-care and autonomy. Also, if medical staff are automatically taking the lead in assessing and planning care interventions, rather than working in partnership with other professionals, users and their families, this might be an indication that they are applying the authoritarian 'quality of care' rather than 'quality of life' paradigm. It requires reflecting upon a (not yet determined) 'long-term care paradigm' which, as a starting point, would include aspects such as inter-professional collaboration, involvement of informal carers or the social and structural environment of the older person in need of care.

The lack of a shared set of values that would be able to shape strategic, organisational and individual choices in LTC is particularly visible where organisations and professionals try to explicitly address problems at the interfaces between health and social care. This is often expressed by the difficulty in creating a common language between health and social care staff. In particular, the newly emerging job profile of case managers in LTC can serve to make such observations more explicit. In this new profession, LTC values can be operationalised in terms of overcoming and integrating specific medical and/or social approaches to care, including user and carer perspectives. However, social care staff often perceive case managers as 'controllers', rather than 'supporters' of a new type of collaboration – even if they are working under the same organisational setting (Faß, 2009: 44). Practice examples have however shown their ability, among other things, to act as catalysts of integrated care – they facilitate collaboration between different health care professionals such as general practitioners, psychologists, nurses and social care practitioners as well as informal carers (Ligthart, 2006; see also Mak, 2011b; Naiditch, 2011a).

Organisational level

Ensuring that LTC values and missions are operational at the core of the caring situation remains not only an individual responsibility, but to a large extent also an organisational challenge. Many shortcomings that emerge from research and practice are organisationally driven:

- Organisational conditions and principles are running counter to the key LTC values and missions (Pleschberger, 2007);
- Trust between hospital staff and primary care and/or social care is lacking (Larsen, 2011);
- Funding of case managers has not become mainstreamed (Mak, 2011b);
- Workforce in LTC for older people are being poorly and not specifically trained (Triantafillou, 2011);

While many people recognise that values such as those addressed by the previously mentioned Charters should prevail, the features above explain why their implementation falls short. But these examples also demonstrate that the identity of LTC is emerging and developing, even if agreed values set out in Charters still do not ensure that organisational cultures are adapted accordingly in practice. The big challenge is, whereas many actors and stakeholders are involved in providing and developing LTC, they need to share the same values and sense, in order to provide optimum care and support to older people and their carers.

Policy level

At the political or systems level, the lack of specific values and the dominance of traditional quality of care and economic paradigms for LTC also creates bottlenecks and restricts reform options to both health and social care, rather than looking for solutions that allow specific characteristics and values of LTC to be appropriately addressed, such as the role of informal care, self-care, dignity and rights of older people.

For instance, in most European countries health reforms over the past two decades have introduced new types of financing mechanisms for hospitals that are based on 'diagnosis-related groups' (DRG). In general, this has resulted in hospital managements' priorities shifting towards reducing patients' length of stay with the aim of improving cost-efficiency within the health system. There is some evidence that this has been achieved, but a wide range of open questions remain to be answered to distinguish specific gains from burdens that have been shifted to primary care (Cylus and Irwin, 2010; Quentin *et al.*, 2010; Busse *et al.*, 2011) or community care. In any case, for older patients the incentives to reduce the length of stay in hospital often imply discharge with important care needs that are not always communicated adequately to their families and/or people and organisations who provide support in their communities. Hence existing services and/or family carers are left with insufficient resources to deal with any ensuing problems, which often results in unnecessary hospital readmissions (Meyer, 2007).

The introduction of 'discharge managers' in hospitals has been one reaction to such shortcomings, but similar to the function of case managers, it is doubtful whether a common language between hospital-based discharge managers, patients, their families and community care services can be easily found if this task has to be shouldered by a single person within the health system. Indeed, conventional procedures, poorly defined competences and resistance of important stakeholders are among those difficulties usually encountered by innovative projects in this area (see, for instance for Austria: BMGFJ, 2005: 164f.). Still, a number of practice examples indicate a way forward in this area (see Holdsworth, 2011c; Hammar, 2011; Wagner, 2011b).

Another example concerns quality assurance where quality is often still monitored and inspected from a quality of care perspective, rather than with a view to quality of life (European Centre, 2010; Capitma *et al.*, 2005). At the same time contracting or commissioning LTC is focusing on individual services, rather than on packages or 'chains of care' (Leichsenring *et al.*, 2010).

Therefore, in some countries efforts have been made to develop a joint vision for LTC at the systems level (Finland, UK) in order to agree upon a joint direction for service development. The question is whether or when such initiatives may really influence daily practice or remain rhetorical.

Putting an LTC identity into practice

This chapter has highlighted how some governments and organisations working in and with LTC have started to develop initiatives, projects, programmes and reforms that show the first signs of 'identity awareness' (Lazenby, 2007) with the aim of implementing LTC-specific values in their daily practice. To make further advancements in this direction we propose to systematically consider the following key issues identified within the INTERLINKS Framework to elaborate on the identity of LTC (http://interlinks. euro.centre.org).

- *To define and agree upon key principles that characterise LTC* – it is certainly a political challenge to find ways to encourage the involvement of citizens or at least a wide range of relevant organisations and stakeholders in a public debate about expectations and complex needs in the realm of LTC concerning the entire population. Existing practice examples show possible ways forward in dealing with this challenge.

- *To reflect upon the values that are shaping political, organisational and individual choices in LTC* – in particular at the organisational level it will be necessary to develop value-based delivery of care further by involving users and carers, and by opening a dialogue between them and professionals about quality of life in the context of LTC. It is even stronger if these examples come from users and carers and their organisations themselves.
- *To adequately address the rights of informal carers and/or family ethics in legislative frameworks* – in some countries, carers have been recognised by legal regulations as a specific target group with rights and entitlements.
- *To address issues of dignity, quality of life and empowerment in policy papers and organisational practices* – a wide range of initiatives have shown the opportunities and positive outcomes of new approaches to care that consider these aspects; these values are now becoming integral to policy and practice. However, as shown above, they are not easy to implement in a consistent way with staff, management, policy, commissioning and monitoring agencies, as well as service users and their carers. But repeating the message and engaging in the dialogue proves to be productive in changing attitudes and cultures.
- *To consider prevention and rehabilitation in long-term care is vital* – while prevention and rehabilitation is often neglected for older people who have been diagnosed as frail or 'dependent', many examples have shown how further deterioration can be prevented or even how allegedly unalterable illness trajectories can be reversed.
- *To embrace the diversity of users and carers (according to gender, culture and social inequalities) and support the specific needs of hard-to-reach groups* – the increasingly multicultural social structure within Europe calls for adequate strategies to take account of different cultural backgrounds and their consequences for coping with LTC. Furthermore, many older people still ignore the various types of help and support to which they should be entitled. Specific attention has to be placed therefore on initiatives that strive to guarantee equal access across health and social care.
- *To explicitly address problems at the interfaces between formal/informal and health and social care, prevention and rehabilitation or the use of migrant care workers.* This requires new operational, quality and funding instruments, as well as coordination mechanisms. However, skills and attitudes of care workers and professionals also need specific attention. Inter-professional collaboration is relatively new in a field in which professional identity is highly valued. As problems of older people are increasingly complex and require multidisciplinary approaches, this needs to become an emerging field in education.

This chapter has shown that it is the specific target group – the service users and their relatives – that drive the need for a specific identity of

LTC. Certainly, older, vulnerable people with multiple long-term needs for care and support may at certain moments also require acute health care or traditional social care, but specifically designed LTC organisations and organisational structures (such as networks, chains of care, partnerships) are able to respond to these needs in an integrated, user-oriented and coordinated way by establishing inter-professional and inter-organisational arrangements. As a corollary, further research is needed to define and identify what makes an organisation, a network or a political initiative competent for turning the LTC values into a well functioning support system for older people and their informal carers: What are the specific operational qualities, the qualities of staff, and the values and missions to make these organisational structures suited for their purpose?

The following chapters of this book will further elaborate on this question and related issues that define and illustrate the INTERLINKS Framework for LTC with the aim of showing ways to further improve policy and practice across Europe.

Part II
Overcoming the Health–Social Care Divide and the Formal–Informal Care Divide

3
User Perspectives in Long-Term Care and the Role of Informal Carers

Michel Naiditch, Judy Triantafillou, Patrizia Di Santo, Stephanie Carretero and Elisabeth Hirsch Durrett

Introduction

This chapter has the goal of re-examining and expanding the results of previous analyses conducted within the INTERLINKS project from the perspectives of two groups of users of long-term care services: first, older people who are frequently faced with difficult decisions about how to reorganise their daily lives and obtain care because of declining health and functional abilities; second, their family, friends and neighbours who form a more or less extended 'social network' of co-care providers, and who informally provide most of the required support, help and care.

The chosen focus aims to examine the place of service users within a recent long-term care (LTC) policy development context that is characterised by an emphasis on the active-client paradigm and also by the discovery or rediscovery of the role played by informal carers. In this regard, LTC reforms over the past decades have not only often disregarded their major role as care providers, but also played down or ignored the differences and potential conflicts of interest between older people and their informal carers, who may have different, or even conflicting needs and expectations. Since the well-being of carers has an immediate impact on that of the older people they care for and vice versa, policies in the field of LTC and policies for informal carers must take into account these links and these contradictions.

So the main goal of this chapter is to critically analyse the extent to which this new paradigm has been successful in achieving the reconciliation of the perspectives of older people in need of care and of informal carers.

In order to tackle this complex issue, this chapter has been structured in the following way. The introduction presents its rationale by identifying two key policy areas related to the user perspective that arose from INTERLINKS research; questions concerning older people's perspectives and informal carers' issues are investigated in turn. For both categories of users, a view of the 'state of the art' based on current literature is first presented, followed by an identification of the main issues and problems, after which the accent is placed on instruments for change in policies and service delivery, underpinned by practice examples that illustrate the INTERLINKS Framework for LTC. To articulate the perspectives of both types of users and illustrate how they are inextricably linked, but may not coincide or even be conflicting, the third section is devoted to the analysis of the impact of cash benefits measures on both users, as these measures are emblematic of the active consumer paradigm measures. Finally we use the last section to draw main conclusions and discuss recommendations regarding how to synergise LTC policies addressing older people's needs with policies to support informal carers. There are three main areas that will be addressed therefore in this chapter. Intrinsic to the discussion will be a consideration of relevant practice and policy issues in this field.

First, although actual conditions of access to and availability of LTC services vary widely across countries, it can generally be stated that the user perspective, that is the opinions, wishes and expectations of older people as well as those of informal carers, is under-represented in policy and practice. Be it at the collective or at the individual level, the voices of both type of users, and thus their priorities, seem to be insufficiently taken into account in designing, planning, managing and delivering LTC services. This statement would appear to partly contradict findings from comparative research at the macro level (Glendinning, 2010; Rostgaard, 2011), that highlight an advance of consumerism and market-driven reforms, which promote the 'active user' in contrast to the 'passive recipient' of long-term care services. One hypothesis regarding this dual judgement is that these reforms have differential impacts on both types of users; this may be because they are based on a paradigm that may only be applicable to some sub-groups of the population of vulnerable older people in need of care, or because there is a barrier between policy statements and their impact on practice.

The purpose of this first section will therefore be to consider the extent to which service provision inspired by the 'active consumer' paradigm has had an impact on empowerment of users. Additionally, the discussion will move to a consideration of how existing gaps in services for the older people in need of care can be overcome.

Second, informal carers have been 'discovered' or 'rediscovered', not only in countries where public policies rely heavily upon their availability as a basic resource to secure the financial sustainability of LTC services for older people, but also in those with a relatively easy access to formal care. Thus the main rationale for a specific policy aimed at supporting informal carers is to guarantee their long-term availability. Informal carers (mainly spouses, children and children-in-law), are not only co-providers of care but may themselves have specific needs for support, as well as specific expectations outside their caring roles (for example employment). Professionals contributing to care cannot be the only source of such support: both for frail older informal carers and for working carers, a wider range of supportive measures may be needed in order to improve the quality of the care they offer, but equally importantly to improve their own overall quality of life and participation in society. To address the issue of analysing such policies, a scheme for the classification of related support measures for informal carers was developed during the INTERLINKS project (Triantafillou *et al.*, 2010). The specific issue to be addressed in this section is to what extent the new 'active user' and consumer paradigm has been influential in the design of these specific policies to support informal carers.

With a focus on informal carers, this second section will consider the extent to which the 'active consumer' paradigm is connected to the implementation and conceptualisation of supportive measures for informal carers. In addition, there will be a consideration of the impact of the measures on carers' short- and long-term well-being.

Third, cash benefit measures can be viewed as the emblem of the consumer choice paradigm as they constitute a primary instrument for the implementation of new LTC policies inspired by this paradigm. They are expected to have a positive impact in terms of greater choice of care provider and services delivery reflected in a higher level of well-being, not only of the older person in need of care but also of the informal carers. Given the previously quoted potential conflicts of interests between users' perspectives, a detailed examination of the contradictions arising within these interests is provided.

This therefore creates an opportunity to critically review the extent to which cash benefits have fulfilled their promise of offering greater choice and enhanced participation of older users as well as to their carers in the third and final section.

Older people as users of LTC systems in Europe

This section examines the question of whether service provision inspired by the 'active consumer' paradigm has resulted in specific difficulties for older people in need of care, despite its claim of increased empowerment of users. Key issues, problems and gaps in LTC systems are first identified from the perspectives of older people themselves, followed by descriptions of the relevant 'instruments for change' aimed at overcoming these problems; relevant practice examples are drawn from those that illustrate the INTERLINKS Framework for LTC.

Response to common LTC issues

The comparative analysis of the background reports of the INTERLINKS partner countries (http://interlinks.euro.centre.org), shows that all European welfare systems are facing the challenge of adequately meeting the increasing needs of older people for LTC. Southern and Eastern European countries (Spain, Greece, Italy and Slovakia) are still a step behind in providing professional services, while even the wealthier countries, due to the economic crisis, are experiencing difficulties (Colombo *et al.*, 2011). Predictably, the countries lacking public services are those where families and other members of their immediate social network provide most of the care and are viewed as the real 'caring agency'. Yet even in Northern European countries with higher levels of public care and assistance, it must be emphasised that informal carers still provide significant levels of help and care. In Sweden, where older people have legal rights to professional care services when needed, there has been a 'rediscovery' of family care, as demonstrated by legislation mandating Swedish municipalities to support caring family members (Triantafillou *et al.*, 2010).

The paradigm change: from 'passive' to 'active'

An important paradigm change in LTC development concerns the promotion of the 'active user' in contrast to the 'passive recipient' of LTC services. Entitlement models, that posited that potential users had a right to services, but were considered as passive recipients of what LTC services authorities were able or willing to provide, have been gradually replaced by 'responsibility-centred models' (Newman and Tonkiens, 2011), founded upon the new paradigm of the 'active consumer'. However, the effectiveness of market-style reforms promoting this new paradigm must be carefully examined in terms of their impact on the older users, their informal carers and LTC professionals. This implies a radical transformation: systems in which access to and provision of social and health services was tightly controlled by professionals and local authorities become systems in which users are seen as free to 'choose'. But what does this mean and how can it be achieved? Are there hidden costs for older people and their informal carers linked to the greater responsibilities given to these 'active customers'? For instance, the introduction of cash benefits in LTC aimed at increasing older people's purchasing power comes with an expectation that they would act as 'responsible people' in choosing between different care arrangements (Arksey and Glendinning, 2007). Cash benefits will be discussed in more detail later, but it begs the question to what extent has this goal been reached? Have users truly been given realistic broader opportunities to exercise choice? If that is the case, what benefits have resulted from this enhanced choice? Has it caused harm and if so, to whom? These issues and questions are explored below.

The need for LTC may arise following a sudden health emergency, or accompany a process of longer-term deterioration of chronic conditions, which gradually erode the autonomy of the older person. While in the latter case support is built up over time, in the former case health professionals have to assume responsibility for immediate care and may have to work with 'passive' and submissive subjects, who – out of necessity – completely acquiesce to decisions taken by professionals. Patients' or relatives' informed consent to interventions, while being an essential legal prerequisite, is too often mainly a technicality, with professional expertise being dominant both in the diagnostic and treatment phases. Patients' own perceptions and expectations are frequently neglected and little consideration given to their social environment. But once the emergency situation is over and continuing LTC support is needed, the situation changes radically; the pertinence and appropriateness of professionals' interventions then becomes directly dependent on social values and on the social context in which care will take place. The involvement of the patient then becomes a goal to be pursued. It first requires a conscious attempt by professionals to move away from the 'hospital/medical cure/intervention model' to adopt a more user-oriented perspective. This necessitates continuous and sustained efforts on the part of community health and social services to 'customise' care plans by taking the social environment into account. Moreover, this perspective requires going beyond the philosophy of the user as a 'rational customer' by acknowledging that it is not sufficient simply to give information to older people in need of care; it must include recognising that an intervention exclusively aimed at satisfying the older person's spontaneous preferences may be as inappropriate as an intervention unilaterally imposed by health and social care professionals (Di Santo, 2004).

Autonomy and self-management in the field of care, in a context of declining health and reduced capacities, are strategies that do not only require the promotion of the empowerment of patients and their families, through information, assistance, monitoring and evaluation actions; adapting professional interventions to the real-life problems users are experiencing is also crucial. Services offered must be explicitly responsive to users' own opinions and expectations regarding their needs, a perspective that implies turning the situation upside down (Wilberforce *et al.*, 2011). Professionals as individuals, but also as members of care organisations, have to communicate and interact with users and with their family networks, not merely to obtain 'patient compliance' but to promote 'patient participation' by helping them to use experiential knowledge as a basis for choosing appropriate and effective care paths and solutions that fit their expectations, (Baxter and Glendinning, 2011) while being compatible with professionals' open-ended views of possible care arrangements. Promoting and supporting patient empowerment comes through interventions whose appropriateness and pertinence are based on up-to-date scientific knowledge. Yet while not neglecting their expertise, professionals have to focus on the resources of users; they must learn to help and support them in identifying their personal potential and strengthening their capacities in order to regain more autonomy in life's choices.

The necessity and effectiveness of this kind of approach, especially in LTC situations, is recognised by the international literature (Martin and Peterson, 2009) and is also reflected in the EU overview report on prevention and rehabilitation in LTC (Kümpers *et al.*, 2010). Prevention and rehabilitation processes within LTC are firmly founded upon a vision of patients being active participants in the process of developing and maintaining their health in partnership with professional advisors. More importantly, these approaches are the only ones capable of insuring a sustainable care path as they take into account patients' expectations and their life settings. Appropriate interventions thus entail 'social sustainability' as they rely on the experiential knowledge of all partners in the social network who can act to improve the patient's health and living conditions (Casale *et al.*, 2006). It is therefore necessary not only to listen to what users know about themselves but also to provide them with the means to become their own 'health experts' and to call upon and include the resources of the world in which they live (community capacity building). This approach obviously includes taking into account the place and role of informal carers and how they interact both with the older person in need of care and with professional caregivers (see next section below) and/or with volunteers.

Older people as active consumers of LTC services

Having looked at the existing gaps for older people in need of care we come now to the issue of how they can be overcome through examining some practice examples.

Including the user perspective in care planning and provision in an emphatic way entails innovative strategic orientations that 'shake up' the system in a way that initially seems to create more complexity rather than reducing it. It requires a network-oriented collaborative approach with individuals; it necessitates strengthening neighbourhood solidarity and other types of community relationships as well as inter-agency collaboration; and it also implies appropriately structuring interventions and using different modalities well evidenced by practice examples described below.

'Active assessment' and user involvement

In many countries, not only are patients regularly confronted with the length and complexity of the process leading to a much-needed multi-dimensional assessment, but they also have only a small chance of their voices and opinions being heard and taken into account regarding the design of their own care plan and care pathway (Scourfield, 2005). Even with home care, interventions taking place within the patient's home may not take into account the views of the users, leading in some cases to significant abuse of older people's human rights (Equality and Human Rights Commission, 2011). So even when the social context is a recognised part of the assessment procedure and care plan, the older person and his or her relatives are often involved right at the end of the process, when decisions about care provision have already been taken and implemented.

Instruments for change: Needs assessment processes. INTERLINKS examples included several innovative assessment procedures that demonstrate how giving users a real opportunity to express their preferences and making them active at the very beginning of the care process contributes to involving them throughout in a meaningful way. Multidimensional assessment of the older person's needs, including the social dimension, entails the use of multi-factorial instruments used jointly by different professionals. Some practice examples that illustrate the INTERLINKS Framework show how taking into account the user's needs and sharing instruments and methods with other professionals help in designing complex but coordinated care pathways (Brichtova, 2011a; Ródenas, 2011; see also Box 3.1).

Box 3.1   Single Assessment Process (United Kingdom)

The Single Assessment Process (SAP) is a UK national initiative aimed at creating a more person-centred approach to care delivery by improving the communication of common assessment information between services and providers, therefore reducing the burden of assessment on the user. Information can be shared between both health and social care services. There are four levels of assessment information that can be included: contact, overview, specialist and comprehensive. The SAP aims to reduce the assessment burden on older people through increased sharing of assessment information among professionals from different provider organisations.

Source: Holdsworth (2011a).

Instruments for change: Stimulating users' resources.   Older people's well-being can be effectively improved by developing their social networks. User-centred approaches imply working with support networks and fostering community solidarity. A first prerequisite for users to take active decisions about their care pathways is the need for professionals to accurately inform and orient them without overriding their opinions. Second, new types of interventions are needed to strengthen links between LTC systems and the community, to promote solidarity through developing connections and information exchange and to mobilise informal local resources. Some practice examples illustrating the INTERLINKS Framework are based on the activation of the family, social and community networks that contribute to increasing the patient's autonomy and well-being through empowerment and strengthening neighbourhood social connections (Mak, 2011a; Repetti, 2011).

Instruments for change: Case managers.   Users attempting to manage their own care need the advice of professionals who are able to orient users' choices while fostering the expression of individual resources in terms of autonomy. Information, assistance and orientation are key words used in all INTERLINKS National Reports, as fundamental elements for an LTC system to be effective and inclusive and to be able to make use of patients' skills and resources (Reilly *et al.*, 2009). Case managers may have different roles and intervene at different stages of the care process. They may act as a member of a care team delivering direct care for a specified number of patients or act as an advocate or broker for the older person on the basis of an elaborated care plan, choosing, organising and coordinating the different professional services. In addition, they may act as users' advocates during the assessment and care plan phase; on the basis of a defined care plan, they may help older people and their families to negotiate a consensus on decisions taken by the professionals. Disagreements about what professionals have planned and the older people's personal strategies and/or the informal carer's agenda may arise. Case managers may have to advocate for the older users' autonomy and help them be 'autonomy-capable' (Sen, 1999), while at the same time taking into account the possibly conflicting views of informal carers. The INTERLINKS Framework contains a number of key issues and practice examples concerning the functions and roles of case managers (see Hammar, 2011; Larsen, 2011; Naiditch, 2011a; Sedlakova, 2011; see also Box 3.2).

Box 3.2   Case managers for people with dementia and their informal caregivers (The Netherlands)

People with dementia and their caregivers may not know how to deal with the impact of dementia or how to find appropriate help. Informal caregivers need additional information and advice, e.g. about how to cope with their relatives' behavioural problems and about the development of the illness trajectory. From 2004 until 2008, an experiment was initiated by the National Dementia Programme (LDP) with 16 pilot regions in the Netherlands, where cooperation in community dementia care would be put into practice and case managers were introduced to coordinate the care for dementia clients. By the middle of 2010, the experiment had been rolled out in 57 regions. The implementation of case managers was the most important consequence of this experiment. The case manager sets up a diagnosis trajectory, coordinates the care provided and informs and advises the client and caregiver, in accordance with their specific needs.

Source: Mak (2011b).

Older people's difficulties in accessing LTC services

Here, it is important to consider the major difficulties older people encounter when accessing LTC services. This section will review how the complexity of LTC systems creates fragmentation to the detriment of care.

Fragmentation and complexity of LTC systems

Whether as passive patients or active consumers of care services, a common characteristic identified in all participating countries was the lack of awareness on the part of professionals, whether at policy, planning, management or delivery levels, regarding the practical problems encountered by older users when confronted with the complexity of LTC systems. However, when care systems are looked at through the eyes and experiences of potential system users, the causes of these difficulties become clear: lack of organisational transparency in LTC systems, fragmentation of facilities and services provided by different organisations and professionals, rigid divisions of labour and competences between professionals and departments, wide geographical separation between service settings, as well as a range of other obstacles hampering the ability of users to comprehend and negotiate the system. Patients and potential service users, with little or no information and support (Baxter and Glendinning, 2011), may have to find their own pathways to accessing services, relying on their own competences as well as on the opinions and advice of other users or relatives. Access to services was identified as a sub-theme in the pathways and processes theme of the INTERLINKS Framework, with two key issues specifically referring to the importance of good practice in the transfer of information to users and carers and between services or agencies, giving priority to the older person's and carer's interests and involvement, and considering their rights.

The high degree of professional integration needed to overcome the fragmentation in LTC services, with its associated problems and barriers at all levels for older users in need of care, was acknowledged as an important difficulty in almost all European welfare systems analysed by INTERLINKS researchers. In Central and Northern Europe these difficulties are minimised to a certain extent because welfare systems are mature, LTC services are more accessible, widely available and comprehensive; in Mediterranean countries however, these challenges are widespread and pervasive at all levels (Lamura *et al.*, 2007a; Larsen, 2011).

Instruments for change: Overcoming fragmentation at the system level. In order to confront the 'fragmentation' issue, international literature points to two main paths: one is prevention and rehabilitation and the other is health and social care coordination, at least at the provider level. Such cooperation may even lead to integration between hospital, residential and community services, between social and health services, and between public and private resources, in order to build the 'service chain' that is indispensable for achieving continuity of care.

Many INTERLINKS practice examples specifically concern actions at the governance level and highlight how various types of division may be bridged, including health/social, and formal/informal care. Examples also point to mechanisms for joint, integrated multidimensional assessment within the hospital; they describe ways to transmit information and set up care arrangements with all services and facilities involved as a necessary condition for guaranteeing continuous care paths.

All the examples, despite differences stemming from different national contexts and types of welfare systems, pinpoint the necessity of implementing some standardised procedures and written agreements between hospital and home care organisations, as well as between different home care agencies active within the same area, so that responsibilities may be defined, practices spelled out and support instruments specified. In many countries as different as Finland, France and Switzerland, practice examples also describe several experimental organisational models and procedures to facilitate older people's long-term care pathways from home to hospital and from hospital to home, with the aim of improving management of the care path. These procedures are based on shared, personalised care plans described in a written document that can be consulted by all actors involved in care planning and decisions, sometimes with the help of technology (see also Gobet, 2011; Hammar, 2011; Naiditch, 2011a; see also Box 3.3).

Box 3.3   Local governance to bridge institutional and professional gaps: Dementia guidelines and informal carers (Sweden)

The first Swedish guidelines regarding dementia care were published by the National Board of Health and Welfare (NBHW) in 2010, presenting evidence-based and evaluated treatments and methods within health care and social care for people with dementia and informal carers. The guidelines are first directed at municipalities (e.g. social care, home help, etc.) and primary care organisations managed by the county councils and aim to provide support to decision-makers within municipalities, county councils and regions, so they can govern the different organisations according to systematic priorities. Other aspects of the guidelines address doctors, nurses and other health and social care professionals on a managerial level. This example shows that it is possible to overcome institutional and professional barriers between different organisational bodies through policy instruments that focus on putting the patient's actual needs at the centre of interventions.

Source: Emilsson (2011).

Instruments for change: Information and counselling for older people and their carers.   In order to foster the appropriateness of care paths and efficient service use, older people and their informal carers must first be given all useful and relevant information in an appropriate way and receive the same coherent and complete information at all access points regarding the kind of care facilities and services they can use. This enables them to better understand procedures before being confronted with them (see Box 3.4).

Box 3.4   Integrated access point for older people (Italy)

In 2006, the regional government of the Veneto region in northern Italy ordered each local health authority (LHA) to set up an Integrated Access Point in every social and health district to facilitate equal access for older people to local health and social services (hospital, home care, residential facilities, etc.), within and across organisations in an inter-professional setting, with a further objective being to ensure continuity of care, i.e. appropriately planned discharge from hospital and integrated home care services. Information and assistance are provided to users about care pathways and by centralising requests for care and support. Establishing an Integrated Access Point requires a new organisational structure that enables staff to activate resources and available services directly. Users are not seen as passive recipients of care, but the goal is to help frail older people and their families exercise their own rights as active and participative clients. Currently, 17 out of the 21 Italian regions have included an Integrated Access Point in regional planning.

Source: Ceruzzi (2011).

Instruments for change: Overcoming inequalities and sustaining diversity. Inequalities in health and access to services, due to social and economic conditions as well as gender and country of origin, mean that disadvantaged older people are not only less healthy but also have less access to services that are needed (Bracci and Cardamone, 2005; Di Santo, 2010). The financial crisis, particularly in Southern European countries (Greece, Italy and Spain) has implications for even further reductions in service provision, leading to increasing inequalities.

It is extremely important for LTC systems to include strategies and policies addressing issues of diversity and equal access; differences in gender, culture as well as social inequalities must be taken into account in order to improve their inclusiveness and universalism. Welfare associations and migrant organisations must be involved alongside other local stakeholders in the participative planning of communication instruments aimed at facilitating access to services by more disadvantaged users, such as the German Forum for Culture-Sensitive Care in Old Age (Stiehr, 2011d).

Discharge management and interdisciplinary work

Current hospital discharge processes also present grave difficulties for older people. Hospital length of stay is decreasing in all European countries, with earlier discharge being frequently poorly managed, resulting in inadequate collaboration/coordination between hospital staff, community services, nursing care agencies, GPs and social workers. Moreover, informal carers are often not formally acknowledged, nor provided in time with the appropriate information necessary to participate in planning and coordinating the home care. Thus in most cases, it becomes the users' responsibility, usually with little support, to constitute their own personal 'chain/network' (Le Bihan and Martin, 2010).

In all INTERLINKS National Reports, hospital discharge is considered a fundamental element for an LTC system to guarantee continuity of care for frail older people. Returning home after hospitalisation can be hard for frail users and extra home care is usually needed. In order to assure continuity of care, it is necessary for hospital professionals to assess the situation together with users and with community professionals. Differences between LTC systems across Europe exist, but INTERLINKS national reports identified a common lack of communication and coordination between hospitals and primary care professionals, leading to poor continuity of care; many practice examples do however show creative attempts to deal with this issue. Services and facilities often depend on different bodies or institutions, thus requiring a stronger coordination between institutions. Transfer of information between different providers can be a problem in assessment procedures. Health care information and appointments are frequently lost in the transition from one agency to another and patients may have to spend more time in hospital than necessary, because coordinating care with the home care services after discharge is slowed by poor communication.

Instruments for change: Improving and integrating health and social care between hospital and home.   Addressing these gaps in the discharge process requires clear lines of communication and teamwork between hospital, home health and LTC providers, including informal carers, GPs, nurses, occupational therapists, physiotherapists, home helps and social workers. Areas of responsibility in patient-centred care procedures must be established in order to eliminate possible gaps in cover and include informal carers in the care team (see also next section on informal carers as co-care providers). Technology plays an increasingly important role in facilitating this process (Gobet, 2011; Naiditch, 2011b; see also Box 3.5).

Box 3.5   Using electronic communication between hospital and municipality: SAM:BO (Denmark)

Targeting frail older people, who receive long-term care services from the municipality or need long-term care after discharge, all 22 municipalities of the region of Southern Denmark cooperate in the SAM:BO project, which aims to shorten the time lag in communication between hospitals and municipalities especially at discharge. Using a commonly agreed electronic tool, the objectives are to:

- Ensure seamless transfer of the older person from hospital to the community;
- Enhance patient satisfaction with health care services;
- Strengthen cooperation between general practice in the municipalities and the hospital;
- Ensure dialogue and coordination between the parties and with the widest possible involvement of patients and relatives;
- Ensure patients and their families participate actively in planning their care.

Nurses from the hospitals and the municipal home care teams are responsible for adhering to the agreement in practice and therapists working in the hospitals and in the municipalities coordinate the older person's rehabilitation by using an electronic care plan. The agreement helps ensure that doctors at the hospitals and general practitioners in the community coordinate the patient's treatment using electronic discharge summaries. Thus SAM :Bo addresses both the means and content of communication between long-term care actors and patients and their families are also enabled to contribute to planning their care.

Source: Møller Andersen (2011).

Informal carers as co-providers and users

As outlined in the introduction, in the context of the 'active user' paradigm, this section aims to examine the rationale behind the implementation of supportive measures for informal carers. After identifying the challenges of informal carers' experiences that arise from the caring role, related measures to address their needs are described. A specific scheme to classify these measures is outlined which attempts to shed light on how their conceptualisation is closely related to the new choice paradigm. This presentation is followed by instruments for change illustrated through practice examples, which describe the extent to which these measures impact on informal carers' well-being.

Informal care provision as a crucial resource for LTC

Informal carers are a dominant resource as co-providers of care for all European LTC systems. Despite problems of data comparability between countries, including definitions of what constitutes an 'informal carer' and of the tasks they undertake, national and European reports consistently show that they constitute the principal source of help and care provision for older people with long-term care needs, representing on average 80% of the total support delivered, whether estimated in kind or in monetary terms (Van den Bergh *et al.*, 2007). According to data coming from SHARE (a prospective representative cohort of the over-50-year-old population in 14 European countries), their contribution largely surpasses that of formal carers even for countries such as Sweden, Denmark, Finland, Norway and the Netherlands, where formal services are widely accessible. This (apparent) contradiction with data coming from the latest Organisation for Economic Co-operation and Development (OECD) report on LTC (Colombo *et al.*, 2011) can be explained by the very broad definition in this data set regarding the type of care and help provided by informal carers, encompassing moral and psychological support, as well as 'keeping company' and activities promoting social inclusion. This definition goes well beyond the predominantly Instrumental Activities of Daily Living (IADL) and Activities of Daily Living (ADL) related tasks provided by formal services that absorb the bulk of public funding and form the basis for most cross-country comparisons.

SHARE data show that the proportion of care, help and support provided by informal carers varies between countries, not only in terms of intensity of tasks, number of hours per day and overall duration, but also in the type of services provided. In countries where the availability of formal services is poor, informal carers are compelled to provide most of the IADL and ADL-related tasks, while in those where formal services are easily available, informal carers can choose to provide less 'hands-on care', as such services are professionally provided. Nevertheless, it is worth noting that the availability of formal services does not lead to a withdrawal of informal care, as demonstrated also in the Colombo report, but rather leads to a shift in the tasks they undertake. These well-evidenced facts lead to important conclusions regarding support policies and measures for informal carers. It is now clear that providing formal services does not 'crowd out', discourage or exclude informal carers from delivering care. On the contrary, by offering real choices, formal services transform the type of care they provide, so that public solidarity does not replace private solidarity but each complements the other (Bonsang, 2009).

Informal care provision and its impact on carer's well-being

Impact of caring on carer's health status

The positive value of caring is reflected in carers' reported primary motivations for providing care, these being mainly emotional bonds of love and affection, followed by a sense of duty or obligation and with only a very small proportion reporting no other alternative than to care (EUROFAMCARE, 2006). This correlates with statistical analysis using SHARE data showing that as long as the overall mix of caring intensity, duration, frequency and type is below a certain threshold, caring, even when combined with working, has a positive effect on the health and well-being of carers (Fontaine *et al.*, 2009). Above this threshold, however, the provision of informal care can have adverse social, physical and mental health effects (EUROCARERS, 2009; Lamura *et al.*, 2008), making it a long-lasting, stressful life event (Zarit, 1998; 2002).

The stress of caring can affect quality of life; UK and Swedish carers, who are relatively well supported by services, report better quality of life than carers in the Mediterranean countries of Greece and Italy, where the level of formal service provision is lower (Lamura *et al.*, 2008). Other factors predictive of a negative impact on the mental health of the caregiver include: greater responsibility for performing care duties, such as the number of weekly hours of care provided (more than 35 hours) (Colombo *et al.*, 2011); the degree of impairment of the care recipient – both functional limitations and mental problems (Gramain and Malavolti, 2004); socio-demographic factors (more advanced age, female gender, lower level of education), the context of the care (co-residence) and the availability of informal and formal support (Garcés *et al.*, 2010; Mestheneos and Triantafillou, 2005). Caregiver stress, through its impact on mental and physical health of the care provider, may also have effects on the recipient of care. In this regard the main consequences documented in the literature are various types of abuse towards the dependent person (Buyck *et al.*, 2011; Carretero *et al.*, 2009; McGuire and Fulmer, 1997).

Caring: social and economic risks

According to Colombo *et al.* (2011), caring enhances short- and long-term risks of poverty for both working and non-working carers. Data show that a disproportionate number of lower income households provide informal care to their older relatives and poverty risk is higher for intensive compared to non-intensive carers, as they tend to have lower incomes. Socio-economic status affects both caregiving and labour market outcomes (Lilly *et al.*, 2007). Socially disadvantaged families are more likely to engage in caregiving rather than working as they have a lower level of employability, and as a result, they end up trapped in caring roles with even fewer labour market opportunities and thus find themselves at higher risk of poverty. This risk of poverty also applies to working carers since they work fewer hours on average than non-carers (two hours less per week) and because they tend to be over-represented in part-time work (Colombo *et al.*, 2011). This difference exists in all countries except in Northern Europe and the Netherlands; a greater reduction in working hours is found in Southern Europe than in Central Europe. The marginal effect of hours of care on hours of work is strong: a one per cent increase in hours of care translates, on average, into slightly more than a one per cent decrease in hours of work, this effect being stronger in Mediterranean countries than in Northern European ones (Colombo *et al.*, 2011).

Thus, the thresholds (with regard to amount, intensity and duration of care) at which the withdrawal from the labour market and/or care burden begin to take place are variable according to country: lower in Eastern and Southern countries; higher in Northern ones and in the Netherlands while the situation for France, UK and Germany can be described as intermediate. Care activities thus have an impact on the carer's career continuity and employment choices: available data show that among a group of carers of elderly dependants, 17 per cent reported that their career advancement had been harmed by their care responsibilities and in the United Kingdom, carers are 30 per cent more likely to hold a temporary job (Colombo *et al.*, 2011).

Finally, long-term lower accrued pension entitlements resulting from a fragmented working career must be added to the poverty risk due to reduced income in the short term. The higher risk of poverty of the informal carer can also be deduced from the reported financial problems caused by caring. The EUROFAMCARE study found that the cost of medications, extra travel, out of pocket payments for health care, adaptations to the home and special food feature among the additional costs of care (Tjadens and Colombo, 2011), and similar data have been found in other studies (Bolin *et al.*, 2008b; Family Caregiver Alliance, 2003; Grunfeld *et al.*, 2004). The implementation of measures intended to combat poverty as well as enabling carers still in the labour market to continue to care are crucial elements of supportive policy.

The care/work conflict and gender

Women carers provide on average twice as many care hours as men, and women tend overwhelmingly to provide personal care, while men concentrate on administrative and logistic tasks, which require less involvement. The burden of care thus falls disproportionately on women (Viitanen, 2007). It has been demonstrated that caring women work on average fewer hours in the labour market than their male peers and, even for those with a limited number of care tasks, their probability of working is lower than for men in the same situation (Viitanen, 2005; 2007). This poses a problem since the age at which women become carers often coincides with the period when their children and grandchildren need their support (Esping-Andersen, 1997; Saraceno and Keck, 2008). This is what Rosenthal (1997) has called the 'sandwich generation', with women being caught between paid work and a double care duty and is directly related to the higher rates of poverty among informal carers. The Lisbon target of enhancing employment rates, especially for women and senior workers (55–64 years), who are the main informal care providers, clearly necessitated the implementation of measures to allow female carers in particular to gain access to or to remain in the labour market, enabling an effective care/work reconciliation policy (European Commission, 2008b).

Governing informal care

Having emphasised the major role of informal carers in the provision of services in all countries as well as identifying the related challenges it creates, it is important to consider reasons for the general lack of specific policies for informal carers. European countries demonstrate a wide variation in how responsibilities for the care of older people with LTC needs are shared between the family, the state, the private and the non-governmental organisation (NGO) sectors. This partly depends on the different welfare state regimes, and specifically on whether the family or the state is given primary legal responsibility for undertaking their financial and material support when needed (Triantafillou *et al.*, 2010). So in countries where care provided informally had been traditionally considered a private affair, informal carers' long-range persistent 'social invisibility' came from the fact that this issue was not considered as necessitating a public debate or requiring regulation at the political level. Indeed informal carers' perspectives on care provision itself indirectly contributed to foster this vision as caring is considered to be a highly valued responsibility.

This general willingness of informal carers to undertake responsibility for care (Eurobarometer, 2007) continues to be inappropriately utilised by decision-makers, who appear to consider them more as a convenient resource than as a group with specific needs. In many cases, the presence and availability of an informal carer has the impact of rationing or curtailing the formal service provider's care tasks in areas where informal carers are supposed to be able to provide care. In Germany for example, the low initial level of funding of the LTC insurance fund was based on the implicit assumption that informal carers would fill the gaps in formal care provision (Triantafillou *et al.*, 2010). In the case of countries (such as Sweden or the Netherlands) where insuring care provision is the responsibility of the state, the social invisibility of carers came from the fact that they were considered as marginal actors when compared to professional carers. In many countries informal carers' voices as stakeholders still remain weak although there exist noticeable exceptions in the Netherlands and UK, where the lobbying power of carers' associations as a collective voice is strong.

The increasing demand for services, together with predictions of the decreasing availability of informal carers have led to the 'rediscovery' of the potential role of families (Da Roit and Le Bihan, 2010); this holds both for countries lacking formal services, as well as for those with a strong network of professional services, which are now being restricted due to the financial crisis. This fear of the unsustainability of informal care networks is linked to ongoing social and cultural changes such as changes in family dynamics and increases in female participation in the labour market, factors that are likely to lead to sharp reductions in the number of relatives willing and able to provide high levels of hands-on care. At the same time, a mix of demographic and epidemiological factors lead to projections showing a growing gap between supply and demand of informal care (Karlsson *et al.*, 2004; Pickard, 2008; Robine *et al.*, 2007). These factors threaten the financial sustainability of LTC budgets, which is especially important in countries where professional services are poorly developed. However, even in countries such as Sweden where the state via the municipalities assumes primary responsibility for LTC according to need, the rising demand for care and the economic downturn have led to a 'rediscovery of family care' (Da Roit and Le Bihan 2010).

Supporting informal carers within LTC policy for older people

As the intensity of the care provided informally already threatens the physical and mental health as well as the socio-economic status of informal carers and thus their long-term availability, most of the proposed measures have the clear aim of securing the continued provision of informal care as a resource for LTC of older people, whether informal carers are on the labour market or not.

When examining the impact of such measures and policies, one main issue is to clarify whether they aim primarily at supporting the informal carer or whether their objective is mainly to ensure 'informal support' of the older person, neglecting the potential drawbacks for the informal carer. When this question was examined during the INTERLINKS project, a potentially problematic issue was identified in that, in many countries, the needs and expectations of informal carers tended to be inappropriately conflated with those of the person they are caring for (Moran *et al.*, 2011). This view relied on the 'natural harmony' that is supposed to exist between the two stakeholders, underpinned by the belief that care relies on altruism and/or reciprocity, provided to parents in repayment for support provided earlier in life. So an assumption is made that choices made by the older person would be in line with the carer's preferences, thus neglecting the complex and frequently conflicting dynamics of the choice process (Glendinning and Kemp, 2008).

The issue of whose needs are or should be targeted in LTC policies has to be clearly addressed. To clarify this complex issue, a scheme for classification of informal carers' support policies and measures was developed during the INTERLINKS project, with the objective of providing a structured method for better assessing their effectiveness At the policy level, this classification can also be used to describe and compare the effectiveness of informal carers' support policies across European countries and assess how these specific policies fit into overall LTC policy addressing older people's needs (Naiditch, 2012).

An analytical scheme for classifying measures relating to informal carer support policies

Broadening the scope of previous work by Glendinning (Eurocarers, 2010), a bottom-up analytical approach was used to develop a scheme for the classification of the various measures designed to support informal carers based on the following principles. Measures (whether in kind or in cash) are first divided into two broad categories:

- *Specific measures* that intend to respond only to informal carers' needs, and
- *Non-specific measures* that address the needs of both informal carers and the older people they care for.

In each case, these measures are then subcategorised as *direct* or *indirect* as follows (see Tables 3.1 and 3.2 for examples):

- *Specific direct measures* are those that help informal carers to *perform their caring tasks*, such as training them in caring techniques; but also training formal carers in how to include and support informal carers in a shared provision of care;
- *Specific indirect measures* are those that enhance the opportunity for informal carers to care by making the *conditions under which they care* more favourable, such as care allowance for all types of carers; care leave; flexible working arrangements for working carers.
- *Non-specific direct measures* are those that primarily target informal carers, such as respite care
- *Non-specific indirect measures* are those that primarily target the older person, such as attendence allowance and all professional services provided to older people

This scheme is used in Table 3.3 (see p. 65f.) to classify the examples of informal carer support measures described in the following sections.

As shown in Table 3.3, different types of action encompassing various policy levels need to be mobilised to respond to informal carers' needs. This section will now continue by focusing on measures in kind and analysing

Table 3.1 Specific measures for the support of informal carers

Specific measures	Examples
Direct (supporting and improving informal care provision/delivery)	*In kind*: Information, training, education, opportunities for the exchange of specific care provision experiences; training formal carers in how to include and support informal carers in a shared provision of care
Indirect (improving the context of informal care)	*In kind*: Legislative recognition of carers' rights, e.g. to an assessment of their own needs; support for lobbying and advocacy activities; flexible working arrangements, care leave, pension and accident insurance
	In cash: care allowance

Table 3.2   Non-specific measures for the support of informal carers and older people

Non-specific measures	Examples
Direct (primarily targeting informal carers)	Respite care, support groups and stress relief through voluntary work initiatives, etc.
Indirect (primarily targeting older people)	*In kind*: All in-home and residential care services for older people, housing accommodation and adaptation, meals on wheels, technical supplies, etc.
	In cash: attendance allowance

them with regard to the choice paradigm, as the in-kind measures relate conceptually to the specific arguments previously discussed.

Instruments for change: Legislative recognition of individual carers' rights (specific indirect measures).   The acknowledgement in legislation of the valuable contribution of informal carers is one of the most important dimensions in promoting explicit recognition of informal care in society. This recognition should apply at the micro level where informal carers' own needs should be taken into consideration when setting up the older person's care plan (see Box 3.6).

Box 3.6   Informal carers' rights to an assessment of their own needs (United Kingdom)

Legislation in the past two decades has recognised the contribution of informal carers and has acknowledged their individual needs. Around 6.8 million adults in Britain are informal carers (referred to as 'carers') and approximately three-quarters of them care for older people. As the population ages, carers may need to provide even more support for longer periods. Their right to an assessment of their needs is the subject of the Carers (Recognition and Services) 1995 Act, Carers and Disabled Children Act 2000 and Carers (Equal Opportunities) Act 2004, covering carers over the age of 16 in England and Wales. The 1995 Act was the first piece of UK legislation to fully recognise the role of carers and the 1995 Act introduced the concept of a carer's assessment, since expanded in the subsequent Acts to give carers the right to an assessment and to give them opportunities equal to those of non-carers in recognition that caring is not a legal obligation. Carers benefit from an entitlement to an individual needs assessment separate to the cared-for person, which means they are also allowed to personally receive services or cash benefits directly from local authorities. This legislation has manifested itself in a range of linked initiatives to support carers which vary across England and Wales, such as increased funding for short-term breaks, emergency support services for carers, and day and respite care for the cared-for person.

Source: Holdsworth (2011e).

Table 3.3 Analytical scheme for classifying measures related to informal carers' support policies and the country in which the example is implemented

Type of support		Measures	Example	Country
Specific direct (*help in performing caring tasks*)	**In kind**			
	Cognitive	Information, advice, counselling	WeDo (Luherne, 2011)	EU
		Training for informal carers	Elderly Care Vocational Skill Building and Certification (Triantafillou, 2011)	EL
		Training for formal carers in how to include and support informal carers in a shared provision of care		
	Emotional, psychological	By professionals Through peer groups	1) Meeting centres for people with dementia and informal carers (Mak, 2011c)	NL
	Social	Recreation/Alzheimer Café	2) Athens Association of Alzheimer Disease and Related Disorders (Kagialaris, 2011b)	EL
	Health	Check-up/medical visit Healthy ageing programme		
Specific indirect (*support and facilitate the caring option*)	Carers' needs recognition	Carers' needs are explicitly assessed during the assessment of the older person in need of care	1) Informal carers' rights to an assessment of needs (Holdsworth, 2011e)	UK
			2) Municipal LTC obligations to support informal carers (Emilsson, 2011a)	SE
	Legislative: Work recognition as carers	Pension rights	Social protection of informal carers (Repková, 2011b)	SK
		Social security benefits (health/ disability pension/sickness/ unemployment/work accidents)		

(continued)

Table 3.3 Continued

Type of support		Measures	Example	Country
	Political recognition	Advocacy groups	'We Care' (Stiehr, 2011b)	DE
	Labour market benefits	Work leave: a) Paid b) Unpaid Flexible work arrangement a) Formal b) Informal	Care Leave Act (Mingot, 2011)	DE
	In cash	Care allowance as: a) Maintenance b) Formal recognition c) Substitutive for formal care	Direct payments for informal carers (Holdsworth, 2011c)	UK
Non-specific direct (*primarily informal carers*)	Respite	Short stay (nursing homes) Day care Home custody (day/ night/24 hours/weekend)	Respite care platform (Naiditch, 2011c)	FR
Non-specific indirect (*primarily older people*)	In cash	Attendance allowance Personal budget Tax exemptions Vouchers		
	In kind	All types of professional home care and residential care services for older people Housing adaptation ICT: simple monitoring, complex monitoring	Integrated access point for older people (Ceruzzi, 2011)	IT

Source: Triantafillou *et al.*, 2010; Naiditch, 2012; http://interlinks.euro.centre.org. Country abbreviations: DE - Germany; EL - Greece; EU - European Union; FR - France; IT - Italy; NL - The Netherlands; SE - Sweden; SK - Slovakia; UK - United Kingdom.

Instruments for change: Enhancing NGO's roles and contribution (specific indirect measures). The visibility of carers' issues in formal policy guidelines and at government legislation level is extremely variable across Europe. Carers' organisations and other NGOs have been pioneers, both nationally and at the European level (Eurocarers; AGE Platform), in raising the issues of effective support of informal carers and the needs of the older people they care for, in particular where state services are insufficient. By providing a collective 'voice' and by lobbying for carers' rights, in the UK for example, they were influential in promoting the UK Equalities Act (October 2010), which considers working carers as an 'at risk group for employment discrimination'. Alzheimer associations have also frequently been successful in implementing cross-sector collaboration to run programmes, which often rely heavily on volunteer support (Kagialaris, 2011b). This is especially important since the behavioural problems associated with Alzheimer's disease are one of the major causes of high levels of stress and burden for carers. The example below (Box 3.7) aims to enhance the quality of the caring environment by giving legal rights to informal carers.

Box 3.7 'We Care' – Representative body of informal carers, relatives and friends (Germany)

'We Care' is a network that aims to represent the interests of informal carers and give them a voice in political decision-making. It is a registered association at national level and organised within regional and local groups, with members who are also professionals and advise and support the requests of informal caregivers. 'We Care' aims to have a systematic impact with regard to the position of informal carers, who are the most important care providers, but often lack more formal supportive conditions concerning recognition of their engagement in the caring process. 'We Care' acts as an interface by disseminating information about existing types of support for informal carers and by directing recommendations for improvement towards political decision-makers. The example demonstrates how a bottom-up initiative of informal carers can operate and achieve results, though under difficult conditions, and represent carers' issues in policymaking.

Source: Stiehr (2011b).

Supporting and improving informal care provision

Measures such as provision of information for informal carers, training in caring techniques and information on how to take care of their own physical and mental health are still scarce or insufficient in most countries, with a corresponding lack of data on impact on both end users of such services. Few countries are engaged in serious efforts to train professional staff in how to systematically assess the needs of and provide support to informal carers or in how to bridge existing care gaps between formal and informal carers. Nevertheless some of the following practice examples highlight projects in which such measures have been implemented. It is also worth noting that many of these measures specifically target informal carers in the field of support to Alzheimer patients, while informal carers supporting severely physically disabled or gravely ill older people with high care needs are rarely defined as target populations. Improving the quality of both informal and formal care provision was identified as a major goal of the INTERLINKS project. For informal carers, programmes of training and education aim to increase their ability to care appropriately and to protect their own physical and mental health. For formal caregivers, programmes that link them with informal caregivers in a shared provision of care can also provide informal carers with opportunities for exchange of information and support.

*Instruments for change: Training of formal and informal carers
(specific direct measure; Box 3.8)*

Box 3.8   Elderly Care Vocational Skill Building and Certification (ECVC) (Greece)

The ECVC project is an e-training programme for caregivers of older people in Cyprus, Greece, Lithuania, Spain and Hungary. It was developed to address the needs of the elderly care sector in Europe for vocationally skilled workers by means of a common level of competencies and certified recognition by the partners, who provide education, which could lead to an accepted EU-level qualification. The Greek programme consists of theoretical and practical training of the trainee care workers in a training institute and two selected residential care homes in the region of Attiki, together with the processes of assessment and certification. The e-training software can also be used by informal caregivers and migrant caregivers. The main benefit for informal caregivers is that training can help reduce the negative consequences of caregiving on their physical and mental health, as well as improving the quality of care they are able to provide. The project demonstrates that the use of computer-based self-training methods can contribute to providing the means to achieve common education and care standards, resulting in higher quality of care, in countries with different care systems.

Source: Triantafillou (2011).

Instruments for change: Combatting the negative impact on carers' health status. The design and uptake levels of in-kind non-specific supportive measures indicate potential conflicts between older people and their informal carers regarding issues of choice. A number of examples that illustrate the INTERLINKS Framework describe practice that overcomes this problem. These include meeting centres in the Netherlands, for example, for people with dementia that consider the needs of informal carers (Mak, 2011c), and respite care (see Box 3.9).

Box 3.9 Respite care platform (France)

A national model of innovative respite care platforms (RCPT) was piloted and tested at 11 sites. The platform aims to provide information and coordinate the provision of a broad set of respite care services for carers of people suffering from Alzheimer's disease, including short stays in nursing homes, day care, home care, day/night 24hr/weekend care, short holidays, recreation, Alzheimer cafés and cultural events. The objective is to efficiently address a range of carers' needs while respecting the needs of the person being cared for and by ensuring easy geographical access and financial equity.

The main benefits for the carers are that their voice is being heard by a competent professional and that they are able to meet and exchange experiences with people in a similar situation, thus reducing isolation and providing continuity of follow-up. Evaluations show that RCPT should focus on the key needs expressed by the main carers themselves in order to achieve the best balance between the carers' expectations and the older persons' needs. This balance appears to be a key element and represents a major challenge in achieving an effective and fruitful use of respite care services. Important issues include a project design built around a single as opposed to multiple local entry points and carefully planned baskets of minimum services to be offered.

Source: Naiditch (2011c).

Instruments for change: Reconciling work and care (specific indirect measures). Following European initiatives from the Council of Europe and/or the European Commission (European Commission, 2006; 2008b; 2011b) some legislation and regulatory documents in the field of care/work reconciliation have been issued at the national political level. They focus on incentives for employers to hire or retain working carers and/or to implement specific care/work reconciliation measures. Associated measures aim at reinforcing the usual 'informal care arrangements' based (when available) on a 'social network' of family carers, friends, neighbours and/or volunteers, which may be effective in the short term, but may lack long-term stability

(Le Bihan and Martin, 2010). In order to fulfil informal carers' complex needs and be compatible with outside employment, support networks must be underpinned by specific as well as non-specific policy measures (Eurostat, 2009), classified as follows according to the analytical scheme and examples illustrating the INTERLINKS Framework:

- *Measures focused on informal carers' roles as workers*: *Specific indirect* measures relate to *schedule and working time issues* (reduction of working hours, adaptable work schedules; opportunities for part-time employment) (Mingot, 2011); other measures concern *adaptations to the working environment or to the job itself* (occupational health; 'well-being at work' measures).
- Measures focused on their role as informal carers; Specific indirect measures include provisions such as care leave (short or more prolonged, with or without pay or benefits compensation, as well as specific direct and non-specific care-related support through information, counselling, practical support for the informal carer through facilitated access to and/ or actual provision of formal services (see Mingot, 2011).
- *Contextual measures* aiming to improve the 'labour context' of carers and in particular to increase the uptake by carers of all types of support measures; this includes measures to increase awareness of care issues and to promote the development of skills among managers and the workforce.

While sufficient data is not yet available, it can safely be stated that the question of care/work reconciliation has been given insufficient attention in many countries, whether by political decision-makers, managers or employers. The implementation of measures in this field is hampered by many managers' strong prejudices, with only a minority of managers believing that a 'carer-friendly policy' will bring positive results (Employers for Carers, 2010). Yet some well-designed studies (for example those carried out in the UK at the initiative of 'Carers UK' (Yeandle and Fry, 2010) have shown that appropriately organised support measures give a positive image of the firm and can lead to positive outcomes in terms of productivity gains, as well as improved retention and recruitment of skilled employees, with less turnover of staff and better relations between individuals and working teams (Buckner and Yeandle, 2006; Employers for Carers, 2010).

However, it seems that current legal provisions are usually not strong enough to successfully combat prejudice, as firms that provide such measures are infrequently rewarded while others not complying with legal requirements are rarely sanctioned. Unwillingness to confront care/work issues leads to a lack of appropriate and well-publicised information on existing support measures, so that employees' knowledge of such possibilities is scant, resulting in low demand. However, even when they do possess the relevant information, employees may be reluctant to request support because of possible drawbacks, such as stigmatisation and negative impacts, not only on their professional advancement, but also on their working relationships with colleagues, resulting in a strained work atmosphere and affecting quality of life in the workplace (Eurofound EWCO, 2008). This clearly challenges the free choice assumptions of the new paradigm. Furthermore some practical problems are put forward to explain the reluctance of firms to implement such measures. They include the unpredictability of the exact time at which short-term care leave needs to be granted. In the case of longer-term leave, the required notice needed by managers to avoid organisational difficulties and possible losses in productivity is also viewed as a problem. So to be effective, the implementation of any type of care/work reconciliation measure not only relies on the customisation and flexibility of the work process, it also entails good labour relations associated with a high level of awareness from managerial staff and the active participation and support of fellow employees in the field. Without these prerequisites it is unlikely that these types of measures can be successfully implemented.

What limited existing data show is that countries where the more generous and highest number of specific care/work reconciliation measures are implemented are those where overall proactive employment and labour policies exist. This positive context for the implementation of specific measures for working carers also facilitates the expression of demands by potential beneficiaries, as they need not fear stigmatisation; this is especially true of senior workers and women (Sweden, the Netherlands, Finland, UK). In these countries, even if some variability exists in terms of economic sector and the size of firm, measures are better known by their potential users since they are better publicised within firms; they are also easier to access as they are articulated with other existing flexible work arrangements, as shown by the following example. In the Netherlands, working carers have a specific right to request fewer hours of work during periods of high-level care activities, whether the care provided concerns children, disabled relatives or older parents, and firms provide opportunities for making up these hours later through flexible working arrangements. It must also be acknowledged that the uptake level of these measures is positively correlated to the high proportion of part-time work among women, as in the Netherlands, Sweden and more recently in Finland and Germany. Here we see again how the choices are contingent to the quality of overall social environment.

Cash benefits for older people and carers: an analysis of their impact under the choice paradigm

Cash schemes can be considered as emblematic of the new paradigm of choice. But their design and potential impact must be analysed within the national context in which they were implemented as they are part of the overall LTC policies (Rostgaard, 2011). In particular, attention should be given to entitlement rules, the nature of controls regarding the appropriate use of related services, as well as their quality monitoring. All these dimensions impact on how these benefits can influence the well-being of older people and their informal carers (Da Roit and Le Bihan, 2010).

Cash benefits and older people as active consumers

Cash benefits can specifically target older people, their carers or both. From a European perspective, they were introduced under different circumstances and at different times in various countries (Ungerson, 2004). In a first group of countries (France, England, Italy, Germany, Austria), they were construed as a step towards shifting care from the status of a private issue to recognising the need for seeing care as a social risk. In this group, cash measures frequently represent the backbone of emerging LTC policies/ systems focusing on the older population. They can take the form of LTC insurance (as in Austria and Germany) or may exist without such a scheme (as in France). In both cases they have the main goal of addressing older people's care needs by enabling them to have access to professional services although they can also be used as a means to compensate informal carers. In a second group of countries, including Sweden and the Netherlands, where there is already broad access to professional services, cash benefits were ostensibly introduced under the new paradigm as an alternative to in-kind services. The primary goal was to increase older people's decision-making power and independence by giving them the opportunity for more choice regarding their care providers and services needed; and also by enhancing their capacity to participate in the care planning process (Baxter and Glendinning, 2011). These goals were less dominant in the first group of countries. However in both of them, competition in the care market was introduced in parallel to enhance the range and diversity of choices and to increase quality of service (Stevens *et al.*, 2011).

So given this opportunity and user-centred ethos, it is important to consider whether the choices inherent within cash benefit schemes are able to provide a real option and for whom. This will be discussed in relation to experiences within Germany, France, the Netherlands and the UK.

Cash benefits for older people: choice of providers, services offered and decision-making issues

The most common form of cash benefit is the attendance allowance (AA), targeted towards older people themselves. But benefits may also take the form of a 'personal budget' which encompasses tax and social discounts as well as vouchers (Glendinning *et al.*, 2009). In the Netherlands additional benefits for which family members are eligible are also available (Decruynaere, 2010).

In Germany, following an assessment, all older people entitled to care services are offered three choices of how their needs can be met, regardless of their financial status and resources: in kind, in cash or as a mix of both. Choosing cash only has a direct impact in terms of coverage/choice mix and produces variable results. On the positive side, the older person is free to use the cash as he or she wants; this corresponds to a privatisation of care services, as the state's responsibility is no longer engaged. However, this choice comes at a price: for a similar level of care needs, the cash option allows older people in need of care to purchase only 60 per cent of the services they would have received had they chosen the 'in-kind option', so that the cash option results in lower financial expenditures for the state (Mingot, 2011). Controls of services purchased with cash benefits hardly existed until 2009 and older people could encounter problems if the benefit did not result in appropriate care being provided. The system also increased social inequality: the wealthiest households tended to choose the in-kind benefit, while those with less financial resources chose the lower and poorly quality-controlled cash option. The 2009 reform tried to address some of these issues in a way that was inspired by Dutch approaches to the same problem.

The French case is somewhat different (Triantafillou *et al.*, 2010). The choice between cash and in-kind benefits is also offered to all older people after an assessment of care needs; should cash benefits be chosen, the option to hire a personal assistant directly or to pay a family carer – except a spouse – is available, with only 9 per cent of caring relatives taking this option. However, as the French care attendance allowance (APA) comes with heavy co-payments scaled according to income (90 per cent when monthly income exceeds €2850), wealthier households prefer to opt out from the APA public scheme and choose another more favourable legal arrangement allowing them to hire a carer directly. In this case the hourly rate paid by the care client will be half of the cost per hour of a professional employed by a recognised agency. This difference, stemming from tax and social charge reductions granted to the older person as an employer, makes this indirectly subsidised option very advantageous. A recent report by the French General Accounting Authority (Cour des Comptes, 2009) showed that this option was 'hijacked' by the wealthiest households, leading to increased inequalities with unequal service use in situations of similar care needs. This drawback is due to the fact that this option and its attendant financial incentives were implemented as part of a labour market-oriented policy aiming not only to eradicate the 'grey' (legal but undeclared) care services market, but also to promote a new broad 'personal services' market that includes recreational and leisure activities as well as care of older people.

In the UK, severe underfunding of public social care services as well as the emergence of a broad range of private providers, including poorly regulated private for-profit home care agencies, led to the introduction of an 'individual budget' pilot programme tested in 13 local authorities between 2006 and 2008 (Wilberforce *et al.*, 2011). Individual budgets were provided flexibly according to users' preferences, so the latter could purchase the services they wanted in one of two ways: either the support of a care manager (usually a social worker) could be provided, with an a priori agreement about expected outcomes negotiated between the care manager and the client, or the individual budget could be managed independently by the older person, with an a posteriori control of its use by local authorities. The scheme had an impact on the way home care providers designed and carried out their care activities in order to respond to the expectations of this new type of client, with day care and established care institutions leaving the market; it also had an impact, perceived as beneficial, on the involvement of social workers within their new task as care managers (Glendinning *et al.*, 2008). Despite these findings, evaluation of the pilot scheme showed that it was not welcomed by a majority of older people especially when they (as well as their informal carers) lacked the support to exert informed and free choices.

However, they did welcome the opportunity to purchase some previously inaccessible services, though no real assessment of the quality of the delivered services was undertaken. Nevertheless, before the end of the assessment process, a decision was taken to extend this approach, now termed 'personal budget', to all people in need of social care and the programme was rapidly expanded. Further assessments (Baxter *et al.*, 2011) yielded similar results and showed that a relatively high level of needs remained unmet. These rather negative outcomes are clearly linked to the rather low level (€6000 annually) of the average personal budget (with only 1.2 per cent of GDP allocated to social care), a crucial feature of the UK care context, as compared with the average annual benefit level in the Netherlands (€17,000 and its 3.1 per cent of GDP budget devoted to LTC; Allen *et al.*, 2011).

In the Netherlands, personal budgets emerged at the beginning of the nineties within a very different context of broad access to good quality professional care and within the framework of a mixed economy of care providers, in which non-profit services were and are still dominant and where care markets are relatively strictly regulated (at least in the field of ADL). Today the personal budget scheme (tested since 1991) is widely implemented throughout the country with 111,000 users in 2009, representing 15 per cent of the overall disabled population, 20 per cent of whom combine this option with various types of in-kind services (Decruynaere, 2010), showing that in-kind services remained the most frequently used option in 2009. Although there is no overall scientific assessment of the scheme, the quoted report acknowledged that satisfaction with the information service was rather low among personal budget older users compared to other indicators. It also showed that the rapid growth of the number of users was largely due to uptake by the younger disabled population, whereas use of the scheme by the older population was hindered by the administrative and psychological cost of budget management.

This comparison illustrates the fact that, in the UK, cash for care is strongly linked to the goal of substituting privately purchased care for publicly run social care services, the latter being not easily accessible and viewed as bureaucratic in their approach to management. This is in contrast to the Netherlands, where personal budgets are seen as supplementing formal care services, either not-for-profit or for-profit organisations, with the goal being to boost quality through regulated competitive bargaining processes.

So it appears that cash schemes targeting older people lead to different outcomes depending on the characteristics of the national context in which they are implemented, though they seem to present similar disadvantages. However, the internal logic of the scheme does bring about similar outcomes regardless of the national context. In both cases, it appears that it is the more educated and wealthy populations that would appear to benefit from any potential advantages, even when professional budget management support is not available. This is not the case for the more deprived segments of the older population, where access to these support services is crucial for any advantages to be evident.

Active consumerism and support for informal carers

As was the case for older people, the new policy paradigm inspired the design of support policies for informal carers, particularly in the field of payment schemes directly targeting them. It developed in a limited number of countries, mainly in the UK and the Netherlands but also marginally in Sweden. Payment schemes were conceived either as a symbolic recognition of the care provided (Colombo *et al.*, 2011), or as additional income support to help older carers combat poverty risks and social isolation, the latter by enabling them for example to pay for leisure activities. This cash was also provided to support working carers of older people with high support needs who are often forced to leave the labour market against their will and lose income, or alternatively compelled to continue to provide care as well as holding on to paid employment (Phillips *et al.*, 2002).

Cash benefits for informal carers

As previously stated, informal carers' risks of poverty are high and their level of social protection usually low or even absent. These risks can be reduced by providing carers with social protection contributions during any periods where they are unable to work, due to their caring responsibilities. Different schemes were developed in the UK, among which is the following example (see Box 3.10).

Box 3.10   Direct payments for informal carers (United Kingdom)

Direct payments were introduced nationally by the Community Care (Direct Payments) Act in 1996 and have been available to carers since 2001. Direct payments are cash payments made by local authorities to carers of older or disabled people living in the community who are assessed as needing help from social services. Carers who can maintain older people at home save local councils money, as they do not have to pay for residential care; therefore initiatives such as direct payments, which support carers who might otherwise give up caring, are viewed as cost-effective.

The aim of direct payments is to increase the choice and independence afforded to carers to meet their needs flexibly and creatively so that they can continue caring. The expectation is that most direct payments will be used to provide the carer with a break as the payment can be used to purchase services or equipment to help them in their caring role and also to help them maintain their health and well-being; it is not a source of income. Direct payments for carers cannot be used to purchase services for the person who is being supported (who may receive for this a cash benefit in their own right).

Source: Holdsworth (2011c).

Evidence-based studies examining cash-based policies in national contexts have failed to provide strong conclusions regarding their cost-effectiveness (Glendinning *et al.*, 2009). In the UK however, a series of extensive and scientific assessments were undertaken examining the successive reforms of the cash scheme (Baxter *et al.*, 2011; Wilberforce, 2011), while in the Netherlands, some administrative reports were published on this subject (Decruynaere, 2010). Although their design and impact (whether expected or observed) are crucially dependent on the national context and the overall LTC policy within which they were implemented, available data and reports suggest that in most countries, care allowances acted more as disincentives to gainful employment. Thus they had negative effects by reducing labour market participation and limiting the country's overall economic potential (Heitmueller and Inglis, 2007). Direct cash payments have failed to significantly help informal carers avoid poverty, since social benefits derived from care allowances are usually lower than those prevailing in the regular labour market.

In general therefore they have been weak contributors to the achievement of an adequate, or even an improved balance between care and subsistence, as well as between care and work. For working carers, except in Sweden and Finland, the low level of care allowances does not reach the minimum market wage, so it is unable to effectively compensate for income lost due to the reduction of hours of outside paid work and consequently provides a restricted choice for the carer. And for carers not in employment, direct payment only marginally contributes to poverty reduction making these carers a category of poorly paid and inadequately protected 'hidden workers'. Only Sweden and Finland provide a variety of cash schemes for informal carers with a level of social protection and income that awards them quasi-professional status, achieved by municipal or recognised care agency contracts with informal carers and implying a high level of control of the quality of services provided (Bureau *et al.*, 2007).

Overall conclusions on cash benefits

Overall it could be stated that in many countries (with the exceptions of Sweden and the Netherlands), the positive effect of lowering public expenditure through a reduction in formal services has been achieved at the expense of creating a new subcategory of 'poor and inadequately protected workers', with the risk of a lower quality of services provided to the cared-for person (Colombo *et al.*, 2011).

But cash benefits also have negative effects. The first effect relates to difficulties in sustaining a future formal care workforce (Matrix Insight, 2012). Although rarely explicitly acknowledged, recognising informal carers as 'paid workers' has potentially negative implications for the formal care sector, by contributing to the deprofessionalisation of this already poorly recognised employment field, already characterised by low pay, low status and poor working conditions. Moreover, it contributes to difficulties in the retention of both the current and future long-term care workforce, by maintaining the low status of care work and its associated poor wage levels for formal carers (Triantafillou *et al.*, 2010). Professional workforce planning policies in the field of LTC will have to face up to a declining formal care labour force (Matrix Insight, 2012), so that issues of migration and legal status, as well as promoting LTC-related professions and jobs, can be dealt with (Rothgang and Engelke, 2009). The important phenomenon of migrant care workers is explored further in Chapter 10.

A second effect of cash benefits in terms of overall LTC policies is that they have contributed to a blurring of boundaries between informal and formal carers. This in turn raises the issue of choice. The increasing use of cash benefits has created a 'mixed care workforce' (informal family carers, migrant workers, personal assistants, formal professional care staff), operating with varying intensity in the planning, organisation and delivery of LTC service provision that makes the already complex LTC system even more difficult to understand and negotiate for users (see section 1). This point reinforces the arguments that older people in need of care, as well as their informal carers, require professional advice and support to make the most appropriate and effective choice of providers and care services. But this has a cost that many countries are not willing to pay for.

Conclusion and recommendations

The concluding part of this chapter on users' perspectives attempts to provide recommendations on mechanisms and policies that would help to foster the synergistic articulation of measures targeting LTC recipients and informal carers to reconcile their potential contradictions.

LTC policies and the new consumer paradigm: fulfilled promises?

When examining the goal of empowering older people and their informal carers through increasing participation and choice in care decisions, our findings challenge the idea that these two policy components automatically lead to optimal outcomes for both types of users. In fact, there seem to be some contradictions between decision-makers' views of the potential benefits of increased choice – and responsibilities – for both carers and older people and how these responsibilities and given opportunities are in fact perceived and experienced by groups targeted by these policies (Newmann and Tonkiens, 2011). A much more muted response is recorded by users, compared with high levels of enthusiasm from policymakers, as well as from some associations representing carers or other users. Evidence from countries with limited access to formal services as well as from those where access is broader shows that choice, when conceptualised in a 'market-based approach', does not necessarily meet the expectations of either the older users or their informal carers. Some older people may indeed experience an increased feeling of control over their lives by being given the opportunity to choose (Glendinning, 2011), as creative solutions may be fostered by the flexibility afforded by cash benefits compared to in-kind services.

Yet when choice is used mainly as an instrument to steer demand in a care market economy, or as a means to lead users to lower their reliance on public service providers, older people do not necessarily feel empowered. What the various examples demonstrate is that 'choice' in many countries is not as free as it seems at first glance, as it is limited by financial and/or geographical disparities and it does not automatically lead to positive outcomes for users. For example, choice between in-kind or cash benefits often turns out to be 'economically determined', leading to care-related social inequalities for both older people in need of care and informal carers (e.g. in France and Germany). In other cases, choice is constrained by the limited availability of the in-kind option due to a lack of care services (e.g. UK and Italy). The cash option for older people and their carers frequently comes with a price tag, usually implying a lower volume of services. This is not necessarily detrimental, as long as the quality of the delivered services remains high, but the cash option is frequently accompanied by reduced levels of quality control.

Consumerism as a form of empowerment for service users may thus have a limited impact on older people, and may actually run counter to the carers' choices and expectations, hindering their opportunities to exert more control over their own lives and to fulfil their goals. In order to make appropriate use of 'choice-enhancing' measures and to effectively benefit from them and reach their full capabilities (Sen, 1999), both older people and informal carers need to be supported by appropriate information and professional advice. As the Dutch example showed (Mak, 2011b), a dynamic synergy can be used in which older people are not left alone to make their own choices (and thus not 'empowered' in a strictly liberal sense), but supported and helped in order to be able to choose a suitable solution to their specific needs. Through policies such as these, choice is actually regulated through a process of professionally supported, shared decision-making, decisions then being based on a higher level of appropriate information and leading to a better quality of service for older people. But in this case, as exemplified by the Netherlands and Sweden, cost reductions are not likely to ensue as reflected by their respective LTC budget. This type of policy implementation requires not only a high level of involvement of all stakeholders, but also an appropriate level of funding (Jones *et al.*, 2011). This combination is unfortunately not found in most European countries today, and the current financial crisis is unlikely to help matters in this regard.

Regarding measures targeting informal carers, examples illustrating the INTERLINKS Framework suggest three main conclusions when designing comprehensive and efficient informal carers' support policies, able to meet their needs and expectations in relation with the choice paradigm. First, if the desired goal of increased employment of women is to be achieved, policies that rely on the present limited involvement of women in the labour force will be confronted with shortages of available informal carers in the future. Men of working age will also have to be more engaged in providing informal care, therefore strong care/work reconciliation policies will constitute crucial elements of any LTC policies and become as important as giving more support to older carers. Here, giving them real choice is clearly at stake.

Second, it is evident that this opportunity could be capitalised upon more easily if greater recognition at the political level was given to the voices of both informal carers and care recipients and their views taken into account in LTC policy design, as well as in the care planning and monitoring processes. Yet the assumption of common interests, frequently presented as self-evident by carers' organisations and by influential groups lobbying on behalf of specific populations such as Alzheimer sufferers and their families, should be critically appraised and more vigorously challenged. Neglecting potential differences and conflicts of interest may endanger both users' and carers' well-being.

Recommendations

Coming from many European countries, examples illustrating the INTERLINKS Framework point to a wide array of measures that have the goal of responding to a range of needs. Yet the only policies to meet the challenge of supporting informal carers by effectively protecting their health and well-being, while simultaneously meeting the needs of those they care for, are clearly founded upon a high level of provision of quality formal care. This is particularly so for home care, as demonstrated in countries such as Sweden, Denmark and the Netherlands (Naiditch, 2012). Even in countries where they are strongly marketed, cash measures are considered marginal to the carers' well-being, when compared to professional services. Also the success of these policies does not stem from the choice paradigm, but from the ability of these countries to implement specific direct measures such as training, and practical support coupled with targeted and highly regulated financial help. In addition, indirect measures must be promoted, such as facilitated access to respite care to make carers' lives easier and thus enhance mutual well-being. All of these features must be in place to help support and improve informal care provision.

Moreover and equally importantly, existing policies in the areas of family, employment and labour, housing, and transport must be maintained as they provide a context favourable to a wide range of specific measures targeted towards working carers. This broad view of support policies for informal carers clearly transcends the health and social care fields and has the goal of facilitating social participation for all types of carers. This approach can contribute to viewing informal care not only as a resource for beneficiaries of the care provided, but as a positive element for societal well-being and as a force contributing to social cohesion and solidarity in society as a whole.

4
Joint Working between Health and Social Care

Kerry Allen, Jon Glasby and Ricardo Rodrigues

Introduction

Faced with a series of demographic, social and technological changes, countries across Europe have found themselves trying to respond to a series of long-term care (LTC) challenges, including: responding to increased numbers of frail older people and older people with dementia; changes in social/family structures (often reducing the availability of informal care); rising public expectations and the need to improve and assure quality; financial sustainability (in a difficult economic context but also long term); staff recruitment and retention; and the need to develop more preventative and rehabilitative approaches.

Against this background, this chapter focuses on key themes, barriers and enablers when countries have sought to promote more effective ways of working across the boundaries between health and social care. In particular, the chapter uses a conceptual framework developed by Glasby (2003, 2007) in order to explore the individual, organisational and societal processes, methods and resources that help overcome fragmentation by promoting dialogue, mutual exchange and joint work between relevant stakeholders. It draws on wider governance literature, alongside practice examples that illustrate the INTERLINKS Framework for LTC (http://interlinks.euro.centre. org) and material from background papers produced during the INTERLINKS project (Allen *et al.*, 2011; see also Chapter 1).

According to Glasby (2003, 2007), governments seeking to promote more effective inter-agency working need to focus on three separate but inter-related levels. Thus, the contribution of *individual* practitioners, though significant, takes place within a local *organisational* context, which itself is influenced by *structural* barriers to improved joint working. Similarly, structural barriers derive at least in part from certain organisational features associated with particular types of health and social care agency and, ultimately, from the individual practitioners working within the organisations concerned. Building on this distinction, later sections of this chapter focus on promising approaches at different individual, organisational and structural levels.

As other chapters in this book suggest, the services that constitute LTC are spread across a broad range of disciplines spanning the health professions and wider social care functions. This breadth of involvement has led most European countries to acknowledge integration and its management as important mechanisms for improving LTC (Axelsson and Axelsson, 2006; Hofmarcher *et al.*, 2007; Leichsenring, 2004b; van Raak *et al.*, 2003). The fragmented nature of LTC services extends beyond health and social care sectors, adding further weight to the case for joint approaches. For instance, Allen *et al.* (2011) observe the complex mix of political responsibilities, funding mechanisms, service providers and purchasers that cut across several areas and administrative levels:

- Across national, regional and local level
- Across hospital and community care settings
- Across public, private and non-profit sectors
- Across formal and informal spheres, including the growing role for migrant care workers and individuals purchasing and designing their own support arrangements

Looking internationally, integrated care and the wider process of integration are used to describe a great number of different services and organisational structures. Commentators have stressed the different extents of integration, drawing attention to a spectrum of integration (Ahgren *et al.*, 2009; Dickinson, 2006; Glendinning, 2003; Leutz, 1999). At one end of the spectrum partners simply share objectives and are aware of each other's contributions towards these common goals. At the other end is full structural integration, where partners share the same mechanisms in their everyday work. In an absolute form this would include synchronised financial and political incentives, joint organisational identities and cultures, budgets, capital and human resources, as well as common tools for all processes such as assessment or monitoring outcomes.

Against this background, this chapter reviews a number of promising approaches found in a variety of European countries at the different levels identified by Glasby (2003, 2007). These include:

- Joint working at an individual level (either joining up services around individual service users or encouraging individual professionals to work together in new ways): case management, multidisciplinary teams and personalisation.
- Joint working at local organisational level (where local organisations have developed new local structures or protocols to tackle specific issues at the fault line between health and social care): hospital discharge and needs assessment.

- Joint working at a structural/social policy/system level (national policies to develop more fully integrated structures): joint finance/financial incentives, structural change and the prevention agenda.

In practice, such categorisations inevitably impose artificial distinctions on a more complex reality – and the different approaches below could probably be placed at slightly different levels depending on the point of view of the observer. However, the key point is that there are different levels of activity/ intervention – and that these frequently overlap and interact. This means that any attempt to promote more joined-up care needs to be equally cross-cutting and multifaceted if it is to stand a chance of being successful.

Joint working at an individual level

Examples of joint working with a specific focus on bringing together services around individual service users are case management, multidisciplinary teams and personalisation. Case management and multidisciplinary teams are widespread across Europe, albeit mostly in the form of isolated pilot activities. Although the term 'personalisation' is most associated with the reform of adult social care in the UK, aiming to give individuals more direct control as consumers of their social care provision, attempts to empower individual service users by giving them access to individualised cash payments or vouchers seem to be gaining increasing importance throughout Europe (see, for example, Da Roit and Le Bihan, 2010; Lundsgaard, 2005; Timonen *et al.*, 2006; Ungerson and Yeandle, 2007).

Case management

Case management approaches target those with complex health and social care needs. Typically a case manager works with the service user to identify their needs and decide on the best provision. Case management approaches are evident in many European countries. However, they were often developed sporadically as local projects, rather than being a mainstreamed mode of delivering LTC.

Case management is an individually tailored approach, helping to navigate a range of services and provider organisations, alongside the service user and often their families and carers. The delivery model can enhance professional awareness of often complex local health and social care systems, as case managers frequently work with primary care practitioners, hospitals, mental health care providers and social services. Thus, the intervention has the potential to increase both professional and patient understandings of health and social care systems and local services. Achieving this potential in practice can be challenging as individual case managers seek to link with a large number of staff and organisations. Still, some local projects demonstrate promising progress in addressing these issues.

The Kompass project highlights an important benefit of case management; it tends to target those with the most complex and highest levels of need (see Box 4.1). As well as increasing access to quality care services, the initiative is designed to reduce demand and workload in acute health settings. The case manager, often a community nurse supported by a small multidisciplinary team, takes responsibility for all care arrangements in complex cases, easing the pressure on acute staff and bridging the gap between health and social care systems.

Box 4.1   Kompass – Case management of lower socio-economic status patients (Switzerland)

Kompass was initiated by the Health Department of the city of Zurich as part of a framework to enhance collaboration and networking between local providers by means of case management for patients with complex health and social care needs. It is available to older people from lower socio-economic groups who are either recovering from acute medical conditions or managing a chronic condition.

The main partners of Kompass are the four largest public hospitals in Zurich, the local and regional health authorities, home care services, social services, churches, psychiatrists and general practitioners. The initial aim of Kompass was to provide greater access to health and social care services by providing case management for patients who are emotionally fragile, poor, socially isolated and who have complex needs due to medical conditions. It addresses the gaps between acute health care and social care, as well as between hospital and home care. It also has the goal of contributing to the reduction of the average length of stay for this group of patients by avoiding unnecessary hospital admissions and improving discharge planning. Initial review suggests improved quality of life for service users. The project constitutes an immediate response to a real need among the Kompass partner organisations. Yet its permanent commissioning has to be confirmed by the City Government.

Source: Gobet (2011).

Another example of case management pilot work in Paris displays the intended benefits and outcomes of this type of intervention, alongside some of the challenges of establishing such services (see Box 4.2). The example demonstrates the highly time and resource consuming nature of initiating an integrated service which revises care pathways and depends on sustained commitment from multiple partners. Capturing evidence of the impact of services on individuals and on the local health system were activities crucial to ensuring the intervention's sustainability and transfer.

Box 4.2   Coordinating Care for Older People (COPA): Teamwork integrating health and social care professionals in community care (France)

COPA is located in Paris as a pilot research action programme promoted by the Public Health Department of St Quentin University. The care of individual patients is managed by a team led by a case manager and a primary care physician. Goals are to allow the target population (frail older people living in the community) a longer, safer home stay while limiting hospitalisation and optimising hospital discharge. The partner network involves health and social care organisations (nursing homes, nursing and home help agencies) located in the catchment area, as well as self-employed health and social care professionals, on a voluntary basis. All partners participated in the development of new care pathways. The example shows it is possible through a specific intervention to locally foster a common culture for planning and implementing new care pathways.

In order to overcome GPs' potential opposition in changing their clinical practice, it was possible to meet the primary care physicians' expectations through a well-designed case management approach, by providing resources to care for their most complex (and thus time consuming) patients. In practice, establishing a new service was demanding on time and needed investment and strong promoters, as well as continued commitment of all professionals involved. In France, as in all European countries, even successful pilot projects need strong support from decision-makers to gain coverage and support and to be rolled out nationally. A well-designed cost-efficiency analysis should be part of the overall assessment as it helps to stengthen the argument for an integrated approach. Although COPA is successful in reducing the overall rate of hospitalisation, it has been only partially replicated in one other French city.

Source: Naiditch (2011a).

In particular, there was difficulty in recording the positive impact on service users. This was thought to be connected to the very frail older target group, with high levels of complex need. Outcomes such as better health status, reduced institutionalisation and delayed mortality were not proved via evaluation. However, it was noted that the control groups used presented less co-morbidity and social isolation. Thus, despite the beneficial nature of case management in targeting those in most need of both health and social care intervention, this very aspect can also pose problems for collating the outcome data that is needed to gain the wider support of managers and politicians. Problems of generating evidence are widespread given the complex and multifaceted nature of older people's conditions and services, as such wider support may be achieved only through longer-standing communication and negotiation processes.

Multidisciplinary teams

The use of multidisciplinary teams across many LTC services such as falls prevention, telecare, dementia care, rehabilitation, needs assessment and case management, as well as across acute, residential, transitory and home care settings has become common practice in many countries (e.g. Denmark, the Netherlands, UK, Finland). Examples of multidisciplinary team members span doctors, nurses, occupational therapists, psychologists, physiotherapists, dieticians, social and community workers and technologists. Development of a more specific focus on the individual is an implicit aim of multidisciplinary teamworking. Incorporating several professional perspectives into one service seeks a more holistic approach to the individual service user. Ideally the joint approach results in a comprehensive care plan covering diverse needs, from benefits information to medical and personal care.

National Reports on governance of LTC (http://interlinks.euro.centre. org/countries) offer several perceptions of the benefits of multidisciplinary teamwork, including individual as well as organisational issues:

- Many professional perspectives allow a fuller assessment of the older person, more likely to cover the real needs of the individual receiving services
- Teamwork across different agencies can reduce duplication and gaps of unmet need
- Interdisciplinary learning within teams increases standards of care, communication and general knowledge of the system among professionals
- Multidisciplinary teams are more likely to provide innovative and patient-based care solutions

The most prominent challenges and obstacles for multidisciplinary and multi-agency teams were identified as: cultural tensions between different professional groups; lack of shared systems/resources to process joint work, e.g. information technology; and different employment conditions and capacity of team members (Allen *et al.*, 2011).

In UK Care Trusts (integrated local health and social care authorities) small multidisciplinary teams are increasingly used to deliver LTC, targeting mental health and learning disability as well as older people's services. UK experience illustrates some of the cultural and practical challenges of setting up joint structures and delivering multidisciplinary care. Integrating separate structures has taken a number of years and raised a series of practical issues (for example, pensions, terms and conditions, etc.) as well as organisational development/cultural issues (for example, the difficulty of combining 'medical' and 'social' models of mental distress in a mental health Care Trust or of bringing together different health and social care professionals in a Primary Care Trust-based Care Trust).

With the increasing incorporation of new technology devices in older people's health and social care, multidisciplinary teams assume the role of delivering telecare services in many countries. The ADaL programme in Finland (see Box 4.3) demonstrates how multidisciplinary teams have been used successfully to implement new technology in residential care settings. As a falls prevention initiative, the use of movement sensors was introduced concurrently with a programme of physical training, provided by physiotherapists. This multidisciplinary falls prevention approach brought the residents' and care staff's attention to the wider reason for the introduction of the motion sensors – to prevent injury and reduce the use of bedrails in line with resident preferences.

Box 4.3   Activating daily life (ADaL) programme with technology solutions (Finland)

ADaL started as a pilot project in a large nursing home. Its main objectives were to improve quality of life for residents by changing nursing practices with the use of technological solutions. The programme consisted of a detailed mapping of the preferences of the residents, individual weekly exercises and the use of movement sensors to alert the nurses to those at risk of falling when trying to stand up from the bed or chair. The main principle was to create a safe environment and an individual model programme for daily living for each resident using technology solutions.

This programme was established by the multidisciplinary team (nurses, physiotherapists) in cooperation with the residents. The results showed that the ADaL programme was able to increase the residents' independence without increasing staff costs. Residents' quality of life increased and there was a reduction in the use of bedrails and bedbound residents.

Source: Hammar and Finne-Soveri (2011).

Personalisation

Personalisation is a broad approach that combines a strong commitment to 'self-directed support' with the use of personal budgets (a post-assessment allocation of money to an individual) and cash payments. On paper, the values which drive personalised services are clearly focused at empowering individuals and providing links to the most appropriate local services. The principles of self-directed support have been outlined by Duffy (2005, p. 10) as:

- Every adult should be in control of their own life, even if they need help with decisions.
- Everybody needs support that is tailored to their situation to help them sustain and build their place in the community.

- Money is most likely to be used well when it is controlled by the person or by people who really care about the person.
- Family and friends can be the most important allies for disabled people and make a positive contribution to their lives.

However, the personalisation approach also has an inherent consumerism rhetoric in many national contexts (Da Roit and Le Bihan, 2010), emphasising the role of users as consumers of care and competition between providers as a tool to bring about increased quality and lower costs. In the UK, the Netherlands and Sweden there are high hopes that more individualised forms of funding could lead to better outcomes for service users, meeting needs more fully and (perhaps) enabling older people to remain living independently for longer. By fully tailoring services to individual needs and circumstances, it is argued, there is scope to provide more innovative, higher quality and better organised support – potentially preventing or delaying future crises in people's health or social situations. This optimism is reflected in political commitment with the advent of personal budgets and the continued expansion of direct payments, in the case of the UK, and expansion of voucher systems enacted at the municipal level but supported by central government grants, in the case of Sweden.

Although the evidence of a direct preventative impact is limited, personal budgets do seem to have had a significant impact on people's quality of life (Glendinning *et al.*, 2008). However, older people reported lower psychological well-being than other younger user groups (e.g. learning disability, physical disability or mental health) perhaps because they felt the processes of planning and managing their own support were burdens. The consumerist underpinning that characterises personal budgets may prove more suitable for some particular groups of users than others. Older age users may require a more enabling and supportive approach in the process of choosing care and shaping older people's services to their preferences and needs.

Across Europe acknowledgement of the importance of patient-centred LTC is highly visible in the majority of countries (see Kümpers *et al.*, 2010; Nies *et al.*, 2010; Triantafillou *et al.*, 2010), but the link between this and individual budgets or payments is less frequently made. Countries with a long history of distributing cash allowances bring further insight into the benefits and dangers of this financial mechanism in practice. For instance, in Austria the experience of a tax-based comprehensive LTC cash allowance (See Box 4.4) has identified several potential tensions for cash-based systems, including:

- The emergence of a 'grey' or 'black' market of care, in particular as the proximity of low-income countries offered favourable opportunities for migrant carers ready to provide 24-hour care in the home of the beneficiary at affordable prices (see Chapter 10);

Box 4.4 Tax-based comprehensive LTC allowance (Austria)

In 1993, Austria was the first European country to introduce a universal coverage of tax-financed LTC allowances as the main instrument to fund LTC. Today, Austria has one of the highest proportions of older people who receive LTC allowances among EU Member States. The scheme aims at partially compensating individual costs occurring due to care needs by a non-means tested cash benefit at fixed rates, aiming to ensure necessary support and increase opportunities for independent living. Benefits are paid in cash according to a needs assessment in seven levels to individual beneficiaries who decide autonomously how to use it. More than 80 per cent of recipients are older people above 65 years of age. Compared to most other LTC schemes in Europe, the Austrian attendance allowance is the most generous in allowing for flexible solutions to pay for care.

However there are strengths and weaknesses of a cash-based approach to finance LTC, and the impact of such an approach on the hitherto clear-cut division between formal and informal care. On the one hand, evaluation studies have shown the scheme's important impact on the visibility of LTC (increased public debate, legal rights), the development of organisational structures and user satisfaction. On the other hand, the allowance has also triggered an increase of co-payments for services, moonlighting migrant carers, and incentives for women to leave paid employment. The allowance and ensuing reform measures had only very limited effects with respect to the persisting social and health care divide – while home care services as such became more integrated, their links with the health care sector (GPs, hospitals) remained restricted.

Source: Leichsenring (2011a).

- Providing incentives for informal carers to leave the labour market;
- The value of the allowance devalued between 1996 and 2007 so that the 'purchasing power' of beneficiaries decreased consistently during that period;
- The allowance does not encourage links with the health care sector (GPs, hospitals);
- Providing users with increased purchasing power was not enough to stimulate increased formal care provision, as supply continued to be strongly regulated and marred by barriers to entry.

These insights give strength to commentators' fears that the positive impact of personalisation could be easily lost if the original values are forgotten amid numerous partners, all with their own motivations. At worst this could result in a situation which '*allows the old system to pay lip-service to the concept while essentially recreating itself*' (Glasby and Littlechild, 2009, p. 86).

This section has focused on how services can be better linked up around individuals. It addressed issues such as how to encourage more cohesive working between health professionals by means of multidisciplinary teams. Against the backdrop of an increasing choice of services, older people in need of care and their carers often have to face complex tasks in arranging their care from various providers across health and social care sectors. This can be particularly challenging where individuals are severely ill, socially isolated, have complex needs or do not have the information about their rights and services to guide these decisions. Case management has displayed an important role in beginning to address these gaps. However this progress tends to be hindered by the complexity of this type of activity, with case managers negotiating a variety of different issues with service users while ensuring links to an equally diverse range of professions. Bringing professionals together into cohesive teams, with specific tasks such as the delivery of telecare, has had some success. But again the time consuming nature of creating teams across different professions and organisations calls for significant and prolonged financial and administrative support to make such ventures work.

Finally, it has been outlined that countries with a history of cash benefits for people in need of LTC display lessons about how to promote a personalised approach without creating further fragmentation of a system. However, a key message from European use of cash allowances is that transferring purchasing power, in and of itself, does not improve access to services or their quality if it is not backed up by related policy measures at the system level.

Joint working at local organisational level

In addition to bringing professionals together in new ways and enabling individuals to join up their own care and support through greater use of cash allowances, many European countries have more formal processes and policies between local organisations in order to tackle specific fault lines in the boundary between health and social care. These often take the form of protocols and address key issues such as:

- How to best ensure patient rights during transitions from acute care settings;
- The role of IT for management of LTC services;
- How to deliver effective multidisciplinary assessments.

Hospital discharge arrangements

Many European countries have started to develop approaches that focus on arranging swift and effective discharge from acute settings, reducing cases of delayed discharge. Often triggered by diagnosis-related group financing, these interventions have often shifted the responsibility for discharge to the non-acute sector. These integrated services concentrate on outlining roles, responsibilities and activities of the different organisations involved in this transition. Hospital discharge has long been a difficult area of policy and practice, often characterised by gaps such as:

- Poor communication between hospital and community;
- Lack of assessment and planning for discharge;
- Inadequate notice of discharge;
- Inadequate consultation with patients and their carers;
- Overreliance on informal support and lack of (or slow) statutory service provision;
- Inattention to the needs of specific groups such as people with dementia or people from minority ethnic communities;
- The risk of premature discharge, with some patients and health/social care practitioners feeling that individual patients have sometimes been discharged from hospital too quickly in order to make room for new patients.

Successful hospital discharge examples such as the 'Express Service' in Bern (Box 4.5) report significant outcomes, such as reducing the hourly cost of care by 40 per cent. As with many hospital discharge services found in Europe, the Express Service provides a rapid response and comprehensive home care rehabilitation service to those ready to leave hospital. Initial resistance to this new model came from home care staff, as a higher degree of flexibility was required to execute the express service. This resonates with experience in the UK, where traditional home care services are being transformed into 're-ablement' services. Ample time and training is essential in order for home care staff to adjust to new ways of working and to find sufficient alternatives, for example intermediate care or home care.

Box 4.5   Improvement of discharge planning through formal collaboration between hospital and home care organisation (Switzerland)

The Express Service of the home care organisation of the City of Bern is intended for older people who require home care after a hospital stay. About 20 per cent of the (regular) hospital discharges are unplanned.

The hospital, GP, informal carers or the patient may announce the impending discharge to the home care organisation, by mail or by phone, through a single entry point. The request for care is forwarded to the home care centre of the district in which the patient is a resident.

The home care professional, who is usually the person who will take care of the patient at home, makes the initial assessment while the patient is still in the hospital. The hospital doctor gives the information on the spot, in the presence of the patient. Having everyone in one place enables more direct and efficient communication between the parties involved during a hospital discharge as gaps and ambiguities are immediately clarified and home care services adequately planned. Thus, the information flow between both institutions is guaranteed and continuity of care is improved by contact with a single nurse. The needs assessment in complex care situations can take place within 2 to 4 hours after notification from the hospital, even on weekends.

Source: Gobet (2011).

There is a broad scope of initiatives which have shown a positive impact on hospital discharge (see also Di Santo, 2011c). For instance, in one Greek hospital extending the use of ICT in everyday clinical practice has been reported to unblock hospital beds, reducing readmission rate and length of stay by 60 per cent (Mastroyiannakis, 2011b). As with many other examples of innovation in LTC, the work of the e-health unit (see Box 4.6) has yet to filter through nationally. Wider government-supported promotion is seen as a missing link in Greece towards fully embedding the improvements realised through the unit.

Box 4.6   E-health unit (Greece)

The e-health unit is situated in the Sotiria hospital in Athens, an 800-bed general public teaching hospital. The unit is both a research and therapeutic unit for the implementation of emerging Information and Communication Technology (ICT) in everyday clinical practice.

It was established in March 1999 as an initiative to modernise the existing rehabilitation services, with the aim of maintaining chronic patients with multiple morbidities at home through effective IT supervision and monitoring. It also aims to prevent disease exacerbations, reduce multiple hospital visits and admissions and support family carers. The services involve no extra costs for the users and target older people, the majority of whom are socially disadvantaged.

This example demonstrates that the use of concurrent technology in clinical practice can effectively respond to the multiple needs of a very fragile population, by empowering them with the opportunity of equal access to high quality care.

Source: Mastroyiannakis (2011b).

However, experiences with hospital discharge have not always produced positive results, even when modelled on relatively successful experiences from other countries. The UK introduced a 'reimbursement' mechanism, modelled on the Swedish example where social services can be 'fined' by local hospitals if people are delayed in hospital for social care-related reasons. In Sweden this was part of a comprehensive reform of social and health care services (*Ädel reform*) that shifted responsibilities and funding for social care to municipalities. Established in a very different institutional setting from Sweden and as only one of several piecemeal practices to reduce 'bed-blockers', the UK 'reimbursement' mechanism has been a contested area of policy and practice. Thus, while some felt that this would help to concentrate the mind of local managers and provide appropriate incentives for swift discharge, others felt that it would undermine existing joint working and penalise one partner for a whole system issue. Delayed discharges fell dramatically in both countries, but this trend was already observed before the introduction of these measures. At least in the British case it proved relatively short lived and the overall effects of this policy, namely in terms of its impact on emergency readmissions and really addressing the causes of delayed discharges, have been strongly questioned in the UK by later research (McCoy *et al.*, 2007).

Needs assessment

For the coordination of health and social care, certain processes and points along the patient pathway have received particular attention. Transition from acute settings is one such point and initial assessment of individuals to access LTC services or facilities is another. Older people can experience a high level of contact with various health and social care providers, leading to repetitive assessments. These unconnected assessments miss important opportunities to create one holistic plan of care for the service user and to share knowledge among provider organisations about what services are being received concurrently. In addition, having one comprehensive assessment has some implicit benefits for the older person, being less invasive and less likely to cause confusion in terms of managing multiple relationships and care plans. Some countries such as the Slovak Republic, Spain and the UK demonstrate a commitment to integrated assessment policies and tools, at different stages of development.

In Spain both regional and national governments have invested in the application of a common protocol for assessment, the Resident Assessment Instrument (RAI) to facilitate coordination between hospitals, residential settings and home services (see Box 4.7). The core of RAI is information analysis and best use of health care data between organisations via software

Box 4.7   Improving the assessment of people with care needs – The RAI system (Spain)

The RAI is used to carry out interdisciplinary assessment of the needs of people who require LTC. The RAI information has been used to describe case mix and staff requirements in different organisations. This tool allows people to be placed in the most appropriate setting, helps to develop common pathways in the LTC system, and improves resource allocation in health and social services provided. The RAI assessment uses the same professional language across health and social services. A case management team defines the pathways and coordinates how the entire portfolio of resources is used.

Source: Ródenas (2011).

tools and databases. The need for shared IT tools and systems to support joint working underpins the development of this assessment protocol.

The Slovak Republic and UK both have national models for joint assessment processes. In both cases there is a legal mandate for a common assessment framework between health and social care providers, but there is flexibility to develop assessment tools locally. A problem encountered in both countries was an initial drain on administrative time due to the demands of coordinating one process between different organisations. In the Slovak Republic in particular it was felt that the availability of central funding to support the switch to comprehensive assessment would have been beneficial (Brichtova, 2011a). In the UK further barriers to successful implementation of the single assessment process are professional resistance to change linked to a lack of common professional ownership of the process (Holdsworth, 2011a).

This section has highlighted hospital discharge and the assessment of individuals for LTC needs as areas in need of greater clarity in terms of the roles and responsibilities of different organisations and providers. Particular gaps are evident around: communication between providers; lack of or duplication of assessment; inattention to the specific needs of people with dementia or from minority ethnic communities; and inadequate consultation with individuals and their carers. Acknowledging these weak points in the LTC pathway, European countries have focused several initiatives to formalise the hospital discharge process, or experiment with techniques that might enhance the experience or efficiency of discharge. Similar attempts to formalise and clarify processes of individual assessment are apparent, consolidating all LTC needs into one comprehensive assessment.

Incentivising swift hospital discharge has proved to be challenging in all countries. A tension occurs where not all partners share the same incentive or capacity to facilitate acute discharges, especially where social care partners have full responsibility without the corresponding means and resources to achieve their targets. There is a growing consensus that intensive rehabilitative home care and incorporation of telecare devices may result in people recovering faster and being able to live in their own homes for longer (Darkins, 2012; see Chapter 14). European experiences suggest that care workers need to be fully supported as their job changes, with the introduction of new technologies, more intensive methods of working and a clearer focus on the outcome of community living.

Joint working at a structural/system/social policy level

Developments at a national level can have a significant impact on the delivery of local LTC services. Policy development, legal changes as well as guidance and tools produced by national organisations not only 'help', but are necessary components to achieving progress. This section explores some of the most used national level developments in European countries; the legal ability to pool budgets between partners, structural reform consolidating LTC administration and national preventative strategies.

Joint finance/financial incentives

Effective practice in LTC can be incentivised by the ability to pool funding and thus create fully integrated service units (Allen *et al.*, 2011; Colombo *et al.*, 2011). Findings suggest financial sustainability of LTC depends on the degree to which health care and other components of social LTC services will be integrated as part of comprehensive reform steps. Different methods of pooling funds are being developed across Europe:

- In the Netherlands, several LTC insurance agencies have started to commission integrated dementia care, through financial incentives to several providers to pool resources; recently a so-called 'care standard' has been agreed between the various parties involved. This care standard specifies how dementia care across services should be organised. It will set a basis for commissioning across sectors;
- In England, use of 'total place' welfare budgets seek to move decisions over benefit payments to a local level, with welfare spending incorporated into pooled area-based budgets.

The commissioning of integrated dementia care in the Netherlands has been seen overall as a success, leading to earlier and better diagnosis, as well as establishing good local networks for dementia care providers. It also served to identify some factors for successful purchasing practice such as: having clear and documented agreements between partners; incorporating national charities and patient representation; and ensuring commitment from local authorities and GPs to ensure greater success (Allen *et al.*, 2011: 80–1). There is also consensus from those countries involved in fully integrated structures that it is a very labour intensive undertaking, often involving some tensions between organisations and professional groups.

Locality-based (neighbourhood) funding such as English pooled budgets have seen some promising indications from individual projects, resulting in positive outcomes for service users (Audit Commission, 2009).

The impacts of integration have been highly commended in localities and include: improved accessibility to intermediate care, occupational therapy, physiotherapy and district nurses; faster rates of assessment, provision of care and installation of home equipment; and reduced use of acute hospital services.

Evidence of efficiencies gained by forming single structures gives incentives to embark upon the route of pooling budgets and forming joint structures. For instance, in one northern city in England a single commissioning unit was created using a pooled budget. Back office savings are estimated to be around €1.5 million per annum (Allen *et al.*, 2009, p. 3). These savings result from shared systems and overheads used by the integrated unit team. The location of the team in shared premises, with a single health intelligence system, single performance management system and aligned indicators and shared outcome goals contribute to more efficient and focused working practice.

Structural change

Situated at the boundaries between health and social care, LTC services draw resources from both health and social care systems. This means that different funding sources, financing and access principles and budgets usually apply to LTC within each country, depending on the nature of services involved.

Health care is usually provided within the framework of 'mandatory' universal health insurance (financed generally by obligatory tax or social insurance contributions), which aims at being comprehensive in the coverage of the population and therefore tends to be free at the point of delivery. Social care services, on the other hand, have often evolved from social assistance laws. As such their financing relies heavily on tax revenues and users' fees calculated on the basis of the income and sometimes also the assets of the users and their families (see Chapter 3); this is particularly the case for residential care. In a number of countries (e.g. England) access to public benefits is means tested. In response to this financial division, many countries have established distinct finance and governance administration systems for LTC. For instance, Germany and the Netherlands have gone as far as having a separated social insurance and assessment processes for LTC. In Spain, Austria, France and Scandinavian countries the financing of LTC takes place within the tax envelope.

Prevention agenda

Many European countries have made the link between investment in prevention and rehabilitation services for older people and the future sustainability of LTC. Embedding prevention and rehabilitation into national systems promises the potential to reduce demand for more expensive acute and residential care. Steps towards achieving more prevention-based approaches can be seen in many EU countries where preventive measures are becoming mainstreamed into policy and legislation, for example in Denmark preventative home visits for people aged 75 and over are legally enforced (See Box 4.8; Kümpers *et al.*, 2010, p. 40) and in England national guidelines on the development of older people's preventative interventions have been published (Department of Health, 2009).

Box 4.8   Legal basis for older people's prevention visits (Denmark)

Since 1998, all Danish local authorities have been required by law to offer two annual prevention-focused home visits to all citizens aged 75 years or older, living independently.

The visits aim to: help older people use their own resources; facilitate the early detection of signs of illness; and provide health promotion advice and signposting to relevant support services.

The visits are usually carried out by district nurses, but an obligatory health check is not included and GPs are rarely directly involved. As an example, the Odense 'Walk and Talk' project seeks to combine preventive home visits with greater health promotion in order to increase physical activity, detect unmet support needs, encourage older people to join with others in walking groups and/or to set up their own and to enable older people to broaden their social networks.

In 2009, home visitors were trained as exercise consultants for older people. The home visit takes the form of a walk, where health and social issues are discussed along the way and signs of tiredness or dementia are observed. Participant feedback shows general satisfaction and funding has been sought to broaden walk and talk to all sections of the local authority. The authority has identified a need to improve their understanding and provide evidence of the impact of such an approach on future service use and the value of preventive home visits as a mainstream service.

Source: Kümpers *et al.* (2010, p. 40).

The vast majority of initiatives in prevention and rehabilitation tend to be small-scale, time-limited pilot projects. Initiatives for older people remain vulnerable to tightening national budgets and 'cuts', because short-term effects are not expected or easy to demonstrate (or policymakers are not convinced of the value of the effects). Therefore, further evidence on often hard to prove impacts of preventative approaches would be required to strengthen national confidence and commitment to such measures, embed preventative services and achieve the resulting cost benefits for the whole system.

National developments in law, infrastructure and policy have responded to the need to increase volume and quality of LTC services, often in the context of financial cuts to public spending. The legal ability to pool budgets and shift money between organisations, especially between health and social care, has opened up opportunities for more genuinely integrated local approaches and services. As with the other multidisciplinary ventures covered in this chapter, those involved report the labour intensive nature of forming new teams and new organisational identities.

The identification of the fragmentation of financial and regulatory administrations has led to the consolidation of LTC functions in some cases. Part of the rationale behind such restructuring activity was ease of national financial and outcomes monitoring which may have helped to close a large gap in what is known about LTC spending in some countries. However, national comparisons have demonstrated that there might not be a single quick fix from which policymakers could learn, as solutions seem to be dependent on cultural and political traditions as well as on existing structures of each welfare regime (Allen *et al.*, 2011).

Making visible specific national agendas for prevention, as well as for older people's services and dementia, have helped to spread good practice and highlight the importance of these areas in some countries. Yet the small scale of investment, complex needs of beneficiaries and short-term nature of many preventative interventions have made it hard to generate clear evidence of impact. Specifically, evidence of the cost-effectiveness of new interventions tends to lag behind the progress being made locally, impeding the ability to make persuasive business cases for new services.

Conclusion

This chapter identified some of the common challenges for joint working in LTC for older people and common trends developed as a response to these. By addressing these through a framework of the individual, organisation and whole system it can be described and better understood how different stakeholder issues at these distinct levels interact. Of course, ideally the experience and outcome for individuals and their carers should guide the development of LTC services in Europe. Yet this individual experience cannot be divorced from its wider context; from the teams who deliver, the managers who invest and the national agencies and policymakers who govern this process.

Looking at some of the common difficulties in joint initiatives reveals the importance of stakeholders at all three levels. Above all there are key messages for policymakers which emerge at all levels of joint working. Whether working as a frontline multidisciplinary team, coordinating a care programme or assessment involving several providers, or pooling budgets to form a new basis for a service, it is easy to underestimate how resource consuming joint ventures are in practice. Not only are the needs of older people complex and varied, but so too is the mix of professionals and providers that constitute LTC services. In practice this gives rise to vast administrative tasks as well as inevitable tensions of organisational cultures and approaches between professions that take time to work through.

Prolonged and visible national level commitment in the form of funding and guidance are central to continued learning about what joint working in LTC can achieve. Evidence of what works in older people's LTC services can be problematic due to the many variables involved for people with complex needs; this is particularly relevant for preventative approaches where proving what outcomes have been achieved adds a further challenge. For those directing policy and investing in services this suggests that a 'realistic' approach to evidence is required (see Chapter 14), perhaps with a high level of ongoing communication and negotiation between policy-makers and those involved in services, rather than a straightforward reliance on more concrete cost-effectiveness results. Difficulties around evidence also point towards the need to support ongoing work through thorough national evaluations, capturing and learning from what has already been achieved. Against the backdrop of limited evidence base, the collation and analysis of practice examples that illustrate the INTERLINKS Framework for long-term care (http://interlinks.euro.centre.org/framework) offers some rare insights into the real experiences, successes and challenges faced by those involved in LTC services working across the health and social care divide.

5
Converging Methods to Link Social and Health Care Systems and Informal Care – Confronting Nordic and Mediterranean Approaches

Jorge Garcés, Francisco Ródenas and Teija Hammar

Introduction

The substantial growth in health expenditures in European countries over recent decades has brought about serious problems for health care management (Burau, 2007; Dixon and Mossialos, 2002; Thomson *et al.*, 2009) and finance (Directorate-General for Economic and Financial Affairs, 2002a, b; Economist Intelligence Unit, 2011), especially in Mediterranean countries. In 2008, European Union (EU) countries devoted 8.3 per cent of their GDP on average (Spain 9 per cent and Finland 8.4 per cent) to health spending, which was up from 7.3 per cent in 1998 (OECD, 2010). This situation can be explained by the concomitance of demographic, social and cultural changes in Europe (Jackson and Howe, 2003; Lee *et al.*, 2010) as well as by the principles that have guided health care policy over the past 40 years. In Spain, for instance, one of the reasons for the sharp increase in health expenditure – with a total outlay on health that increased from 5.3 per cent of GDP in 1980 to 9.2 per cent in 2009 – is the increasingly arbitrary distinction made between health and social care systems and the lack of long-term care (LTC) services, which is typical not only for Mediterranean welfare systems (García-Armesto *et al.*, 2010; OECD, 2011).

The differences in service infrastructures, and problems in home care and discharge practices are common in many EU countries, for instance with respect to shortcomings in the flow of information and in the continuity of care between hospital and home care as well as between health and social care (Ala-Nikkola and Valokivi, 1997; Gérvas, 2008; Gobet, 2011a; Payne *et al.*, 2002; Witherington *et al.*, 2008). In addition, there is a lack of clarity concerning responsibilities for patients after discharge from hospital to home care and a lack of integration of services based on the patient's needs (Atwal, 2002; Bryan *et al.*, 2006; Payne *et al.*, 2002). Furthermore, hospitals have increased their efficiency in terms of reducing the average length of stay (Cano *et al.*, 2008). As a consequence older people are more frail when being discharged, thus increasing the risk of readmissions if out-patient and community care services are lacking. In this context it has been shown that the first two weeks at home after hospital stays is a particularly critical period (Anderson *et al.*, 2005). Frail older people are therefore a target group that calls for special attention when planning and implementing hospital discharge management, community health facilities or home care services after a hospital stay. Appropriate service packages should contain multiple services provided by various professionals (Anderson *et al.*, 2005; Hedman *et al.*, 2007; Landi *et al.*, 2001; Modin and Furhoff, 2004; Trydegård and Thorslund, 2001; Tepponen, 2009). Processes and procedures should also consider that, in the current situation, older people often feel that their perceptions concerning their health problems or their self-care ability are not sufficiently taken into account when their care is being planned and when services are put into place (Janlöv *et al.*, 2005).

New discharge and home care practices such as case management and multidisciplinary teamwork have been developed to address the above-mentioned shortcomings and also to cut or restrain the costs of health and social care. Perhaps the most common innovation in job profiles has been the development of discharge managers whose task consists of facilitating the transition between hospital and home care services (Nies *et al.*, 2010; Table 5.1). In France, a medical doctor may obtain the status of 'coordinating physician' in care homes following relevant training. In Finland, case management methodology consisting of two professionals working closely together has been used to improve home care and discharge practices. This 'case management pair' is familiar to the patient and informal caregivers as well as with the service organisation in general. With support from the multidisciplinary team (information exchange) the pair is able to effectively integrate the various health and social services as well as informal caregivers' support. By offering adequate care and services at the right time it is possible to prevent or at least slow down the deterioration of clients' functional ability (FA) and health-related quality of life (HRQoL) and improving care efficiency (Hammar *et al.*, 2009). In Spain, new service packages for older people with lower costs have been developed by using case management to link the network of health and social resources with the support of a multi-disciplinary team (Garcés *et al.*, 2011).

All of these pilots were able to show evidence for various aspects of improvement. For instance, in the French project 'Coordinating Care for Older People (COPA)' the rate of hospitalisations was reduced and the quality of life of older people was enhanced through case management (Naiditch, 2011a). The results of the Swiss Kompass project (Gobet, 2011a) showed that by implementing the case management methodology it is possible to shorten hospital stay among people with lower socio-economic status, with complex

Table 5.1 Selected examples of case management practices in Europe

Author (Country): Title of programme	Aim	Target group	Type of case management (CM): professional background, training, duties, caseload	Setting	Results
Naiditch, 2011e (France): Governing the building process of care innovation for Alzheimer patients and their families: the MAIA national pilot project	Re-engineering care delivery, new pathways, optimising resources	Alzheimer patients and their caregivers	Multidisciplinary teamwork CM: social worker, nurse, psychologist, occupational therapist.Caseload average: 15–30 patients.	National pilot project with strong policy support in the specified area, 17 pilot sites	New pathways Involvement of families was poor
Naiditch, 2011a (France): Coordinating care for older people (COPA): teamwork integrating health and social care professionals in community care	Enable older people to stay at home longer, improve coordination of services and develop new pathways and care processes	Frail older people (65+) living in community	Multidisciplinary teamwork (CM, primary care physicians) Long-term CM and temporary CM: nurse. Training. Caseload: 106 patients/CM Tasks: assessing the patient, work with physicians, taking main responsibility for follow-up, cooperation with informal carers, organising in-patient visits and hospital discharges with hospital team.	National pilot programmes in catchment area (Paris, 16th district)	Lower rate of hospitalisation No change in mortality or rate of overall hospitalisation Enhanced quality of life with lower rates of depression No effect on carer's satisfaction nor involvement in care choices

Mak, 2011b (The Netherlands): Case managers for people with dementia and their informal caregivers	Coordinate the care for clients suffering from dementia, improve cooperation	Older people with dementia and their caregivers	Multidisciplinary teamwork (CM, geriatricians, psychologists, dementia consultant) CM: nurses, social workers, experience with psychogeriatric clients + training courses. Caseload: up to 70 clients/CM. Tasks: CM responsible for planning, information and advice; coordinates the care provided	National dementia programme of the Ministry of Health, Welfare and Sports, 16 frontrunner regions	Positive effect on client's wellbeing and informal caregiver's burden. Delayed admission to a nursing home. Positive effects on health care cost.
Gobet, 2011a (Switzerland): Case management for patients of lower socio-economic status experiencing complex somatic and psychosocial problems (Kompass)	Facilitate and improve networking, promote collaboration	People with lower socio-economic status with complex somatic and psychosocial problems	Multidisciplinary teamwork (psychiatric nurses, psychologist, social worker, student social worker) Caseload: 111 patients/CM (approximately 30% of the supported persons were 65 or older)	Pilot project, one area, in the City of Zurich	Shorter hospital stay

Source: Adapted from http://interlinks. euro.centre.org.

somatic and psychosocial problems. In the Netherlands, a specialised case management programme has been targeted on a group of people with dementia and their informal caregivers – with a remarkably positive effect on client's wellbeing and informal caregiver's burden (Mak, 2011b).

Increased multidisciplinary working, the blurring of professional boundaries and the emergence of new professional roles contribute towards an improvement in how the complex needs of older people are met (Triantafillou *et al.*, 2010; Kümpers *et al.*, 2010; Nies *et al.*, 2010; Allen *et al.*, 2011). There is a trend to have professionals work across different settings and perform activities that are not strictly within their traditional professional profile. Coordination of care sometimes facilitates a joint venture on the professional level, with the aim of following up and evaluating (reviewing) care provided by different organisations and professionals. Though still quite exceptional there is also a growing tendency to assign 'boundary spanners' with a cross-functional liaison role. Their task is to integrate different professions, such as in health care settings, where they combine the roles of case managers and primary care nurses. In this context, case management has increased significantly in recent decades in Europe and at an international level because this mode of working has been identified as an effective integration strategy by a wide range of research studies (see for instance Billings and Leichsenring, 2005; Challis *et al.*, 2001; Engel and Engels, 2000; Hammar *et al.*, 2009; Kowalska, 2007; Scharlach *et al.*, 2001; WHO, 2002). Case management aims at matching supply and demand for people with complex needs – those with functional impairment and a high risk of institutionalisation, for example – through the creation of a network of services over time and across systemic boundaries, as well as by empowering clients and their relatives to use the support in a way that best meets their needs. The coordination of care delivery can avoid or reduce the loss of information and duplicated treatments, hence optimising the use of care services.

The implementation of case management as a method to steer hospital use is based on patients' clinical and social data and a professional dialogue with relevant stakeholders to agree upon the appropriateness and timeliness of admission and/or the referral to other, more appropriate resources (Smith and Terrisca, 2000; Walsh and Clark, 2002).

Based on a description and analysis of two distinct cases from Finland and Spain, this chapter aims to show that more efficient alternatives to the above scenarios for people in need of care can be achieved by coordinating both health and social perspectives of patients' care and by enhancing the involvement of patients and their caregivers in decision-making. The chapter will emphasise that the main principles that move integration of care forward are shared visions and aims, shared practice and resources, and also shared risks across care pathways, requiring all actors to identify and define their place and tasks along this pathway. The authors point out that these features were identified as converging principles for the implementation of case management in LTC, independent from the systemic context. Still, as the contextual framework has to be considered as a decisive factor for initiating, implementing and monitoring case management methods, the first section provides a brief overview of case management trends that are evident despite the presence of stark differences between the Spanish and Finnish health and social care systems.

Spain and Finland – different health and social care systems, but similar strategies for improving coordination

In Spain, health care is an individual right and free of charge for all citizens, whereas access to social care requires a specific needs assessment and is based on a means-tested payment. As a consequence almost 68 per cent of people requiring LTC in Spain are cared for by health services. The usual scenario for people with LTC needs, that is, people requiring assistance for the activities of daily life (ADL), is treatment through health care services, especially hospitals, which are particularly costly and, when occupied unnecessarily, incur an opportunity cost for patients on waiting lists. In this context, in December 2011 the Ministry of Health, Social Policy and Equality published the 'White Paper on Care Coordination in Spain' that reviewed regional best practice and made recommendations to better link the health and social care systems.

In Finland, as well as in other Nordic countries, health and social care services are mainly publicly provided and/or financed, even if a growing sector of private service provision (not-for-profit or for profit) has been observed over the past two decades (Vuorenkoski et al., 2008). The Finnish municipalities, i.e. local authorities (n=342 in 2010) have a legal responsibility to arrange health and social services for their citizens including services for older people according to the Primary Health Care Act (66/1972) and legislation concerning the 'Status and Right of Social Welfare Clients' (812/2000). The municipalities can provide services alone or together with other municipalities. The government (state) participates in funding by means of a state subsidy to municipalities. Client/patient fees for health and social care services are paid by the service users themselves. Further to this, a municipality may purchase services from private service providers (profit, non-profit) or provide the client with a service voucher, which can be used to purchase services from the private sector (part or all of the cost).

Spain and Finland are situated in different parts of Europe and have different types of health and social care structures, including how they are funded. Both countries, however, show similarities such as increasing health and social care costs due to population ageing (Table 5.2).[1]

The Finnish policy on ageing aims to ensure that the greatest possible numbers of older people live independently in their homes, supported by informal caregivers and health and social services. Therefore, services

Table 5.2 Population, health and social care structure, cost and funding in Spain and Finland

	Spain	Finland
Population 2010 (European Union, 2012)		
Total (million)	46.1	5.4
Population 65+ as % of total population	17.0	17.6
Population 80+ as % of total population	5.0	4.7
Structure of health and social care	Decentralised health and social care system; Regional Ministries of Health are fully responsible for providing health and social care	Decentralised health and social care system; municipalities responsible for arranging health and social services. Ministry of Social Affairs and Health sets general policy objectives and provides 'guidance through information'
LTC provider mix in service provision (Allen et al., 2011)		
• Public	high	high
• Non-profit sector	low	low (NGOs)
• Private	low (health) medium (social)	medium
Number of people in need of LTC (in 1,000) according to national eligibility criteria, 2010 (European Union, 2012)		
Total	2,485	437
of which:		
• formal care	673	174
• cash benefit or informal care	1,812	264
Health care spending as % of GDP (European Union, 2012)		
Demographic scenario, 2010	6.5	6.0
Demographic scenario, 2050	7.8	7.1
LTC spending as % of GDP (European Union, 2012)		
Demographic scenario, 2010	0.8	2.5
Demographic scenario, 2050	1.4	4.7
Percentage of LTC expenditure by source of funding around 2008 (Allen et al., 2011; OECD, 2011)		
Total (around 2008), public expenditure of LTC as % of GDP	0.3	1.8
• Public	67.2	76.9
• Private insurance	1.7	0.0
• Private out of pocket	30.4	17.8
• Others	0.7	5.3

Sources: Allen *et al.* (2011); European Union (2012); OECD (2011); http://interlinks.euro.centre.org.

supporting those living at home should be boosted to reduce the proportion of people in institutional care. This reduction is also necessary from an economic perspective, because in Finland 69 per cent of total public spending on LTC finances residential care; in Spain this percentage amounts to 59 per cent. In spite of policies that promote 'ageing in place' and the expansion of care at home, a tremendously high proportion of public spending is nevertheless still required to cover costs for only 4.8 per cent of older people living in Finnish care homes and for 4.1 per cent of people living in residential care in Spain (Allen *et al.*, 2011). In Finland, the Ministry of Social Affairs and Health and the Association of Finnish Local and Regional Authorities (2008) have jointly published guidelines – the 'National Framework for High-Quality Services for Older People' – to define quantitative targets of service provision. For instance, of those aged 75 years or above, 91–92 per cent should be living at home either independently or with assistance, of which 14 per cent should receive regular home care and 5–6 per cent of households with older people should receive care allowances. Those targets have not been reached yet, but with 11.4 per cent of older people (75+) who received home care services in 2009 (THL OSF Statistical Yearbook on Social Welfare and Health Care, 2010) the Finnish home care supply is one of the highest in Europe. In Spain, only 4.2 per cent of older people (65+) received support through care services at home (Allen *et al.*, 2011); the average age of these people is 78 (Institute of Older People and Social Services, 2009).

The Sustainable Socio-Health Model (SSHM) in Spain and the integrated home care and discharge practice for home care clients (PALKOmodel) in Finland

The Sustainable Socio-Health Model (SSHM), as promoted by the Polibienestar Research Institute (University of Valencia), consists of a joint reorganisation of health and social care services to respond to people requiring LTC (Garcés *et al.*, 2011). This new model is based on three principles: social sustainability; quality of life and dignified dying; social co-responsibility. These principles include the criteria of sustainable health care in terms of affordability, quality, appropriateness and accessibility (Garcés and Ródenas, 2012). The model was first applied and tested in 2004 and 2005 in the Valencian Community (Spain) with the older population in need of LTC. At that time, the Ministry of Health estimated that around 65,000 older people above the age of 65 were living at home with LTC needs in the Valencian Community, out of approximately 4 million inhabitants (Valencian Government, 2004). The model focused, among other things, on the creation of case management teams and the implementation of new care pathways with the aim of achieving significant savings.

The PALKOmodel was developed as part of the PALKO project (1997–2007) and was conducted by the National Research and Development Centre for Welfare and Health (Stakes) from which it received its main funding,

Table 5.3 Comparisons of SSHM and PALKO models

	Sustainable Socio-Health Model (SSHM)	PALKOmodel
Aim	To design and monitor new care pathways in primary care to improve the efficiency of social and health care for people with LTC needs	To standardise practice and make written agreements between hospitals and home care services, and also within home care, to define practice, responsibilities and support tools. Describing and documenting the patient's whole 'care chain' to and from hospital and home. Assigned a case manager pair to home care patients.
Target group	Older patients (≥ 65 years) and his/her carer	Older people (≥ 65 years) who were discharged from hospital to home care with home care services
Teamwork and CM (professional background, training, duties, caseload)	Multidisciplinary team: a physician, a nurse and a social worker. The team received specific training on the use of referral protocols, resource management and the use of assessment method. One case manager had to deal with 30–35 clients, but the decisions on each client were taken in a consensual manner by the three professionals who were part of the team.	Multidisciplinary teamwork in home care and in hospital was strengthened. The home care team (home nurses, home helps/aides, a physician, a social worker, physiotherapist) identified a named working pair (i.e. a case manager) within the team. This case manager pair usually consisted of a home nurse and a home help, but the pair could also be a home nurse and a physiotherapist or a home help and a social worker. The structure of the pair varied and was based on patients' needs. No specific training.

Interfaces, network	Public health care (primary care centre, palliative care unit at an LTC public hospital, public mental health unit, ambulance service or non-pharmaceutical complementary benefits), public social care services (public nursing home, home help, remote care, technical aids or removal of architectural barriers) and private care services (a day care centre).	Public health care (hospitals, health centres, home nursing/home care, ambulance services), social care services (home help, social worker), private and third sectors, older people and their informal caregivers.
Steps	The patient and his/her informal caregivers were assessed at home by the case management team. Following the assessment, the team designed the care plan with the patient and the informal caregivers, and informed the doctor, nurse and/or the social worker who had previously dealt with the patient about the programme with the pathway proposal to be agreed. Following the agreement, the resources were activated and the intervention started. The team monitored the process and became the reference point for both the patient and the caregivers in relation to all administrative processes.	The case manager pair planned and integrated home care services together with the patient, with the informal caregivers and with other service providers (private, third sector providers). The pair participated in planning the patient's discharge from hospital to home care with hospital staff to guarantee proactive discharge planning (home care staff consultation and meetings). Teams processed obstacles between interfaces concerning information (transfer, contents, tools) and cooperation, and planned an improvement process. For example hospital–home care teams standardised practices and made written agreements between hospitals and home care services.

Table 5.4 Study designs and effects and cost-effectiveness of the SSHM and PALKO models

	Sustainable Socio-Health Model (SSHM)	PALKOmodel
Design and settings	A municipality. 152 older patients, of which 101 were randomly assigned to the intervention group (receiving case management) and 51 to the control group. The sample reflected total population demography.	Cluster randomised trial (CRT), 22 municipalities with 35 patients. The municipalities were randomised to 11 trial and 11 non-trial municipalities. Population in study municipalities represented 14% of the elderly population (65+) in Finland and 14% of home care clients.
Inclusion and exclusion criteria of participants	Inclusion: older patients (65+), referred by doctors, nurses or social workers of the primary health care services, using a referral protocol. The patients need 15 points or higher in social discriminators and 10 points or higher in the health discriminators. Exclusion: low scores on the referral protocol, some psychiatric diagnosis or dementia.	Inclusion: older people (65+) who were admitted to hospital and discharged with home care services. Exclusion: primary admission diagnosis was cancer, dementia or some psychiatric diagnosis as well as those patients who were unable to answer the Short Portable Mental Status Questionnaire (SPMSQ-test, Pfeiffer 1975).
Data collection methods	Patient interviews at home (first week and after 6/9 months), medical and social records and care registers.	Patient interviews at discharge (n=669), 3-week and 6-month follow-ups, medical records, and care registers. Questionnaires to home care staff at 3 weeks and 6 months.
Quantitative measures/ outcome variables	Diseases were assessed according to the International Classification of Diseases (ICD-10). Functional ability (FA) assessed through Barthel (Mahoney and Barthel 1965) and Lawton and Brody Indexes (Lawton and Brody, 1969). Informal Caregiver's Burden using Zarit Burden Interview (Martín *et al.*, 1996; Zarit *et al.* 1980). Use (number) and cost (unit cost) of health and social care services. Patient satisfaction as well as satisfaction of her/his informal caregiver with the pathway.	Functional ability (FA) using Finnish version of the Activities of Daily Living (ADL) (Jylhä, 1985). Health-related quality of life (HRQoL) using the Nottingham Health Profile (NHP)[2] and the EQ-5D[3] instruments. Use (number) and cost (unit cost) of health and social care services. Cost-effectiveness was calculated for changes in HRQoL. Changes in home care personnel's job, work satisfaction and the quality of services.

Time of intervention	6–9 months	1–1.5 years
Results and findings	More than a quarter of patients could be supported better and/or more cheaply through community care services, rather than by a hospital stay.	No improvements in FA and HRQoL, except in physical mobility at a 3-week follow-up in trial group.
	Identified 33 possible packages of services (pathway that integrated personal and domestic care).	No differences in the use of hospital care, readmissions and deaths.
	Decreased the use of acute care hospital) and increased the use of social resources such tele-care applications or day centre.	The use and cost of home care services, visits to a laboratory or physician decreased in the trial group.
	Exclusive use of health care resources was lower in the group of participants in the case management programme than in the control group.	With the NPH the PALKOmodel seemed to be a cost-effective alternative, but with EQ-5D the intervention was likely to be more cost-effective only at a lower willingness to pay for improvements in EQ-5D.
	With the new pathways an additional 16.4 per cent of patients could be treated in hospital at no extra cost.	The model improved the process in terms of a clarified and improved flow of information, defined responsibilities and standardised practice. It also helped integrate services and increase the proactive way of working.

with co-financing by the participating municipalities, the Academy of Finland and the Finnish Ministry of Education. The PALKOmodel is a generic prototype of case management practice that was tailored to Finnish municipalities' needs. The intervention consisted of defining criteria and developing a method for integrated care, including the real-time transfer of information, cooperation at the interfaces of organisations, coordination of services, as well as the creation of awareness concerning shared objectives and cooperation between all those involved in home care, including older people and their informal caregivers. In addition the PALKOmodel consists of elements which promote patients' and informal caregivers' participation in decision-making (Hammar *et al.*, 2007; Perälä and Hammar, 2003). It was assumed that a case manager pair who were familiar with the patient, the service organisation and with the required information and support from a multidisciplinary team were better able to effectively integrate various home care services based on the patient's needs. Table 5.3 shows a comparison of both models' structural features.

Study designs and effects of the models

The effects of both models were studied using randomised trial designs (Table 5.4). In SSHM the patients were randomised to trial and non-trial groups and the intervention was applied to the patients. In the PALKOmodel the municipalities were randomised, the intervention was focused on staff practice and their effect on patients was thought to be due to this changing practice (Bland, 2004; Drummond *et al.*, 2001).

According to the results, both models contributed to decreasing the use and cost of services. In the Spanish model, new service packages were developed while the Finnish model focused on standardised practices at the interfaces and on better integration of existing services (Table 5.4). The Spanish project identified 33 possible packages of services. For example, if the patient was in a short-stay medical unit (STS) with a high need of care for basic ADL, and a high clinical complexity, but had no psychiatric disorders and a family carer available, she or he could be included in a pathway that integrates personal and domestic care of high intensity in their home (Carretero *et al.*, 2007), along with the temporary support of a 'hospital at home' unit.

In Finland, before the introduction of the PALKOmodel no common agreement between home care and hospital staff existed on how to inform each other if a patient was admitted to hospital from home out of regular hours. As a consequence, if this occurred, the home care nurse or a home help would often be unaware and would devote considerable time and energy in trying to locate the patient. Particularly in rural areas of Finland, where distances between the home care agency and the patient's home tend to be very long, such incidences could take up to three or four hours including locating the client's home travel and search efforts. During the

PALKO project, hospital and home care staff made an agreement on how to act in this kind of situation to avoid unnecessary work (Hammar *et al.*, 2007; Hammar *et al.*, 2009; Perälä and Hammar, 2003; Toljamo and Perälä, 2008).

The implementation of the PALKOmodel and SSHM did not demand extra resources or new stakeholders but was funded through current resources and implemented during normal working hours. There were some small costs incurred as a result of the intervention that were attributed to the loss of working hours during the training process (meetings, seminars). There were some differences between the two models regarding staff requirements, however; the PALKOmodel did not require the appointment of additional staff (such as a liaison nurse) as the case manager pair was appointed from within the existing home care team, while the SSHM model required the creation of new teams or a new professional figure ('continuity nurse').

Comparison of results

According to results, home care staff thought that the content of the PALKOmodel was appropriate and the model worked in practice in a meaningful way. Patients were able to be cared for using less service input (visits) through a reduction in overlapping or unnecessary work which decreased costs. The results of the PALKO project suggest that, by developing the discharge and home care practice according to the PALKOmodel, municipalities should be able to offer services to older people with LTC needs more efficiently (Hammar *et al.*, 2007; Hammar *et al.*, 2009; Toljamo and Perälä, 2008). The case management in the SSHM model promotes health and social services delivery at home, management of integrated services packages and the training of multidisciplinary teams (Garcés *et al.*, 2006). These teams are better able to handle the difficulties reported by health care professionals; these difficulties prevent access to benefits and services that are available in the system, but are not easily accessible.

Unfortunately, both models have only been implemented in a restricted number of municipalities, though many public administrations and organisations have expressed their interest in developing services according to ideas emanating from these models in Spain and Finland. No follow-up studies have been made, neither have professionals in new municipalities been trained to use the models because of the lack of resources for a new study. SSHM and PALKO material can be used free of charge, so municipalities are able to develop their services accordingly. In fact, there are municipalities that use this concept when developing their home care and discharge practice.

Discussion and conclusions

In this chapter two new models to support and improve the integration of health and social care for older people in two different EU countries were described and analysed – the Sustainable Socio-Health Model (SSHM) in Spain and the PALKOmodel in Finland. The basic hypothesis of both models was that it is possible to improve efficiency in structures and processes, for example by decreasing overlapping work and reducing the average length of stay in hospital, by avoiding unnecessary readmissions and by improving the patient's quality of life at home. To achieve improvements in structures and processes a case management methodology with multidisciplinary teamwork was applied in both cases to permit the design of personal care pathways supported by health and social care services. The results supported the basic hypotheses. The efficiency of care improved in both models. In Spain, unnecessary hospital stays were reduced and patients' needs were satisfied more appropriately by referring them to other, less costly services that provide equal or better quality of life (Garcés *et al.*, 2004; Ródenas *et al.*, 2008). The Finnish PALKOmodel helped integrate services and increase the proactive way of working. Consequently, patients were able to be cared for using less services (Hammar *et al.*, 2009). Similar findings have been found in other European projects using case management as a method to overcome bottlenecks at the interface between health and social care services (Gobet, 2011a, b; Mak, 2011b; Naiditch, 2011a, e). However, much depends on the specific design of the intervention as well as of evaluation research. For instance, the French COPA project (Naiditch, 2011a) reported a decreasing rate of hospitalisations of older people via the A&E Department room – a result that could not be convincingly demonstrated in the Spanish or Finnish projects.

There were many positive features within the interventions of the PALKOmodel, SSHM and the other European examples cited above.

First, in all models different elements to support integration of health and social care services were successfully applied. The structure (care team, case managers) and processes of care (teamwork, needs assessment, transfer of information, coordination) were improved to achieve a better integration of services and, as a consequence, more appropriate care for patients/ users for less cost was achieved. For example, the Swiss Kompass project (Gobet, 2011a) has approached the concept of case management similarly to the Spanish project by creating interdisciplinary teams consisting of five professionals (two psychiatric nurses, one psychologist, two social workers). The 'case manager pair' was only used in Finland. Furthermore, tools for better integration of services have been developed, for example new care pathways in the Spanish model and written agreements between hospital and home care in the Finnish model. New pathways and care processes have also been created in other countries, for instance in France within the COPA project (Naiditch, 2011a). In Finland, strategies to improve home care services had previously attempted to implement structural changes, such as by combining home care units and functions, and health and social welfare departments at the systems level. Although most Finnish municipalities have merged home help and home nursing in an integrated home care unit, it has been found that old structures are difficult to change and merging home help and home nursing services alone does not guarantee an integrated service provision. As a corollary it may be stated that structural changes are a necessary precondition, but they need to be complemented by integrative processes and tools.

Second, the various models have been established in cooperation with the practitioners in health and social care to define patient pathways through the systems. It was demonstrated that detailing the patients' care and service networks can reveal obstacles in the care chain that can be overcome to improve current practice. In the Spanish model, the patient pathway was described from hospital to home and decisions were taken together, while in the Finnish model a key issue was to enable all actors involved in in-patient care to oversee a patient's whole care process as a continuum that starts at the home of the older person and continues seamlessly from hospital to LTC and back. Additionally, in the French COPA project (Naiditch, 2011a) a participative development approach was used to involve practitioners more keenly in developing the work patterns.

Third, the political context and the need to improve efficiency with limited resources were the main drivers to develop more effective services with a view to increasing demand for health and social care services. Spain and Finland have different types of health and social care structures, including their funding (see Table 5.2). For example, in Spain, home care cannot build on a long history such as in Finland, because in the context of a Mediterranean welfare regime it is mainly the family taking care of older people. The still scarce provision of home help services is generally part of social services departments. Notwithstanding these differences, the application of case management methods produced similar results in terms of usage and cost of services.

However, there were also some weaknesses in the selected interventions described in this chapter. Rubenstein *et al.* (2007) have indicated that experimental care programmes might fail to demonstrate major effects due to the limited services available, the need for intense interventions and the lack of a systematic follow-up. Another reason may be poor implementation of the intervention. The implementation periods of the presented Spanish and Finnish models were quite short which might have weakened their effects, but this was important in order to avoid attrition through patient death or decreasing morbidity. A further aspect concerned behaviour change. To change working practices and workers' attitudes demands time and resources. A very important feature concerns the involvement of practitioners by allowing everyone to take part in developmental work. While participation was achieved to some extent in the Spanish model, the concept of case management had not been fully consolidated, which was partly due to the voluntary character of the intervention and the fact that no economic compensation for staff involved had been planned. As a result, decision-makers established the management nurse as a new professional figure, thus distorting the original, integrated approach of coordination by multi-professional teams. Furthermore, a participative management approach was needed to underpin cooperation and the control of interfaces between different professional groups and different management levels. Managers have to take leadership and responsibility for the development including practical arrangements (see also Naiditch, 2011a).

The approaches, tools and results of the case management models that were implemented in the Spanish and Finnish health and social care environment have produced relevant insights for the development of services for older people in other countries and contexts. Starting from a description and analysis of current practices and problems, changes to improve practice were implemented to make both models suitable for all user groups in different settings and organisations. There is therefore the scope to extend this to other countries. With respect to the case management method in general, similar findings have been highlighted in other interventions such as the Evercare programme. In line with the findings of the Spanish study, the Evercare programme showed that case management facilitated the coordinated use of health and social resources from primary care (Gravelle *et al.*, 2007). Similar approaches to address key issues in LTC delivery can be found in many other European countries (see Table 5.1). In the Netherlands, case management has been used in psychiatry for a long time, showing that it has a positive effect on clients with chronic problems and prevents admission to institutions. The case management method has therefore also been used successfully in integrating dementia care. In this programme, defined procedures and multidisciplinary cooperation led to earlier and better diagnosis and more appropriate care for the patient. Patients and informal carers considered a case manager as very important for continuity and integration of care (Allen *et al.*, 2011; see also Mak, 2011b).

While both the problems of care coordination and the types of solutions by applying case management methods have been similar across Europe, it is interesting to note that many interventions have remained temporary projects only. For instance, while in the Spanish case the implementation of case management on a systems level happened in a clearly distorted way with the main aim to reduce costs within the health system, the Finnish model remained restricted to the original pilot municipalities. Not only in these cases, lack of funding and support by policymakers and public authorities were considered the main argument for stopping innovative approaches – notwithstanding positive evaluation results showing the effectiveness and efficiency of the intervention. The other important issue concerns the interfaces between professionals and older people and their caregivers. Although for instance the Finnish PALKOmodel includes several elements which promote patients' participation in decision-making (Perälä and Hammar, 2003) and the Dutch case management project (Mak, 2011b) involved older people and their caregivers more meaningfully, real cooperation and participation are still rather scarce. Given the dominant role of informal care in LTC and the fact that the user and his/her family are co-producers and co-financers of LTC, this is still a serious shortcoming.

The reasons for immobility in the area of LTC must therefore be found in the wider context of LTC policies that, as such, are still in an early stage of development – not only in the Mediterranean countries. A wide range of pioneers have contributed to the promotion of approaches, concepts, methods and tools to better link health and social care services in Europe, but these efforts have not yet produced a distinct identity of LTC systems (see Chapter 2), LTC professionals and generally acknowledged LTC methods. Case management is therefore often used as a quick-fix, a remedy to mend gaps and to manage the increasing complexity of provider organisations in regulatory and economic terms, rather than in terms of improving the quality of life of users and their carers. Still, case management projects and programmes could also be seen as one of the cornerstones for developing sustainable LTC systems on national and European levels. On the latter, the exchange of experiences and mutual learning as promoted by the INTERLINKS Framework of LTC (http://interlinks.euro.centre.org) could be further developed in the context of the European Commission's 'Open Method of Coordination' (OMC) in the area of LTC.

Notes

1. Kraus *et al.* (2010) have indeed suggested that, when it comes to LTC use and financing, countries as different as Finland and Spain (including Austria, England and France) may end up in the same cluster as their national characteristics show a similar pattern.
2. The NHP is composed of 38 assertions (yes/no) from which 6 dimensions can be derived. The values in each dimension vary from 0 (best) to 100 (worst) (Koivukangas *et al.*, 1995).
3. The EQ-5D is a generic HRQoL instrument consisting of five dimensions: mobility, self-care, usual activities, pain/discomfort and anxiety/depression. After a weighting procedure, a general index value for HRQoL, varying between 0 (dead) and 1 (best), is derived (Ohinmaa and Sintonen, 1999).

6
Integration as 'Boundary Redefinition Process'

Pierre Gobet and Thomas Emilsson

Introduction: initial stance and intentions

The system of long-term care (LTC) is intended to provide a set of health, social and personal care services to older people requiring support in their daily life. In this book it is approached as a field under construction, faced with increasing needs that cannot be adequately met at current level of service provision. These challenges do not only require a quantitative expansion of services, but also necessitate a qualitative reorganisation; integration, coordination and networking are considered the most promising principles in this regard (Billings and Leichsenring, 2005; Nies and Berman, 2004; Stein and Rieder, 2009; Williams and Sullivan, 2009; see in particular Chapters 1, 4, 5 and 15 in this book). These issues are also prominent within the INTERLINKS Framework for LTC (http://interlinks.euro.centre.org); the examination of the 135 key issues and of the numerous practice examples that illustrate trends towards more integrated or coordinated planning, organisation and delivery of LTC services clearly reveal this. The issue of integration and the question of the way it can be conceptualised are of particular interest here.

Several definitions of integration have been proposed. Konrad (1996, p. 6) views it as 'a process through which two or more entities establish linkages for the purpose of improving outcomes for needy people'. Kodner and Spreeuwenberg (2002, p. 3) propose to define it as 'a coherent set of methods and models on the funding, administrative, organisational, service delivery and clinical levels designed to create connectivity, alignment and collaboration within and between the cure and care sectors'. Funding pools, joint needs assessments, discharge agreements or case management for example, are part and parcel of this set of methods. At the same time, these integration and coordination tools serve as indicators of quality, assuming that their systematic use is a valid measure of an integration quotient, and that a higher level of continuity of care is an indicator of the efficiency of service provision (Åhgren and Axelsson, 2005; Vondeling, 2004).

This chapter proposes a different take on integration, which is understood as a 'boundary redefinition process' (Denis *et al.*, 1999, p. 105). Referring to the boundary approach developed in the social sciences during the past two decades (Lamont and Molnar, 2002; Pachucki *et al.*, 2007), integration is considered as a reshaping process resulting from the 'boundary-work' (Gieryn, 1983) performed – intentionally or not – by stakeholders through their activities. The practice examples gathered to illustrate the INTERLINKS Framework for LTC may be viewed as instances of boundary transformation activities. Since they include boundary-work, they contribute to redraw the outlines of the LTC field and its components, thus giving them a new content. According to the approach set forth in this chapter, integration should not primarily be viewed as a way of configuring closer ties between entities. It rather suggests that integration modifies the relationship between entities by transforming the entities themselves.

The nature of boundaries

Boundaries commonly mark the end of an object and the beginning of its outside environment. In this perspective boundaries do not exist for themselves; nor are they viewed as having their own reality. The observer's attention is focused on the outlined entity itself. In this 'centre-fixed approach' (Medick, 2006) objects of the social world are defined in a movement going from core to periphery.

The way boundaries are perceived has changed in the context of European construction and economic globalisation; acknowledging the fact that modern societies reshape an already socialised world – a phenomenon captured by the concept of 'second or reflexive modernity' (Beck, 2003). Boundaries have become the raw material of a structuring process that takes place in a space already organised by previous social activities. As such, they must be seen as possessing their own substance. The boundary approach in the social sciences formally takes note of this new state of affairs by elaborating specific theoretical underpinnings focused on boundaries themselves. Moreover, they become a heuristic and an epistemological principle, things having to be considered from their boundaries. As Chris Rumford puts it: 'We need to think from borders, not just study them' (Rumford, 2006, p. 166). Objects are thus apprehended from the outside, going from the periphery to the centre.

The essay 'things of boundaries' (Abbott, 1995a) is one of the first to illustrate this change of perspective. The author postulates the temporal priority of boundaries: 'social entities come into existence when social actors tie social boundaries together in certain ways. Boundaries come first, then entities' (Abbott, 1995a, p. 860). As an example, he uses social work, a field he sees as arising from a differentiation process between tasks that are subsumed under social work and tasks that are not (Abbott, 1995b).

So, boundaries can be approached as 'sites of differences' (Abbott, 1995a, p. 862). Charles Tilly gives a definition based on the same premises. 'Social boundaries', he says, are 'any contiguous zone of contrasting density, rapid transition, or separation between internally connected clusters of population and/or activity' (Tilly, 2004, p. 214). Two main ways to analyse boundaries may be distinguished:

- The interactionist approach views boundaries as the *result* of differentiation activities called boundary-work carried out by social actors (Lamont and Molnar, 2002; Pachucki *et al.*, 2007). Two types of boundary-work generate two types of boundaries. Symbolic boundary-work originates in the activities of categorisation and classification by which social actors differentiate objects, people 'and even space and time' (Lamont and Molnar, 2002, p. 168). Symbolic boundaries separate people in distinctive groups that compete in the production and institutionalisation of their own principles of classification. On the other hand, social boundaries are concrete, objectified social differences. They arise through social boundary-work that consists of establishing unequal access and distribution of resources and social opportunities among groups of people. The literature on boundary-work focuses on the properties of boundaries, considering their permeability, salience, durability, visibility and on their dynamics; it asks how these are 'created, imposed, defended, bridged, subverted and transformed' (Pachucki *et al.*, 2007, p. 332).
- On the other hand, the systemic approach identifies boundaries as the *source* of differentiation processes; it sees them as difference markers arising in the course of the functional differentiation mechanism to which every system is subjected. As 'membranes' (Eigmüller, 2006, p. 65), they do not only shape things by delineating an inside and an outside, they also regulate the relationship between these two spheres. To fulfil this function, boundaries operate in 'separating' and 'joining', 'segregating' and 'blending' or 'buffering' and 'spanning' (Karafillidis, 2009, p. 111). Boundaries thus simultaneously dissociate and connect. As structural elements, they are able to condition the activities of social actors when they interact with the social system.

For Eigmüller (2006) these two views reflect the 'dual character of boundaries' since they are both products of actors' decisions and producers of social order. Despite the epistemological difficulties of the undertaking, she suggests integrating them. In her understanding, 'boundaries are institutions that indeed emanate from the interaction of actors, but at the same time form units that act irrespective of these actors, their actions and their interests, thus themselves structuring social orders' (Eigmüller, 2006, p. 73).

The definition of boundaries then has to be operationalised so that empirical investigations can be carried out. In this regard, Tilly proposes to distinguish between 'mechanisms [...] that cause boundary change', 'mechanisms [...] that constitute boundary change' and the 'effects of boundary changes' (Tilly, 2004, pp. 215f.). Yet the term 'mechanism' may be misleading, because it could lend a deterministic character to processes affecting or stemming from boundary change. That is certainly not Tilly's intent. In this context, the word mechanism indicates a relational process rather than a cause or a causal sequence. For this reason, Abbott suggests replacing it by the term 'event' that Tilly also uses as synonym (Abbott, 1992; 2007; Tilly, 2004).

Events are bundles of activities performed by social actors pursuing a given goal in a circumscribed social time and place. They are relational processes in the sense that their meaning 'is shaped by [their] surround' (Abbott, 2007, p. 19). When events contain boundary activities, they form or transform boundaries, changing their own environment by shaping new entities or transforming the ones that inhabit it already. On the other hand, entities are any kind of distinguishable social constructs that actors form through the boundary-work contained in the events they are engaged in. Through their boundaries, entities acquire their own profile, their own identity.

Building on Abbott (1992; 2007) and Tilly (2004, p. 215) four types of effects on boundaries can be distinguished. These are inscription-erasure, activation-deactivation, translation and resettlement. A boundary is *inscribed* when it is drawn for the first time. On the contrary, it is *erased*, when it disappears for good. An existing boundary is *activated*, when it becomes more salient as a structuring element. It is *deactivated* when its effect becomes less perceivable for social actors. A boundary is *translated*, when it gets a new 'run' and it is *resettled* when it is placed in a new location.

Each effect is the result of a particular type of boundary-work. *Boundary-raising* activities lead to boundary inscription, while *boundary-removing* activities result in boundary erasure. A boundary is activated through *boundary-closure* activities, while *boundary-crossing* activities lead to their deactivation. Finally, boundary translations are induced by *boundary-shifting* activities and boundary resettlement by *boundary-relocation* activities.

Boundary-work in practice

The chapter uses as its database 94 practice examples (policies, implemented practice, pilot and ongoing projects) identified between May 2010 and October 2011 in 13 European countries by the 15 partner organisations of the INTERLINKS consortium to illustrate key issues of the INTERLINKS Framework for LTC. The themes of the Framework were identified on the basis of an organisational development model (Glasl *et al.*, 2005); sub-themes and key issues were then illustrated by practice examples (see Chapter 1). For present purposes, the practice examples will be considered as bundles of activities carrying boundary-work likely to shape entities. In other words, they will be treated as events.

In order to consider events in all their dimensions, Abbott suggests (Abbott, 2007, p. 8) submitting them to the following five questions:

- What general issue is addressed by the event (theme)?
- When and where does it take place (scene)?
- Who took the initiative of the event and who is targeted (agents)?
- How, with what means is the issue treated (agency)?
- Why should the issue be addressed and with what intentions (rationale and purpose)?

These questions also formed the basis of the template used to analyse the practice examples that illustrate the INTERLINKS Framework for LTC (Billings *et al.*, 2011).

According to the theoretical and methodological considerations presented above, the analysis will provide initial answers to the following research questions:

- What kind of entities are formed or transformed by the events (practice examples) reported by INTERLINKS?
- What types of boundary-work are contained in these events?
- What particular entity profiles result from the boundary-work identified?

The 94 practice examples examined contribute to shaping four entities corresponding to the characteristic structural components and settings of the LTC sector. Twenty-eight of them contribute to defining the role of the older person in need of care; 11 play a part in constructing the role of the informal carer. Both are associated with boundary-raising and boundary-crossing activities. Thirty-two examples determine the construction of the identity of formal long-term care organisations. They include boundary removing, crossing and raising activities. Finally, a specific characteristic of long-term care consists in the fact that it tends to transform the contours of the private home into a kind of informal LTC organisation: 23 of the analysed practice examples play a role related to this specificity; they include boundary-shifting and boundary-relocation activities.

Shaping the identity of the older person in need of care

Older people with care needs are not simply frail, vulnerable and isolated. A definition of their identity is the result of two kinds of boundary activities. Boundary-raising activities distinguish the specific features of the older people that define them as a social group. Boundary-raising activities also contribute to instituting frailty and vulnerability as a new, distinctive social risk requiring collective investment. Boundary-crossing activities, on the contrary, have the goal of promoting the integration of the older person in the community by fostering exchanges with members of other groups and generations. The theme 'identity of long-term care' therefore has a prominent place in the INTERLINKS Framework for long-term care (see Chapter 2).

Boundary-raising activities mainly contribute to establishing older people in need of care as autonomous individuals, emphasizing their particular characteristics as people with their own rights and responsibilities. At the European level, the rights of older people in need of care are set forth in the 'European Charter of the rights and responsibilities of older people in need of long-term care and assistance' (EUSTaCEA, 2010). The same kind of initiatives have also been taken on national and regional levels, as illustrated by an example from Germany (Box 6.1).

Box 6.1   Charter of rights for people in need of long-term care and assistance (Germany)

> The Charter was developed between 2003 and 2005 in one of the working groups of the national 'Round Table Care', a conference which brought together 200 experts representing all realms of LTC including local governments (represented by their national association), federal states, bodies responsible for care institutions, charitable associations, associations of responsible bodies, nursing home supervisory bodies, LTC insurance funds, advocacy groups for older people, researchers and foundations. Work was performed under the joint lead of the Ministry of Health and the Ministry for Family, Seniors, Women and Youth and integrated a variety of stakeholders and perspectives from the realms of health and social care. The Charter was developed to strengthen the role and rights of people in need of LTC by summarising basic and indisputable rights of people in need of assistance, support and care, which are an expression of respect for human dignity and are anchored in national and international legal texts.
>
> *Source*: Stiehr (2011a).

Defined in this way, older people in need of care are considered primarily as members of a broader group of increasing social and political relevance – a group that may progressively impose itself as a true stakeholder in the LTC field.

Yet boundary-raising activities do not only contribute to strengthening the formal protection of the older people in need of care. They further establish the older people as having needs that are distinct from other groups of health and social services users. In some situations, as that of migrant workers, the biography of the older person determines the contours of his or her requirements. Generally however, medical and nursing problems, e.g. chronic diseases and in particular dementia (Box 6.2), represent key determinants for defining and assessing the care needs of older people.

Box 6.2   Alzheimer Café (The Netherlands)

The Alzheimer Café is an accessible venue (there is no indication or registration necessary) for people with all types of dementia, their partners, family, friends, caregivers and other interested parties. They offer the warmth of a café in combination with providing information in an informal way. The guests share their experiences and information. They talk about the possibilities for assistance or search for practical solutions, about their experiences and adventures associated with the dementia.

The opening of the first Alzheimer Café was in 1997. Subsequently, 187 Alzheimer Cafés, in different regions in the country, opened their doors. Most Alzheimer Cafés take place on a regular evening every month from September till June. The meeting lasts approximately one and a half hours. Each meeting has a different theme. In most Alzheimer Cafés they follow the course of the dementia process. The café is run by volunteers. There are no participation charges for visitors. Experts and speakers who give a presentation provide this for free or at the expense of their own organisation.

Source: Mak (2011a).

The specificity of the needs of the older person invites a reconsideration of the conception of the care provided. The development of acute geriatric medicine as a new field of medical care gives an example of this specialisation process: based on an holistic approach to the older person, the geriatric assessment gives as much importance to physical, psychological and social resources as to deficits. It may also give rise to the creation of new job descriptions or professions, as observed in Germany (Dieterich, 2011b). But the needs of older people are not only specific; they are also transversal and cross the traditional professional and institutional work divide. Appropriate care provision is therefore confronted with the challenge to find means for bridging professional and organisational boundaries.

If older people in need of care acquire a specific profile through boundary-raising activities, they are often portrayed as outsiders or as passive 'care recipients'. However, when it comes to LTC, they are seen as having rights, duties, needs and resources like all other community members – although their concrete expression may differ. Some boundary-crossing activities thus seek to span the generational divide and explicitly define the older person in need of care as a full member of the community. Examples such as the 'Neighbourhood Solidarity' project in Switzerland have the potential to tighten the mesh of the social and institutional net by supporting older people through active networking among stakeholders and community members (Box 6.3).

Box 6.3 Neighbourhood solidarity (Switzerland)

The project was initiated in 2003 by a local non-profit organisation. One social worker, one trainee and one methods specialist called 'change agents' have been especially recruited for it. In a first step, they assess the needs of older people and other community residents by means of community forums. In the second phase, they help set up projects likely to deal with the identified problems by associating older people in need of care with other community members who may be able to help solve the targeted problems.

Source: Repetti (2011).

The fundamental rights perspective means that older people are viewed as being able to choose and implement their own life goals as well as possessing all the resources needed for the life they wish to live. Their identity is thus built upon conceptions inherited from the Enlightenment, i.e. a view of the individual as a fully autonomous and independent being, guided solely by reason. It goes without saying that this model only imperfectly reflects the actual conditions of existence of older people in need of care. Dependent on care, they may not always satisfy the implicit conditions for recognition as a fully-fledged 'person'. Thus, they may run the risk that their basic human rights may be contested, albeit in a latent manner. Boundary-crossing events pursue a new paradigm in which true personhood and dependency are no longer mutually exclusive. In order to reach that goal, another conception of the human being is put forward: instead of deriving their status as 'people' from their own characteristics as bearers of inalienable attributes like reason and autonomy, people are defined from outside, through the network of relationships in which they develop. In this way, whether needing care or not, each human being is defined as dependent on others to become a person. In the LTC field, dependency is then defined alongside autonomy as a second universal anthropological feature (Gilligan, 2001; Tronto, 1993).

Configuring the identity of the informal carer

The figure of the older person in need of care is drawn jointly with that of other stakeholders such as formal care providers, public policymakers and, since the end of the 1990s, informal carers.

The figure of the informal carer appears as a mirror image of that of the older person in need of care: while the latter is viewed as weakened and needing to be empowered to affirm his or her autonomy, the informal carer is construed as an essentially autonomous person whose frailty must be recognised and accepted. To date, a wide range of practice examples can be identified that contribute to shaping the specific identity of informal carers, mainly by boundary-raising and boundary-crossing activities. While boundary-raising activities have the goal of guaranteeing appropriate legal and financial protection, as well as establishing informal carers' special needs, boundary-crossing activities aim to endow them with quasi-professional competencies through professional knowledge transfer.

Supported by boundary-raising activities, the term 'carer' has entered the area of legal regulations in many European countries. For instance, in the United Kingdom, the role of informal carers in providing care achieved prominence for the first time in 1995, when the 'Carers (Recognition and Services) Act' came into force. This legislation was a response to the observation that people who provide more than 50 hours of care weekly are frequently 65 or over and that their own health may be at risk. It entitles carers providing a predefined amount of care to request an assessment of their own needs at the time of the assessment of the care receiver's needs. In 2000, the 'Carers and Disabled Children Act' extended the range of carers who may solicit such an assessment; it must now be provided even if the care receiver refuses help from the formal care services. It also imparts more competencies to local authorities: they can supply carers directly with money or services and support them with short-term break vouchers. Since 2001, informal carers have had access to direct payments by virtue of the Community Care (Direct Payments) Act adopted nationally in 1996. Carers receive these cash payments if they are assessed as needing help by social service teams. Direct payments are not meant to be an income, and they must be used only to support the caregiver. Practically, direct payments are intended for the funding of short breaks. The 'Carers (Equal Opportunities) Act' introduced in 2004 represents the last step in strengthening the protection of English carers to date. As people cannot be compelled to provide care, the Act aims to insure that carers have the same opportunities as non-caring citizens enjoy. It states in particular that the carer's wishes relating to work, leisure, education and training are to be considered during the assessment (Holdsworth, 2011a; Holdsworth, 2011b).

In Sweden, the first law to support informal carers was implemented in 1998. It stimulated research about people involved in informal care activities, improving the knowledge of their specific situation. Their support also improved markedly as several programmes aimed at reducing the burden of care were introduced. Since then, the rights of informal carers have been consolidated (Box 6.4).

Box 6.4   Municipal LTC obligations to support informal carers (Sweden)

The responsibility for formal health care and formal social services is divided between three levels in Sweden: the policies are defined on a national level, while on a regional level the counties are in charge of the health and medical care and on a local level the municipalities have to insure the provision of LTC. The shift in recognition of informal care was made visible in the recent change in the Social Services Act. Until the summer of 2009 the law stipulated that municipalities should offer support to relatives that provide care on a regular basis to their kin who have a chronic need of care. The law is now significantly more binding for municipalities, as they not only *should* but *shall* support informal carers. Informal carers' rights for support are thus now treated as fundamental rights. This gives the citizens the right to apply (and appeal if the support is rejected by the municipalities).

Source: Emilsson (2011).

Many of these boundary-raising activities were initiated and supported by self-defence organisations to improve the protection of informal carers and to strengthen their identity among other stakeholders. For instance, in Germany, the development of lobby organisations representing informal carers at the local or national level has triggered the creation of an umbrella organisation called 'We Care' in 2008. The network had 100 member organisations two years later. Besides informal caregivers, researchers and senior advisors participate in its activities that focus, among other things, on avoiding poverty due to caring duties and the reconciliation of work and care (Stiehr, 2011b). Inspired by 'UK Carers', organisations now exist in Scotland, Finland, the Netherlands and other European countries; at the European level an organisation officially called 'European Association working for Carers', better known as 'Eurocarers' (www.eurocarers.org) has been founded.

Generally, boundary-raising activities such as the ones presented above are a part of broader social protection programmes that recognise and guarantee specific needs and rights. Measures implemented in this context include cash or in-kind benefits, for instance the implementation and extension of respite care services (Box 6.5).

Box 6.5   Respite care platform: Organising a range of respite services in the community (France)

It is well evidenced that carers of older people with Alzheimer's disease experience a heavy care burden which has detrimental consequences on their physical, psychological and mental health status and is an important factor in their quality of life and potential risk of social isolation.

Built upon providers that were already managing respite care services such as day care centres or carer support groups, this platform aims at providing information and at coordinating the provision of a broad range of respite care services for carers of people suffering from Alzheimer's disease: short stays in nursing homes; day care, home care; day/night 24hr/weekend care; short holidays, recreation, Alzheimer Café, cultural events, etc. The objective is to efficiently match various needs of carers while respecting the needs of the person being cared for and by ensuring an easy geographical access and financial equity.

Source: Naiditch (2011c).

Boundary-crossing activities contributing to shaping the identity of the informal carer are focused on promoting the reconciliation of work and care (Box 6.6) and the transfer of knowledge from the formal to the informal care sector (Box 6.7). This second type of measures aims to reduce the burden of informal carers and to improve the quality of their work, in particular by means of training and counselling.

Box 6.6   Care Leave Act (Germany)

The Care Leave Act was created to facilitate the reconciliation of employment with caring for relatives. The regulation stipulates that since 2008, employees have been entitled to take leave for a period up to six months to care for close relatives with care needs without giving up their employment or having to fear dismissal. However, the leave does not include wage continuation.

Source: Mingot (2011).

Box 6.7   Care Companions (Germany)

The training of volunteers as 'care companions' through a 120-hour training programme is focusing on the empowerment of family carers, offering information about direct and indirect supportive measures and building bridges towards professional service providers. 'Care companions' are functioning as confidantes who understand the specific problems and burdens that the care situation entails. Situated at the interface between formal and informal care delivery, this initiative has developed a specifically interesting way to combine volunteering, the needs of informal carers and older people in need of care.

Source: Stiehr (2011c).

The view that dependency is a fundamental trait of human existence seems to be of particular relevance in shaping the figure of the informal carer who is defined as a 'dependent' person even though he or she might not be in need of care. Informal carers as 'giving' individuals need to be supported to continue doing so. Yet the attention they deserve is largely based on economic, rather than anthropological or ethical rationales. Depending on the care setting, informal carers may be perceived simply as an unpaid labour force, as co-workers who contribute to the LTC alongside formal carers, or as co-clients since they have to be supported in order to strengthen their ability to provide care (Triantafillou *et al.*, 2010; Twigg and Atkin, 1994).

Beyond these distinctions, informal carers are primarily producers of informal work. Three distinguishing features of informal care work can be noted: it is a private activity, carried out by a person who is not specifically trained and who is not paid. Informal workers are thus not institutionally, vocationally nor financially bound. This makes the provision of informal work difficult to control and to steer, and thus fundamentally uncertain. In that sense, the boundary between informal and formal work becomes increasingly blurred as informal care work moves towards partial institutionalisation – it can be partially financially compensated under certain conditions, and can become partially professionalised through specific training. Therefore, the question of how unpaid care activities should be shared among community members is constitutive of the informal carer identity; this issue also plays a fundamental part in the future development of LTC systems and in the gendered division of labour in the field of care.

Drawing the outlines of the formal LTC organisation

The formal LTC organisation is a central entity of LTC provision. More than one out of three practice examples that illustrate the INTERLINKS Framework for LTC is dedicated to its construction. Some examples entail boundary-removing activities that are key to *formalisation*, the process by which LTC organisations turn into economic production units following the model of industrial manufacturing and general service provision. Others illustrate boundary-crossing activities, showing that LTC organisations are not defined by themselves, as independent organisms, but primarily as constitutive elements of their institutional environment. Finally, several examples promote boundary-raising activities binding in various ways the characteristic traits of formal LTC organisations to form a coherent whole.

The development of guidelines, job profiles and curricula for care home directors may serve as an example of boundary-removing activities supporting this progressive integration of LTC organisations into the economic fabric. For instance, since about 15 years ago, the 'European Association for Directors and Providers of Long-Term Care Services for the Elderly' (E.D.E.) has promoted specific courses for directors of care homes; in many countries, these posts still have not been defined in terms of specific education and job profile requirements. Based on the E.D.E. guidelines, the Greek Care Homes Association (PEMFI) – as one of E.D.E.'s members – elaborated a document summarising the general attributes of a true LTC formal organisation: constantly striving to improve facilities offered to residents, to ameliorate the fit of services to their individual and common needs and to improve the recruitment, the retention and the support of staff, in particular through the promotion of specialised training (Triantafillou, 2011). This association, as well as similar organisations in other countries show that LTC organisations are beginning to position themselves as relevant stakeholders in defining and developing the identity of LTC.

The implementation of management tools (see Box 6.8), still often restricted to one type of structure rather than for quality assurance across the spectrum of LTC settings (Nies *et al.*, 2010), gives another example of this formalisation process.

Box 6.8   RAI-benchmarking: An instrument for leadership and development (Finland)

The use of the Residents Assessment Instrument (RAI) allows a not-for-profit third sector social and health care organisation to conjoin needs assessment, quality control and performance measure. This multifunctional tool also helps to pinpoint the strengths and weaknesses of each care unit and, by calculating its case mix, to estimate the workload of its team members. Furthermore, RAI allows for a comparison of costs and quality between care units by providing a web-based benchmarking database.

Source: Hammar *et al.* (2011).

The economic normalisation of production in the LTC field is also fostered by important efforts made during the last decade to adapt the quality management systems inherited from industry to the specificities of LTC. This was partly due to the implementation of New Public Management strategies in the areas of health and social care. Changes included new (for-profit) providers entering the quasi-markets as well as detailed requests for information from public authorities on services they purchased. But quality development also became an intrinsic interest of LTC providers seeking to become more transparent for potential users, residents and their friends and families, in particular in the area of residential care (European Centre, 2010). As a result, a wide array of methods and instruments has been constructed, from specific quality management systems (Leichsenring, 2011a) to specific outcome indicators to measure quality of care and quality of life, but also other contextual quality specifications (Leichsenring, 2011b).

Events exemplifying boundary-crossing activities illustrate bringing areas, organisations and services together that are normally separate although without erasing their differences. Classical examples are those contributing to the integration of health and social care. Social and health care organisations may move toward an integrated LTC organisation by coordinating activities they generally carry out separately. The single assessment process, in which personal information needed for both health and social care is collected jointly and reported on the same document, is a good example of such an integrated activity (Holdsworth, 2011c). The establishment of shared access points for older people is another way to promote the integration of both sectors (Box 6.9).

Box 6.9   Integrated access point for older people (Italy)

In 2006, the government of the Veneto region established integrated access points for the first time in Italy. Each access point is run by a team of four full-time social workers who are responsible for arranging the necessary services. They inform the older person in need of social and/or health care about the existing facilities, initiate the multiprofessional assessments by filling in the corresponding forms, request services and transfer information.

The integrated access point aims to make every citizen aware of his or her rights and of the opportunities that the service network offers, thus promoting and enabling access. The user is seen as an active and participative subject, involved in the planning and implementation of personalised service arrangements, rather than being a passive recipient of services planned and delivered by others. Currently 17 out of the 21 Italian regions have included an integrated access point in their planning.

Source: Ceruzzi (2011).

Boundary-crossing events are not restricted to the field of integrating social and health care activities. They may also result in closer ties between LTC and primary care or between private and public initiatives. Among the reported boundary-crossing activities, the majority focus on facilitating speedy discharge from hospital by improving the integration between acute and LTC sectors. This type of activity has been implemented in a wide range of countries, e.g. in Austria, Denmark, Finland, Sweden, Spain, Germany, Italy, France, the UK and in Switzerland. For instance, in the City of Bern (Switzerland) an agreement concluded between the home care organisation and a growing number of hospitals provides for the former to be informed of all discharges, even when they are unplanned. The needs assessment is conducted in the hospital by the home care nurse, who ensures the follow-up at home, if her workload allows it. The hospital doctor gives the information on the spot, in the presence of the patient. The nurse is a member of the regular home care team (Gobet, 2011b).

Finally, boundary-raising activities that draw the contours of the LTC organisation as a whole, thus shaping its specific identity, have to be considered. These are conceived as integrated networks rather than as sharp delimited corporations providing specific professional services. The LTC organisations identified in various European countries share common traits: they are territorially rooted and their mission is to provide care to the inhabitants of the surrounding region. Fundamentally, they follow two kinds of integration processes. In horizontal integration, LTC organisations are usually composed of functionally complementary, but legally independent organisations brought together in an umbrella organisation they manage together. The Canadian PRISMA model (Hébert et al., 2009) has been a paradigmatic example of this type of integration. However, networks can also be run directly by the different stakeholders within a cooperative governance scheme (Box 6.10). These structures are set up as 'non-centralised multi-organisational systems' (Dupuis and Farinas, 2010, p. 557); the creation of meta-institutions is not necessary in this kind of organisational setting.

Box 6.10   'Living comfortably in Menterwolde': Integration of health and social services in the local community for people who need care (The Netherlands)

The centrepiece of this organisation is a multifunctional building, which includes 18 apartments for older people (of whom 70 per cent need care) and 26 studios for people with a learning disability. The residents come from the immediate neighbourhood. The centre runs a restaurant and a playground for children which are also open to the local population. It also provides support to frail people in the neighbourhood who live in their own houses.

Several care and welfare organisations cooperate to provide the necessary health and social services. The organisations involved are the housing corporation, the welfare organisation, the care organisation for older people and two services for people with a learning disability. In the meetings, end users are also involved.

Source: Overmars-Marx (2011).

The alternative, vertical integration consists of building an organisation providing the broadest range of services as possible under a single management structure. This integration model, implying the merger of several organisations giving up their independence to constitute a new entity, is similar to the Social/HMO classically found in the US (Newcomer *et al.*, 1990). An example reported from Finland shows how such an organisation can be implemented in a European context (Tepponen and Hammar, 2011a).

Formal LTC organisations are growing into ordinary production units. They tend to adopt standardised management tools, performance measures, quality control and workload estimates. Most of these tools are adaptations of instruments initially developed for industry. Yet formal LTC organisations are also expected to satisfy comprehensively assessed needs. Therefore, they must bring together various sectors of care work, encompassing professionals and informal carers, and specialised and untrained care workers. Boundary-crossing activities thus do not only shape the identity of LTC organisations, they actually represent one of the key characteristics of the services offered. Formal LTC organisations are and will remain primarily defined by their environment, by their ability to connect to other elements and thus to link parts of the system that would remain separate without them. The formal LTC organisation is not characterised by clearly defined outlines, but rather by how far it is able to reach out and bridge different kinds of organisational and professional cultures.

Reshaping the private home as a non-formal LTC organisation

Home is usually a place where people can withdraw from the outside world, where they are shielded from the social environment and find a place for privacy. One would thus assume that home is not a place where external help is required on a permanent basis. Yet all policy papers and legal initiatives in the area of LTC have the goal of enabling older people in need of LTC to remain at home as long as possible, to expand home care services and to support informal carers. Boundary-shifting activities are processes by which the frontiers of the private home are extended so that dependency does not preclude staying at home. Two kinds of events stand out, illustrated by practice examples. The first are dedicated to the development of intermediate care and/or home care, while the others are related to the implementation of care technologies. Beside boundary-shifting activities, boundary-relocation activities also contribute to modifying the contours of what is usually called home. They are found when activities normally carried out in one location take place in another.

As an example of boundary-shifting activities, intermediate care refers to a wide range of services including rapid response, step-up and step-down care-home beds, supported discharge and residential/day rehabilitation. Initially conceived of as an array of services linking primary care, hospital, nursing and home care, it is now focused on helping to avoid unnecessary hospital stays, to prevent inappropriate admissions to residential care and to facilitate hospital discharges. A prominent subgroup of intermediate care programmes is dedicated to rehabilitation (Box 6.11; see also Müller, 2011).

Box 6.11   'As long as possible in one's own life': Home-rehabilitation (Denmark)

Implemented in 2008 in one of the 98 Danish municipalities, this 'home-rehabilitation' programme aims to prolong the period of self-care, to avoid the 'revolving door' effect and minimize the need for compensatory help of people over 65 who go home after a hospital stay. Once home, the patient is immediately referred to a physiotherapist who works on maintaining or, if possible improving capacities. This package of home training and rehabilitation is also proposed to people applying for home help for the first time. By restoring and improving self-care, the measure explicitly represents a paradigm shift in care provision. With the new philosophy, passively receiving care is replaced by a principle of engagement in one's own care as the key to independence and self-determination.

Source: Campbell and Wagner (2011).

In the United Kingdom, a comprehensive intermediate care system has been developed since 2000 as part of a prevention package for older people. Provided on the basis of a single assessment framework, involving professional collaboration, intermediate care is limited in time, lasting generally one to two weeks, and has the specific goal of enabling people to resume living at home (Allen, 2011c).

Boundary-shifting activities are also effectively founded upon new care technologies; these are seen as having particularly promising potential for improving the ability of people with care needs to organise and facilitate their life at home (see also Chapter 12). Apart from home alarm systems (Brichtova, 2011b), a range of devices has been developed and implemented to varying degrees in a wide range of countries, e.g. to improve monitoring of people with cognitive impairment (see Box 6.12) or to improve remote contact with health professionals (Tepponen, 2011; Cordero, 2011). The wide array of examples shows the vitality of a rapidly growing sector. Although many of these devices are still not rolled out or implemented across countries, they may profoundly change the everyday life of older people in need of care living alone at home.

Box 6.12   ICT solutions and new welfare technology facilitating integration (Finland)

The system implemented in 2008 at district level in the south of the country combines different innovative and intelligent scalable care technologies to improve the quality of life of older people with pre- and mild dementia living in their own homes. So, for example, it indicates whether a patient who has got up at night has gone back to bed within half an hour by means of bed sensors. Falls may also be reported. Additionally, it facilitates tracking of patients when they go out.

The system works automatically in the background at older people's homes. Its steering platform is directly accessible on the Web. Thus, it allows well-trained informal caregivers a remote monitoring of the cared-for person's activities.

Source: Tepponen and Hammar (2011b).

Boundary-relocation activities consist of moving the delivery of care usually performed elsewhere into the person's own home. Hospital at Home projects have initially been developed to reduce the length of hospital stay of cancer patients by relocating care to other settings, in particular to the home of the patient. For instance, in France the 'hospital-at-home service' has now been extended to all patients over 65 in need of nursing and social care with unstable acute medical conditions (Box 6.13). The implementation of palliative care allowing terminally ill people to stay at home with their families as long as possible or to ensure appropriate transfer from hospital care to home care (Ruppe, 2011) also contributes to a relocation of the care delivery.

Box 6.13   Hospital at home for older people (France)

Initially developed to reduce the length of hospital stay of cancer patients, this initiative concerns people over 65 in need of nursing and social care with unstable acute medical conditions. Operated through a single access point, organisations providing this kind of service may be independent or part of a hospital structure, while functioning with a high degree of autonomy. The multiprofessional care team is led by a medical doctor who sets the care plan, the nurse being responsible for carrying it out.

Source: Com-Ruelle and Naiditch (2011).

Boundary-relocation activities may also be carried out by transforming formal institutions into private housing as has been done in Denmark in order to deinstitutionalise care work (Box 6.14). They may also be exemplified by shifts in the opposite direction, as in the example of the sheltered housing scheme developed in Kent which entails relocating the private home in the institutional context (Box 6.15).

Box 6.14   Bridging the gap between nursing home and community care: The Skævinge Project (Denmark)

This event is rooted in the decision taken in the 1980s to deinstitution-alise nursing homes by turning them into private housing for dependent people with a high degree of user involvement. For this purpose two main measures have been taken. First, residents, who had previously received pocket money only – the rest of their pension being kept by nursing home management – received their full pension and paid nursing home bills themselves. Second, the implementation of self-care principles required new orientations in staff training in order to change professional attitudes toward older people. Conducted between 1984 and 1987, the project has had a lasting influence on Danish health and social policy.

Source: Wagner (2011).

Box 6.15   Better Homes Active Lives – Extra-care housing scheme in Kent (United Kingdom)

Extra-care housing schemes are a mix between independent living in the community with home care support and assisted living. This form of very high-level sheltered housing is designed for people who can care for themselves, but whose housing is not suitable for self-care. While seemingly living in an intermediate housing facility, the residents have the same housing rights as 'normal' adults living in apartment buildings.

Source: Holdsworth (2011d).

The examples presented show that the home of an older person in need of care is not a secluded and private retreat. It is hardly a space character-ised by intimacy, but rather a frequently busy workplace where services are given and received. Thanks to new care technologies, the older person may not even be truly alone but watched by remote control devices when nobody else is present. This development of LTC may be interpreted in both a positive way, because long-term care at home is being facilitated, but also in a negative manner as a threat to the existence of the boundary between private and public, between leisure and workplace, transforming what is usually seen as a private space into an informal care organisation.

Entities' transformations and stakeholders' action

In this chapter, practice examples that illustrate the INTERLINKS Framework for LTC have been used to outline European trends in constructing and structuring the area of long-term care. The exploratory analysis shows that LTC in Europe is getting progressively structured around the figures of the older person in need of care, the informal carer, the formal and the informal LTC organisation. These four entities are not fundamentally new. Yet they are about to be profoundly reshaped by the boundary-work reported by the selected examples. All four acquire their specific contours through the blurring of well-established differences:

- The figure of the 'older person in need of care' only emerges when autonomy and dependency are no longer viewed as mutually exclusive concepts;
- The development of the figure of the informal carer (including family, friends, volunteers and migrant care workers) puts distinctions between productive and unproductive work as well as between paid and unpaid work into question;
- Formal LTC organisations are struggling to reconcile standardisation and individualisation as well as 'caring' and 'managerial' tasks, while
- The emergence of the informal LTC organisations is predicated upon the fact that the boundary between private and 'public' space has lost most of its significance.

As structuring elements of the LTC field, these entities condition the actions of stakeholders. But what structural effects can be expected from constructs designed to bridge differentiations that are considered in all other contexts as fundamental as well as unavoidable? The analysis suggests that the success of a restructuring of LTC may necessitate at least four conditions.

First, it should designate dependency as one of the fundamental traits of the human condition. If it does not, the resources of the older person in need of care may be permanently overstretched. Moreover, autonomy could increasingly be perceived as a constraint rather than an ideal, and self-care may resemble a disciplinary measure rather than a desirable means to (re)gain independence.

Second, the resocialisation of care work through the support of informal care requires overcoming the profoundly gendered division of care work as well as the gendered distribution of paid and unpaid work and even the general values of 'productive societies'.

Third, through the deinstitutionalisation of care work and its corollary, the 'colonisation' of the older person's private living space, home may become a total institution, in which life can be planned and organised from morning to night. This issue will therefore become as important for home care professionals as it is today in acute health care as well as in residential care facilities.

Fourth, care provision cannot be assimilated to an industrial process or a classical service. To be appropriate, personal care has to be situation-related, individualised and comprehensive. Furthermore, the interpersonal relationship between care receiver and caregivers is a constitutive part of care provision, care then being addressed as a co-constructed solution to a given problem (Du Tertre, 2009) rather than as a product. Viewed in that way, care is clearly an exotic category in the conceptual landscape of everyday economics. Yet, as its importance increases in the future, it may announce the coming of a 'service model' that could progressively supersede the 'Taylorised' factory model with a new mode of production yet to be defined (Du Tertre, 2009).

Conclusion: different approaches to integration

Integration is usually conceptualised as a process through which new methods of working together bring actors and/or things closer to one another and allow them to become more tightly bound to each other. In this perspective, a care system is integrated when dysfunctional barriers are overcome and smoother system function is attained.

The analysis presented above intends to show however that the ongoing integration in the LTC field has a much deeper impact than simply establishing or improving connections and links between existing entities. When seen as the result of boundary-work, integration is viewed as changing these entities themselves at the interface to their environment, giving them a new profile and thus a new content.

As a boundary redefinition process, integration is meaning-related. It may take place without being explicitly intended by the actors. Inversely, actions deliberately carried out in order to achieve integration may miss their mark. Integration, then, does not conform to a mechanistic procedure. There is no definitive 'recipe' for integration.

Boundary activities do not derive their true meaning from the activities deployed but from the interrelations to the entities they transform. A boundary activity that is not related to an entity is meaningless. Inversely, entities that are subjected to boundary activities acquire another meaning, because they relate in a different way to their surroundings.

Further, we notice that boundary-crossing and boundary-removing activities are accompanied by boundary-raising activities. Integration thus seems to imply differentiation as well as merger, as Leutz suggested when he posited 'your integration is my fragmentation' as one of the five laws for integrating medical and social services (Leutz, 1999).

In an integrative perspective, LTC is conceived as a transversal object rather than a newly defined specialised care field. It is therefore better addressed by defragmenting care knowledge, know-how and provision. Integration always invites stakeholders to question the rationales underlying the division of care work. However, there are different ways to conceive of integration depending on the model chosen to analyse LTC. When LTC is viewed as a network, integration involves rearranging links and ties in order to improve connectivity. In a systemic approach, integration is seen as difference management, through which the system and its elements regulate their relationship to their environment. However, if integration is conceived as a boundary redefinition process, as in this chapter, another analytical metaphor can emerge. Rather than a network or a system, LTC is conceptualised as an expanse or an area. Its analysis can then be topological: it is structured by professional, institutional or organisational landscapes characterised by blurry limits and its outside border is conditioned by the shape of the elements that surround it.

Part III

Innovative Cases in the Construction of Long-Term Care in Europe

7
Prevention and Rehabilitation within Long-Term Care: Applying a Comprehensive Perspective

Susanne Kümpers, Georg Ruppe, Lis Wagner and Anja Dieterich

Introduction

In the light of demographic changes, governments and actors in health and social care systems are compelled to search for sustainable strategies to cope with the care needs of ageing populations (Colombo and Mercier, 2011) and a shrinking workforce in health and social care (Tjadens and Colombo, 2011). Therefore, an important element of policy debates and research is the focus on the potential of prevention and rehabilitation (P&R) for older people in order to postpone or prevent care needs. Prevention for older people so far is conceptualised mainly for healthier and younger older people, as can be observed in the European Union (EU)-financed project healthPROelderly (2006–2008) and the European Year for Active Ageing and Solidarity between Generations (2012).[1] For those with existing long-term care (LTC) needs, the issue of prevention and rehabilitation have so far been given less attention in European countries; respective opportunities seem to be underestimated (Kümpers *et al.*, 2010). Against this backdrop, approaches to prevention and rehabilitation for older people with varying kinds of long-term care needs must be explored and monitored and be placed on the health and social care agenda in the future; there is a need for development in research, policy and practice.

Of central concern here are topics with respect to prevention and rehabilitation for older people in long-term care in Europe which relate specifically to a comprehensive understanding of LTC, as defined by the World Health Organization (WHO):

> Long-term care is the system of activities undertaken by informal caregivers (family, friends, and/or neighbours) and/or professionals (health, social, and others) to ensure that a person who is not fully capable of self-care can maintain the highest possible quality of life, according to his or her individual preferences, with the greatest possible degree of independence, autonomy, participation, personal fulfillment, and human dignity.
>
> (WHO, 2000, p.6)

This definition reflects the broader and long-term nature of LTC. While LTC regularly includes elements of health care, it additionally has to consider other broader aspects that are relevant for daily living, such as social inclusion, autonomy, and issues of diversity and equality. Regarding prevention and rehabilitation, the conventional definition within Mosby's medical dictionary is followed (Box 7.1).

Box 7.1 Definitions of prevention and rehabilitation

Primary prevention: 'A programme of activities directed at improving general well-being while also involving specific protection for selected diseases, such as immunisation against measles.'

Secondary prevention: 'A level of preventive medicine that focuses on early diagnosis, use of referral services, and rapid initiation of treatment to stop the progress of disease processes or a handicapping disability.'

Tertiary prevention: 'A level of preventive medicine that deals with the rehabilitation and return of a patient to a status of maximum usefulness with a minimum risk of recurrence of a physical or mental disorder.'

Rehabilitation: 'The restoration or partial restoration of an individual to normal or near normal function after a disabling disease, injury, addiction, or incarceration.'

Source: Mosby's Medical Dictionary (2009).

LTC for older people and P&R measures are not mutually exclusive; on the contrary, they are closely related and intertwined concepts. It is therefore a central concern of this chapter to demonstrate that LTC as a comprehensive care concept does not only integrate specific preventive, health-promoting and rehabilitative measures at different levels, but represents a preventive and protective system in itself. It must be emphasised that integration and quality of services as well as coordination and transition processes for the individual client must be firmly in place.

However, in studying systematic LTC provision and pathway-related P&R strategies, it became obvious that there are significant differences among INTERLINKS partner countries within political and professional debates as well as at the level of service delivery (cf. Kümpers *et al.*, 2010).[2]

Although a wide variety of opportunities can be identified across Europe due to different cultural, political and social preconditions in different countries, there is a common EU endeavour to bridge gaps and to improve coordination between services. This is in order to increase efficiency as well as to *prevent* people from unpromising and sometimes deteriorating LTC processes (European Commission, 2008a). As the term suggests, people often live with and in LTC for years. Quality of life – in itself a multifaceted concept (Kelley-Gillespie, 2009) – should therefore be an important frame of reference, having crucial implications for the P&R potential of LTC. This is especially pertinent in view of the ongoing individualisation of European societies and the trend towards loosening family ties and informal care, coupled with the threatening European economic crisis.

Within the INTERLINKS project, seven national research teams from Austria, Denmark, Germany, Greece, the Netherlands, Slovenia and the UK cooperated in exploring the thematic field of P&R in LTC as one central topic for modelling sustainable LTC systems. After an initial literature review, research teams from West, North, East, South and Central European countries identified relevant common issues for policy and practice, investigated the state of the art in their countries, looking at political and professional debates, scientific evidence, legislative and institutional contexts and good practice examples and analysed strategies regarding service development and delivery (Kümpers *et al.*, 2010). As is the case for the whole INTERLINKS project, the target group is older people (65+), most of whom regularly and for an extended period of time depend on different kinds of LTC services. These services are based on formal (professional) or informal (private) arrangements, and enable older people to cope with physical, mental and social restrictions, and to manage activities of daily living (ADL) and instrumental activities of daily living (IADLs).

This chapter discusses topics that relate to the current state of the art regarding prevention and rehabilitation for older people in long-term care across countries in Europe, which have emerged specifically from a comprehensive LTC standpoint. In the following, these topics are explored from a general perspective, rather than to compare countries with each other and to specify their differences.

The first part of the chapter deals with a brief discussion of the scientific state of the art of P&R in LTC. It shows that evidence on LTC does not yet cover the full range of dimensions from a P&R perspective that are required from a broad and systemic perspective on LTC. The ensuing section therefore describes two types of interventions as opposing points on a continuum, from isolated interventions to systematic P&R strategies, in order to explain the characteristics of a comprehensive approach. Following this, the next section discusses aspects of governance that promote or hinder comprehensive approaches, such as incentives that foster competition sometimes at the expense of cooperation.

The second part features some specific topics that were identified as being inherent to a comprehensive approach of P&R within LTC. These topics also highlight areas where there are vulnerabilities and risk for older people at the interface of P&R and LTC, such as the autonomy of older people with LTC needs, the position of older people with LTC needs and a low socio-economic status and/or an ethnic minority background, and social inclusion of older people. P&R in dementia care are discussed as an exemplary topic of central significance to all LTC contexts in ageing societies.

Selected practice examples that were identified to illustrate the INTERLINKS Framework for LTC will be described to demonstrate the P&R potential of approaches in different countries. The concluding section will sum up and critically evaluate the findings, especially with respect to the reported practice examples and recommendations for policy, research and practice.

The scientific state of the art in LTC as the starting point: too narrow in focus

In order to provide an overview of current research and practice concerning P&R in LTC for older people, a European literature review was conducted to identify the key findings and gaps at European level (for a comprehensive summary, see Kümpers *et al.*, 2010). The overall purpose of the exercise was to map key themes and gaps within the European literature in order to guide national data collection and reviewing. The analysis was divided into three literature types: European scientific literature (texts and publications drawn from eight key medical, health and social care databases),[3] other European-level databases such as the online library of Cochrane, data from WHO and the Organisation for Economic Co-operation and Development (OECD), and literature from EU research and development and public health projects (CORDIS database, website of DG Health and Consumer Protection/Public Health).

The results indicated a rather fragmentary landscape of evidence. There was a strong focus on research and evidence on some prominent types of intervention such as fall prevention measures and programmes, a range of physical training interventions, some specific geriatric assessment tools and preventive home visits, although with a focus on the healthier, younger age cohorts of older people. With respect to more systemic and complex interventions, some positive impacts were demonstrated with regard to multidisciplinary, nurse-led teams across settings for older people with complex needs, and dementia in particular.

One poignant reflection from the literature analysis was that it revealed the scarcity of findings regarding research specific to LTC. Summing up, existing evidence tended to be restricted to a 'medical model' of P&R. Behavioural and health care-related approaches formed the bulk of interventions studied, while social and environmental aspects were scanty; also issues of diversity and equal access or involvement and participation of users and carers in shaping the services remained largely unaddressed. In addition, governance and management debates on topics such as quality assurance and financing of LTC did not seem to consider the inclusion of P&R. Thus the literature did not reflect the broad understanding of P&R in LTC as used in INTERLINKS and outlined in the introduction as well as in the following sections of this chapter.

Two layers of P&R in LTC: isolated interventions versus integrated care as a system-related P&R strategy

Parallel to the European literature review, National Reports from INTERLINKS also indicated that there was a limited and narrow view of how P&R in LTC was understood and applied, neglecting scientific knowledge in this area. In several countries a traditional medical paradigm within health care services still counters the development of comprehensive, patient-centred and needs-led LTC services for a growing number of older patients and clients. Linear disease-oriented interventions following a traditional medical model tend to predominate, at the expense of those that are associated with multilayered, process-oriented and integrated approaches. In most of the countries, this medical dominance leads to crisis-focused, acute interventions, which relegate LTC to a low priority on the political agenda.

One reason might be the underlying tension between the labour-intensive, bureaucratic and economically burdensome nature of LTC provision, and the recognition of the need to respond to the demands of an ageing population by embedding P&R in mainstream services. Coupled with negative stereotyping of chronically ill older people, and the perception of LTC as a 'dead end', this dichotomy between economically driven services and recognition of growing demand leads to lip service being paid to long-term investment in P&R for chronically ill older people, and at the same time to a distinct lack of evidence of sustained, evaluated interventions.

Looking at the types of interventions that can be attributed to P&R within LTC for older people, they can roughly be depicted in two layers – *isolated* (i.e. short-term and/or one-dimensional) and *system-related* (integrated) P&R interventions (Table 7.1).

Isolated interventions comprise a range of focused and conventional measures of P&R, such as vaccination programmes, preventive medical check-ups, lifestyle recommendations (e.g. anti-smoking campaigns, counselling for healthy nutrition, exercise, etc.), medical rehabilitation, physiotherapy and occupational therapy, and falls prevention campaigns including the use of hip protectors and other devices. Interventions of this kind are well implemented in public health policies in most countries. A common feature of these examples is the focus on one specific problem or disease and a strong orientation towards biomedical functions and concepts. While many of these initiatives are valuable and provide evidence of effectiveness, some are not particularly dedicated to older people *within* LTC (Strümpel and Billings, 2008) and others, such as hip protectors, seem to be quite limited in their acceptance and efficiency (Birks *et al.*, 2003) – perhaps also due to a rather prescriptive and, thus, top-down orientation. In Austria, for example, the free preventive medical check-up once a year for all Austrian citizens above the age of 18 (including special examinations for older adults) is used anually by only 10.5 per cent of

Table 7.1   Characteristics of isolated vs system-related P&R measures

Isolated P&R measures	vs	System-related P&R
Independent elements	<>	Context/pathway/user-related
Predominant professional (medical) paradigms	<>	Multidimensional/Multidisciplinary
Short-term effect oriented	<>	Long-term oriented effects
Evaluation possible/evidence-based	<>	Evaluation/difficult evidence base
Long-standing and scientifically accepted measures	<>	Mainly (pilot) projects

Source: Kümpers *et al.* (2010).

men and 9.5 per cent of women in the 75+ age group (Hauptverband der Sozialversicherungsträger, 2006).

However, the advantage of mainly short-term oriented and one-dimensional interventions in P&R is that they lend themselves more easily to evaluation and the generation of evidence-based knowledge, which explains their long-standing professional and scientific approval.

System-related or integrated interventions refer to the direct and indirect preventive/rehabilitative potential inherent in organisational elements, such as elements of coordination and organisation, transition processes between services or the quality of services themselves along the individual LTC pathway of a client. So far the concept of 'integrated care' (see definition in Box 7.2) has often emphasised advantages such as the efficiency, quality and timing of services, smooth transition and administration processes and, thus, higher satisfaction of consumers and provider.

Box 7.2   A definition of Integrated Care

'A coherent set of methods and models on the funding, administrative, organisational, service delivery and clinical levels designed to create connectivity, alignment and collaboration within and between the cure and care sectors ... (to) enhance quality of care and quality of life, consumer satisfaction and system efficiency for patients with complex problems cutting across multiple services, providers and settings.'

Source: Kodner and Spreeuwenberg (2002).

However, the preventive and protective potential inherent in integrated care strategies – integrated care for older people in particular – has either been underestimated or has not had sufficient attention so far (exceptions exist in Denmark; see Wagner, 1997). Effects are however evident, especially from the user perspective. On the one hand, well organised discharge management might directly prevent a person from physical deterioration, if follow-up services or rehabilitation can be provided at the right time and place. Mobile health and social care services might also directly prevent people from unnecessary and even harmful hospital admissions. On the other hand, positive effects of social policies, such as disability-friendly public spaces, might increase social participation, quality of life and satisfaction of older people and in this way indirectly motivate them to take care of their own health and to make use of health-related entitlements (Greaves and Rogers-Clark, 2009), such as free preventive medical check-ups. However, evidence from psychoneuroimmunology, psychosomatic medicine and health psychology demonstrates that an increase in quality of life as well as feelings of satisfaction can additionally have a direct and positive impact on physical health status (Dockray and Steptoe, 2010; Pressman and Cohen, 2005).

A pilot project in Austria, for example, concerned with outreach geriatric remobilisation (see Box 7.3) shows well the preventive and protective effects of integrated and multiprofessional care provision for older people. It explains how a multiprofessional team organising appropriate rehabilitation at home at the right time can not only help to avoid and protect a patient from

Box 7.3   The hospital comes to your home – Outreach Geriatric Remobilisation (Austria)

This project has been designed to remobilise patients with severe multi-morbidities in their own homes and to reintegrate them into society, in order to increase their self-sufficiency and reduce their need for care, thus avoiding the cycle of hospital readmissions ('revolving door medicine').

One of the basic concepts of Outreach Geriatric Remobilisation is that it takes place in an environment that is familiar to patients (i.e. their home). Patients are mobilised where they spend their everyday lives and therefore learn how to cope with the different obstacles they may encounter in their daily routine. In this way remobilisation is tailored to the person's familiar housing environment and can involve the assistance of family members, friends, or carers. Thus, necessary treatment and rehabilitation becomes part of normal daily routine. Additionally, patients do not have the burden of travelling to a clinic for treatment and barriers to remobilisation in the living space can be reduced. Problems are solved where they occur. Thus, patients can be encouraged to live an autonomous life as far as possible and social inclusion can be supported, as the conditions of a familiar environment can be adapted to suit individual needs and to achieve higher quality of life despite certain health problems.

Source: Müller (2011).

unnecessary (re-)hospitalisation, but also allow for increased social participation and improved quality of life.

A role model for this pilot project in Austria is the German home care service, *Domiciliary Remobilisation (MoRe)*, which basically has the same intentions and can be reimbursed by the Social Health Insurance, but which is currently offered at only five locations in Germany. The rather slow spread of domiciliary remobilisation in Germany is explained by low financial attractiveness and enduring traditions in geriatric rehabilitation (Kümpers, 2011a).

Many aspects of integrated LTC, such as improvements in coordination or collaboration within or between institutions and service providers for older people, were not established on preventive or rehabilitative grounds. However they have had an indirect impact as they permit individualised care pathways to be created and thereby take the individual situation and needs of a patient into account.

As research results from INTERLINKS show, a sustainable implementation of preventive and rehabilitative integrated LTC pathways is still faced with recurring difficulties in most countries. Challenges in development and implementation emerge from questions about scientific evidence, and from the financial, political and organisational requirements of complex restructuring. The following highlights some key points in this respect:

- There is a need to provide sufficient scientific evidence and recognition at policy level for the preventive and rehabilitative potential of whole-systems approaches and long-term effect-oriented measures. However, proving that something is prevented is difficult. To evaluate complex interventions provides methodological challenges (Stame, 2004); they cannot sensibly be measured relying only on methods and criteria that are routinely applied to clinical processes. As a consequence policymakers find it difficult to buy into large-scale preventive initiatives.
- It is important to remove fragmented financial budgets and adverse economic/legislative (dis-)incentives which are disruptive for a seamless and 'preventive' approach to service provision. However, outcomes of P&R are often very long term, while political timescales often require more immediate indications of success. Investing in P&R arguably requires a degree of additional funding, but whether it will reduce future demands remains contested.
- Outdated models of paternalistic and disempowering social care must be eliminated in favour of establishing P&R interventions that take all or at least most settings into account. However, P&R are conceptualised in diverse ways, and different organisations/professions are working with different ideas about what these ways of working entail and what should be the desired outcomes.

Developing a more preventive approach has been a stated aim of many governments over the past decades; however, strategies for comprehensive implementation are rare. Greater acknowledgement of this priority at a strategic level and a clear vision of the possible activities, services and benefits of the approach are needed to enhance mainstreaming. It is clear that a sea-change in attitude would be required if future policy is to bring about real change. In the current period of economic crisis, governments are being forced to think creatively for the future and they might learn from project results.

Competition and collaboration as promoting or hindering comprehensive approaches

There is debate about the evidence of the impact of steering instruments, such as increasing market regulations which stem from new public management concepts, on the efficiency of health and social care development and delivery. Greater competition often brings with it a series of linked challenges concerning the best way to coordinate services and prevent excessive fragmentation. However, the discussion about competition and collaboration as opposing forces has moved to a discussion about how to shape and combine elements of both (Ham, 2012). According to INTERLINKS research (Kümpers *et al.*, 2010), the impact of competitive economic measures on preventive and rehabilitative care quality has hardly been explicitly quantified. However, where incentives induce competition with a certain specified goal, such as to enhance local integration, overall care quality is expected to increase (Ham, 2012). This would result from commissioning services strategically so that care providers can collaborate and form partnerships (e.g. England, Denmark, the Netherlands), thus increasing the quality and efficiency of the output. However, this requires that there is, at the very least, a local steering and commissioning organisation in place.

If incentives predominantly strengthen competition on economic gains only, reductions in quality are likely to occur. For example, if social care providers compete for the same clients within the same or overlapping catchment areas, as reported from Germany for example, this sets up an obvious disincentive to simultaneously collaborate and network, since out of necessity their primary objective is to safeguard their economic existence by sustaining or enhancing their market share – at the expense of competing services (Bode, 2005). Therefore, competition within an under-regulated local care environment can seriously impede collaboration, knowledge transfer and joint local responsibility (Dulac *et al.*, 2012). Vulnerable older people from disadvantaged groups are particularly at risk here as they do not have the capacities to organise their own care themselves (Bode, 2005).

The Swiss example 'Kompass' (Box 7.4) shows an attempt to provide individual compensation for local gaps in service integration.

Box 7.4   'Kompass' – Case management for patients of lower socio-economic status experiencing complex somatic and psychosocial problems (Switzerland)

Kompass was developed within the framework of the 'Health Network 2025' programme in Zurich. The Health Network aims to facilitate and improve networking among service providers and to promote collaboration between them. Kompass provides case management to people from lower socio-economic groups, including older people, recovering from acute medical conditions or suffering from chronic diseases. It addresses gaps between acute health care and social care as well as between hospital and home care. The interdisciplinary team includes psychiatric nurses, psychologists and social workers. Team members have to handle a wide variety of issues including medical, psychiatric and social problems as well as domestic violence, abuse, crime, or extreme self-neglect. The main benefit of Kompass is to provide a single reference person for the whole range of problems and needs.

Professional care staff improve coordinated access and monitoring of health and social services, thereby contributing to a better quality of life while simultaneously improving economic efficiency through shorter hospital stay.

Source: Gobet (2011a).

In addition, efficiency pressures in marketised 'care industries' similarly affect home care and care homes (Oldman und Quilgars, 1999). They stress quick performance of care tasks, sometimes characterised as 'care Taylorism' (Houten, 1999), often at the cost of personal and relational qualities of care processes (Malone, 2003). Since the respective systems of care delivery with 'Taylorised' care tasks are based on a reductionist concept of social care, relational and social aspects of care delivery tend to be neglected (Bode, 2006; Klie, 2009; Pfau-Effinger *et al.*, 2007). An interaction of New Public Management approaches and the weakening of comprehensive care philosophies is discussed in this respect (Bode, 2006; Klie, 2009). This development can act as a barrier to the development and implementation of more subtle interactive approaches with preventive and rehabilitative potential, such as activating self-care. The Dutch organisation 'Buurtzorg' (Box 7.5) developed and managed by professionals, however, can be seen as an initiative that runs counter to this kind of general development.

Countries with strong local governance capacities seem to provide a better context for the development of collaborative 'integrated care' approaches to P&R in LTC (e.g. Denmark). In Germany, however, the weakness of local governance regarding LTC is often lamented (Klie, 2009).

Box 7.5 'Buurtzorg' – Care in the neighbourhood: better home care at reduced cost (The Netherlands)

Buurtzorg ('neighbourhood care') is an innovative approach in the Netherlands to deliver home care. The organisational model of Buurtzorg is to have care delivered by small, self-managed teams, consisting of a maximum of 12 care professionals, that are located in the neighbourhood (hence the name). Organisational costs are kept as low as possible, partially by using ICT for the organisation and registration of care. From the LTC process point of view, Buurtzorg introduces a built-in attempt to contact and integrate with other local carers and with informal caregivers. Also, Buurtzorg aims to deliver care to a client for as short a period as possible, by involving and reinforcing the client's resources.

Buurtzorg responds to the client's care needs, tries to find solutions together with the client and his or her informal carers and other formal carers involved, arranges care around social life and supports self-decision of the client about what is necessary. This may explain the very high user satisfaction scores and reduced costs.

Source: Huijbers (2011).

A local integrated care approach has to promote a comprehensive and holistic view of complete care pathways and general health outcomes and has to encompass multidisciplinary perspectives (Ham, 2012). Preventive and rehabilitative potentials, often hidden behind conflicting interests of independent actors, can be discovered and addressed more easily. Also, opportunities for social participation at a local level are communicated better and therefore can be made accessible within functioning formal and informal networks (Falk, 2012). One element of this might be that professional (and other) knowledge is transferred more smoothly, if professional stakeholders conceptualise their tasks from a 'whole-systems perspective' (Kümpers *et al.*, 2006). An example of this is given by the Geriatric Network and Geriatric Academy Brandenburg in the North East of Germany, which provide a bottom-up structure for cooperation and knowledge transfer among health and social care professionals with respect to geriatric patients at the regional level (Box 7.6).

However, a strong local steering and widely integrated system of LTC that promises the delivery of appropriate quality of care and integrated P&R might imply a restriction of choice as providers might form a monopoly. The tensions between an integrated system (see also Box 7.7) as against the advantages and inconveniences of a fully developed (and therefore fragmented) market should be explored further in future research.

Without doubt, the role of a supportive institutional framework for collaboration equipped with some power at a local level (formal or informal), must be emphasised. Innovations such as preventive and rehabilitative approaches in LTC need organisational coherence and coordination rather than fragmentation. This kind of (social) innovation requires new professional goals, routines, values and forms of multiprofessional/inter-organisational cooperation (Ham, 2012). Such a restructuring process was realised more than 25 years ago in a small Danish municipality (Box 7.7) but, unfortunately, rather than spreading all over Europe, the model has been endangered by recent municipal reforms.

Box 7.6   Geriatric Network (GeriNet) Brandenburg (Germany)

GeriNet is an 'umbrella-network' (at federal level, Land Brandenburg, Eastern Germany) of several regional networks for geriatric care, with hospitals with geriatric wards, GP practices, home care providers and therapists participating. At the centre of each network is a geriatric hospital or ward. Main approaches are regional network-building and the development of an inclusive quality concept for comprehensive geriatric care, part of which is the provision of joint geriatric training for various health professions, and the regular attendance of professionals in quality circles. Aims are to improve continuity and quality of care for older people at a regional level by increasing geriatric knowledge across professional and organisational boundaries, including prevention and rehabilitation. Medical as well as nursing, health and social needs are dealt with. The holistic view on the needs of frail older people helps to adopt a preventive and rehabilitative attitude. Family carers are addressed as 'external members' of the geriatric team, simultaneously having support needs of their own.

However, because of missing financial incentives to engage in outpatient health care for older people, participation, especially by general practitioners, is unsatisfactory in some of the regions.

Source: Kümpers (2011b).

Box 7.7   Bridging the gap between nursing home and community care: The Skaevinge project and prevention of hospitalisation (Denmark)

The Skaevinge project shows how changing an organisation and its culture can prevent hospitalisation. The project is approximately 25 years old. In 1986, Skaevinge municipality introduced the concept of 'integrated care'; traditional barriers between nursing homes, home care, etc. were removed. Via a bottom-up approach, citizens, staff groups and politicians developed more inclusive attitudes together towards older people and their families, based on shared and horizontal decision-making. Prevention was the rationale underlying all measures in the Skaevinge project. This included a focus on early intervention during the course of illness (both before and after hospitalisation). The aim of early intervention was to implement flexible arrangements before and during illness in order to prevent crises and breakdowns. In practice this was carried out by improvement and adjustments of acute care facilities in the health centre, and while these tasks normally were the responsibility of the hospital, the Skaevinge municipality assumed co-responsibility for the assignments as well. Loneliness could be prevented and social networks maintained by different strategies used to facilitate social relations, being based either on professional concepts or on ideas developed by the older people themselves.

New laws in housing (deinstitutionalising) were implemented; moreover, former nursing home residents who previously had to spend all their pensions on nursing home fees regained control of their pensions. Ten years later, evaluations showed that new ways (self-care) of delivering services and caring processes, as well as new values, had emerged and had led to an innovative service philosophy.

Source: Wagner (2011).

Facilitating institutional frameworks are needed to enable practice development of new forms of stakeholder-driven working arrangements. Scanning the National Reports, a variety of steering mechanisms are operational. These are either state, local authority (England, Denmark) or governance functions implemented in regionalised insurance institutions combined with municipal responsibilities (The Netherlands).

Vulnerable topics at the interface between prevention and rehabilitation and long-term care

Having highlighted the requirements, barriers and opportunities of comprehensive P&R interventions in the above sections, the discussion now turns to specific areas where there are vulnerabilities for older people at the interface of P&R and LTC. In the following paragraphs, autonomy, equal access and diversity, and social inclusion as well as dementia care will be discussed in the context of P&R within LTC, and existing barriers to their implementation will be highlighted. Again, examples of good practice for comprehensive P&R in LTC will show opportunities for overcoming these barriers.

Autonomy

The meaning of autonomy has different interpretations that are important to make clear in order to connect them to the concept of LTC. The original denotation refers to self-determination – the focus here – while sometimes it is interpreted as being independent from help and support. This latter meaning cannot generally apply to people with LTC needs (Kümpers and Zander, 2012).

The ideal of autonomy supports care policies that focus on the resources, abilities and decisions that older people make themselves with regard to their care needs; these factors should stimulate professionals to support their clients' functioning in daily life. Therefore autonomy can have a preventive and rehabilitative effect. Older people's freedom to live autonomously in spite of LTC needs may be limited in practice by restrictions in the kinds of options available to them, due to physical and mental health problems, financial constraints, environmental barriers, limited support services and the necessity of adapting their lives to fit in with other family members. Nevertheless, the ability to compromise and prioritise the importance of different elements to an individual's well-being and quality of life has been identified as a major factor contributing to 'successful ageing' (Kelley-Gillespie, 2009). Both informal and formal caregivers can help to maintain autonomy in older people by supporting them in making often difficult decisions about their own care. A Dutch project aimed at enhancing autonomy and empowerment of people with LTC needs shows how to integrate issues of care, quality of life and community life (Box 7.8).

Box 7.8 Living Comfortably in Menterwolde: integration of health and social services in the local community for people who need care (The Netherlands)

Comfortabel Wonen Menterwolde (Living Comfortably in Menterwolde) is an infrastructure in the local community which provides a range of public health and social services. The De Gilde building is a multifunctional meeting centre which contains 18 apartments for older people, of whom 70 per cent have a need for care, and 26 studios for people with a learning disability. The strength of the project is the cooperation between health and social services, which not only focus on care facilities, but also try to find solutions for living independently. In the meeting centre, several organisations provide services such as home care, a playground for children, activities for older people and people with a learning disability, such as computer courses and cooking. A restaurant for local citizens is also available. The activities focus particularly on social interaction. The health and social services which are provided in De Gilde support people who live in the apartments, but also vulnerable people in the neighbourhood who live in their own houses.

The main benefit of the Living Comfortably in Menterwolde project is that older people and people with disabilities who have care needs can remain independent in their own neighbourhood in a meaningful way. Empowerment of vulnerable people is an important part of the project. People who are in need of care become more independent, empowered and integrated into the community. People experience care close to their homes, they have opportunities to meet each other and, importantly, feel safe.

Source: Overmars-Marx (2011).

The current trend to focus on ideals of autonomy and self-care (Holstein and Minkler, 2003) reflects a change of attitude in policymakers, professionals and users, emphasising the greater responsibility of the individual and family, and has led to the provision of funds instead of services. The position of the consumer is arguably also strengthened by counselling and advocacy (Klie, 2009), in order to promote informed consumer choice. To be effective however, this kind of policy has to be guided by sustained responsibility and accountability on the part of public services, otherwise concerns may arise with regard to ignoring vulnerable groups and 'blaming the individual' if care needs are insufficiently met (Holstein and Minkler, 2003). Other constraints such as local availability of appropriate services, lack of personal information and lack of financial means to pay for services, as well as inappropriate support for those with mental health needs, may also act as restrictions to a person's free choice and have to be considered. These matters differ, not only between European countries, but also between regions within countries.

A project in Bremen, Germany, was set up to help socially disadvantaged and isolated older people with emerging LTC needs to manage their lives and to prevent them from dropping out of social networks and failing to get access to services they were entitled to (Box 7.9).

Box 7.9 Social work with older people – Frequent home visits for socially isolated older people/people with low levels of care needs (Germany)

This pilot project was initiated by the state government of Bremen involving local municipalities, care insurers, charities, churches and volunteer organisations. Project aims were to reach older people with a low level of care needs who were (in danger of being) socially excluded and disadvantaged and having difficulties managing their lives. Thus the main target group were older people without mobility problems living in isolation or with insufficient support from their families or neighbours, unable to overcome isolation by themselves.

Health and social care needs were assessed simultaneously and a personal plan set up with the user. Depending upon the plan, assistance from professional carers and/or volunteers could be made available, visiting once or twice a week. Volunteers might potentially reduce any barriers to integration, encourage external engagement, and help them to access social care. After the first year of the project, a group organising regular outings was initiated by some of the service users themselves. Also mutual visits among service users became more common. An evaluation signalled satisfaction and an improvement of quality of life in service users.

Source: Weigl (2011).

Such approaches to empowerment offer solutions to enhance autonomy and independence of disadvantaged individuals without delegating responsibility for care away from the health and social care systems. However, these approaches so far only exist at a few locations.

Access and diversity: challenges to equity

In most European countries the number of older people from ethnic minorities is growing. Their specific needs are often neglected in mainstream services. Also, aspects of gender or specific needs of socially disadvantaged populations are seldom taken into account when conceptualising health and social care services. In particular, older women with care needs who are dependent on the respective LTC system are at risk of both social exclusion and having any care needs unmet (Theobald, 2006). Despite this, there is a growing awareness of the diversity in older people and discrete services are being set up specifically to deal with the needs of these population groups. In the UK, there is enhanced political attention on discrimination that occurs due to age and ethnic background; in the Netherlands there is an emphasis on integration of older migrants in voluntary work, and this is seen as a measure of social inclusion and health promotion; in Denmark, a series of pilot projects concerning older migrants and health were funded (prevention, participation in service development, dementia care); and from Germany, individual projects for older migrants are reported. However, in none of the countries is there extensive work in this area, so that the need for further development in research, policy and practice is clear – on the one hand for services for socially disadvantaged groups and on the other, for an increase in diversity-competence within generic services.

Equality of access to services for vulnerable and heterogeneous groups of older people is a key policy issue throughout Europe. In a report by Huber *et al.* (2008) titled 'Quality in and Equality of Access to Healthcare Services', which is based on findings from eight European countries, it has been demonstrated that difficulties in accessing health and social care services are in most countries the result of various compounding and interacting factors such as poverty issues, organisational and geographical barriers as well as health beliefs or limitations in health literacy. Poverty and cost-related issues sometimes appear as underestimated barriers, as EU countries generally provide universal public health coverage and other networks of social security. However, as frequent users of social and health care services, older dependent people experience high financial burden due to cost-sharing requirements, private out-of-pocket expenses or insufficient exemption rules. In addition, in many countries, informal payments are expected in return for speedier, better or more personal services (Huber *et al.*, 2008). Limited access of older dependent people to integrated health and social services directly as well as indirectly results in a loss of P&R opportunities for this target group. A Greek initiative shows an opportunity for helping older people living in poverty who would otherwise have no access to appropriate care (Box 7.10).

Box 7.10   Archdiocese of Athens Social Services Department – Christian Solidarity (Greece)

While social and charitable work has always been expected of and provided by the Church, today the current programmes of the Church's Social Services Department target all vulnerable groups, including older people, with respect to their rights to benefit from social protection regardless of race, sex, age, religion and ethnicity. Easy and open access to the neighbourhood church and knowledge of the social circumstances of the older members of the congregation ensure that appropriate and timely support can be available to meet their changing needs, through a variety of professionally organised services, which rely heavily on volunteers.

The department's staff includes social workers and psychologists; their work with older people addresses acute socio-economic and health crises involving not only isolated older people, but also those whose families are no longer able to meet their needs for care. Apart from spiritual, psychological or mental support and guidance, actions of parishes for those in need include: regular financial support; free meals; residential care accommodation and home care services for older people (with high quality of services, but at affordable cost); free clothing and footwear; summer holidays in special facilities built or adapted to serve older people. Older people are encouraged to participate as volunteers and as leaders, and, while enjoying the services that are provided, they also enrich, support and nourish the variety, quality and quantity of the work of Christian Solidarity.

Source: Mastroyiannakis (2011a).

As a common theme, spatial disparities in access to services emerged from several National Reports. Chronically ill older people in rural, border or inaccessible areas often have less access to P&R services. The aforementioned report by the European Commission (Huber *et al.*, 2008) identifies the lack of mobility, organisational issues and adequate transport as well as limited service infrastructure in rural areas as important reasons for such disparities, which can have a significant impact on health outcomes. A recent German study showed that regional disparities interact with scarcity of resources of groups of older people with low socio-economic status and influence their chances for maintaining autonomy and social inclusion (Falk, 2012; Heusinger, 2012).

Enabling older people in LTC to stay socially included

There is evidence that social isolation of older people has a negative impact on their health (Cattan *et al.*, 2005; Holt-Lunstad *et al.*, 2010). Prevention of social exclusion and isolation therefore needs to be highlighted as an issue in comprehensive approaches to LTC, as it is neglected in many common LTC strategies and research. Social inclusion or exclusion of older people facing LTC needs is linked to several aspects of vulnerability, deteriorating health and being in need of LTC, as demonstrated in an Australian qualitative study (Greaves and Rogers-Clark, 2009). Isolation and the impossibility of social participation is furthermore a frequent situation for people with LTC needs, linked to restrictions to mobility, and social participation is often central to what they miss the most (Heusinger, 2012). However, intervention knowledge and evidence is still limited (Dickens *et al.*, 2011), and the importance of, and in particular the preventive and even rehabilitative effects of 'social inclusion' is barely acknowledged in common P&R strategies for older people, especially for those with LTC needs. Therefore, social inclusion must be tackled as a cross-cutting issue in LTC policies.

Existing research (Dickens *et al.*, 2011) indicates that in order to target social isolation of older people, group-based interventions are superior to one-to-one interventions. However, group interventions for this target group are especially demanding in relation to mobility and psychosocial factors. An example in the French-speaking part of Switzerland shows an approach that is focusing on the needs and wishes of older people, simultaneously of those with LTC needs, their informal carers and healthy older people. At the local level, it succeeded in (re-)organising the community, fostering social inclusion and supporting older people with LTC needs living at home (Box 7.11).

Box 7.11   *Quartiers Solidaires* – Neighbourhood Solidarity (Switzerland)

The Neighbourhood Solidarity concept started as a local project in 2003 in Lausanne. The underpinning philosophy of the programme was to help vulnerable older people to remain at home to improve the quality of their life, foster the integration of older people in their neighbourhood and enable informal carers to cope with the difficult situations they are facing. To achieve these goals, finding new ways for 'reinforcing neighbourliness' was deemed essential.

After investigating the needs and resources of the inhabitants of a neighbourhood and assessing structural resources in the area (such as social and medical care structure, shops, benchs, meeting places, etc.), a social worker organised community forums for older people and their relatives that aimed to identify problems and assess related needs. The forums are used as a 'think tank' for finding solutions. Projects selected through the forum are developed by the social worker together with neighbourhood volunteers (for example, in one neighbourhood, every Monday three retired people organise meals for 30 older people or a café is opened and manned by retired people one morning per week).

An evaluation showed that solidarity and participation of older people in neighbourhood life had increased: social contacts of beneficiaries have improved and older people are more involved in social activities (Genton, 2008). Individual welfare and the overall quality of life of the community were improved. Moreover, better interaction between inhabitants and professionals was promoted.

Source: Repetti (2011).

The examples in this section show that, depending on the respective drivers for developing specific projects, 'social inclusion of older people' might be either a primary focus of initiatives or it may become an indirect and secondary consequence of an initiative.

Special topic: Dementia

Dementia is a major topic in LTC. First, in ageing societies the prevalence of dementia is rising, in both absolute and relative terms. In residential care it will be the main task in the coming decades. Coupled with this, dementia care poses specific challenges for professionals as well as informal carers and requires special efforts regarding professional qualification, training, knowledge transfer and support structures. In responding to this double challenge, some activities at the European level focus on how to prepare health and social care systems. For example, the Alzheimer Europe project 'European collaboration on Dementia – EuroCoDe', financially supported by the European Commission under its programme for community action in the field of public health (2003–8), compiled knowledge about dementia care regarding social support, psychosocial interventions, risk factors and prevention (Audit Commission, 1997). The Council of the European Union (2008) decided on 'Council Conclusions on public health strategies to combat neurodegenerative diseases associated with ageing and in particular Alzheimer's disease'. Regarding LTC, Council Conclusions did not relate explicitly to P&R;

however, they made a point about integrating health and social care, and about promoting secure and friendly environments and social inclusion for dementia patients and carers, thereby raising issues which were identified as having preventative and rehabilitative effects. In the 'Dementia in Europe Yearbook 2008' by Alzheimer Europe, strategies for primary prevention[4] to delay the onset of dementia are particularly featured. These do not relate specifically to P&R within LTC. Recommendations for social support for dementia patients and carers, however, list multiple interventions suitable to promote patients' and carers' health and well-being including information, advocacy, psychosocial counselling, flexible respite care, which taken together support patients and carers and their interactions in order to prevent precarious care situations and avoidable deterioration in quality of life.

Recommendations are also directed to empowerment and self-determination – as dementia especially endangers these aspects of quality of life. Furthermore, the recommendations underline that in this field services also have to be orientated to cultural and social diversity. Within the 'General Framework for Care and Support' (Alzheimer Europe, 2008, p. 29ff.) some topics are highlighted with respect to P&R in LTC for dementia patients and their carers. First, as an element of secondary prevention, the importance of early diagnosis is referred to as opening access to subsequent services and treatment (including medication) to 'use all possible means to maintain both the quality of life and functional capacity of people with dementia' (Alzheimer Europe, 2008, p. 33). Furthermore, a general rehabilitative approach shaping the complete care pathway of people with dementia is postulated, referring to the rationale of the International Classification of Functioning, Disability and Health (ICF) (WHO, 2001). An example of a comprehensive approach addressing people with dementia and their carers with both preventive and rehabilitative elements is provided by the Dutch Meeting Centres for people with dementia and their informal carers (Box 7.12).

Box 7.12   Meeting centres for people with dementia and their informal caregiver(s) (The Netherlands)

Meeting Centres were developed by the Vrije Universiteit of Amsterdam in the 90s. They integrate different types of support and offer a wide range of activities, collaborating with health and welfare services. People with dementia can participate in recreational and social activities, such as reading newspapers and receiving reminiscence training. Informal caregivers can visit informative meetings, discussion groups and get assistance with practical, emotional and social problems. Consulting hours, social events and excursions are available for both client and informal caregiver.

Thus the support programme integrates practical, emotional and social support for both client and informal carer. People with dementia using the centres show fewer depressive symptoms and have a higher self-esteem than clients in regular day care. Furthermore, admission to a nursing home is postponed. Carers experience more support, have an increased feeling of competence, and feel less burdened.

Source: Mak (2011b).

As dementia affects all aspects of a person's capacity, care pathways have to be shaped to use all opportunities to recognise, support and (re)train lost and remaining skills and functional capacity (Alzheimer Europe, 2008, p. 31). In this sense, the main elements that have been already identified with respect to general requirements for P&R reappear within LTC in dementia care.

Conclusions

This chapter aimed to explore the state of the art and the current develop-ment of P&R within LTC, starting from what is currently understood by the term 'LTC'. It has been confirmed during the INTERLINKS project that this topic needs further development – at research, policy and practice level. However, remarkable differences between countries have been seen with the UK, Denmark and the Netherlands all demonstrating significant elements of person-centred and comprehensive approaches. Central European countries such as Germany and Austria however are still in discussion about such approaches rather than having them extensively implemented, and Eastern and Southern countries are still in the process of searching for systemic development of their LTC in general (Kümpers *et al.*, 2010). However, as was mentioned above, this was not the focus of this chapter.

The previous sections sought to bring to the fore the various dimensions that P&R as a cross-cutting theme for older people within LTC may have. After reporting the scarce evidence base for comprehensive approaches within LTC, the 'nature' of comprehensive approaches was discussed. In order to identify comprehensive and sustainable strategies of P&R, characteristics of 'isolated' versus 'system-related' P&R were contrasted as ideal types along a continuum. As barriers to the sustainable implementation of preventive and rehabi-litative LTC pathways, the following recurring difficulties were identified: it is a complex and difficult process to provide sufficient evidence and recognition at policy level for the preventive and rehabilitative potential of whole-systems approaches and sustainable measures; it is challenging to overcome outdated models of paternalistic and disempowering care approaches; and it seems almost impossible to eliminate fragmented financial flows and adverse economic/legislative (dis-)incentives which are disruptive to a seamless and 'preventive' approach to service provision.

Thus, preconditions for an integration of P&R aspects into a comprehen-sive LTC approach are multiple and complex. They extend to questions of governance. Governments' efforts to promote the integration of P&R in LTC need to handle steering instruments carefully. Market instruments can counteract collaborative processes needed to shape comprehensive care pathways. Therefore competitive instruments need to be oriented towards enabling or at least not undermining the cooperation needed for comprehensive LTC pathways. Forms of local governance also impact significantly on opportunities for integrated LTC pathways and shape the ability of users to participate in these matters.

However, even within adverse or at least ambiguous circumstances, several projects or initiatives showed promising approaches able to counteract or even use these conditions in an unforeseen constructive way: The Dutch example 'Buurtzorg' establishes a seemingly successful professional-based and comprehensive outpatient approach to home care in a highly competitive and bureaucratised environment (Huijbers, 2011). With a more complementary character in relation to mainstream services, German and Austrian concepts of geriatric rehabilitation at home are used effectively – given the successful local implementation – to remedy service gaps by putting multidisciplinary teams into service in complex care situations.

Coming into perspective at an individual level, autonomy, responsibility and accountability for living with one's own choices may be seen from different angles, depending on the opportunities and choices one may have in planning, coordinating and receiving preventive and rehabilitative LTC services. At the policy level a trend towards increasing and supporting self-care and autonomy is noticeable. Cash allowances or in-kind services are provided to different degrees across the EU countries. Besides the positive perspectives of enhanced individual choice and decision-making, debates are ongoing as to whether the transfer of authority from the state to the individual may reflect a political effort to camouflage the tendency of shrinking public responsibility and accountability, leaving the individual to cover his own care needs – whether he or she has the required capabilities or not – including P&R within LTC. This goes together with a public debate, which in part tends to become normative with respect to concepts of ageing such as *active, productive* or *healthy ageing*, valuable concepts since they have replaced explicitly negative ones related to ageing. However, they might develop problematically, if the autonomous decision of older people to reduce activities is to be judged negatively, as suggested in a recent statement made by the Federal Government of Germany (Bundesregierung, 2010), and having LTC needs could be associated with unsuccessful ageing (Holstein and Minkler, 2003).

The significance of equal access and diversity will increase more, since the European crisis affects the Welfare States across countries, which had already been weakened in the decades before, influenced by the neoliberal paradigm. With regard to ethnic minorities, migration waves do not decrease, and in many countries migrant cohorts reach old age. The poverty rate is already higher for older people, especially for older women, and can be expected to further increase (Zaidi, 2010). Therefore researchers as well as professionals should not lessen their attention to exclusion and poverty, especially in the vulnerable group of older people with LTC needs. It became clear from the selected practice examples illustrating the INTERLINKS Framework for LTC that many initiatives are working successfully at the local level, providing tangible support to older people; however, regarding their scale they must be considered as a drop in the ocean.

Focusing on social participation and inclusion as an essential element of preventive and rehabilitative strategies – especially in their more comprehensive and holistic interpretation – turned out to be even more difficult. A small number of examples could be identified that point to and continue this topic. However, when searching for distinct strategies and interventions, information became often rather poor, and this was even more so when claiming scientific or at least other forms of evidence. This can be translated into a strong recommendation for research, policy and practice, to enrich and further develop the scope of P&R within LTC, including explicitly the topic of social inclusion.

Dementia represents a major challenge to P&R within LTC. All kinds of interventions to maintain self-care capacities and quality of life have to be further developed. Especially with dementia, traditional ideas have to be challenged to open and adjust P&R services to the needs of patients and carers. The Dutch example showed simultaneously pragmatic and efficient approaches empowering people with dementia and their informal carers by focusing on central challenges to cope with the consequences of the disease.

It does not come as a surprise that most of the good practice examples briefly introduced within this chapter seem to be more or less interchangeable regarding the paragraphs they are linked to. As a concept of 'comprehensive approaches to P&R within LTC' was shaped and given depth within the chapter, the relatedness and interaction of the separately discussed topics became compelling. It is also obvious that for most vulnerable groups, access to services and social inclusion are most at risk and need to be protected by all European societies.

Notes

1. www.healthproelderly.com/index_en.php; www.age-platform.eu/images/stories/EN/EY2012_Coalition_Roadmap.pdf
2. Slovenia and Slovakia, for example, had initial difficulties tracing LTC as an existing and consistent concept of service provision in their countries at all. Against this background, P&R within LTC was an even less tangible topic. Other countries, such as Austria, Germany and Greece are currently at a stage of development where there are stated intentions to overcome gaps between the social and the health sector and between different organisational structures by mainly upgrading or implementing intermediate structures, such as day care centres, short-term care

or specialised acute care facilities. Countries such as Denmark, the Netherlands and the UK are already oriented towards a needs-based, person-centred, horizontal provision of continuous services. In the Netherlands and Denmark for instance, patient pathways have turned the provision of everyday health and care services at all levels into work standards. In the UK, 'intermediate care' does not refer to specific organisational structures but rather to a developing concept among professionals and policymakers, covering a network of P&R services. In Denmark, it was two decades ago that conventional nursing homes were substituted with multidisciplinary coordination and 24-hour community care services oriented to the individual needs of clients (Kümpers *et al.*, 2010).

3. Health Management Information Consortium (HMIC), Medline, Excerpta Medica database (EMBASE), Cumulative Index to Nursing and Allied Health Literature (CINAHL), Social Care Online (SCIE), Applied Social Sciences Index and Abstracts (ASSIA), Proquest and Enhanced Business Source Complete (EBSCOHost).

4. Strategies of primary prevention as described here focus especially on cardiovascular risk factors partly linked to behavioural topics such as exercise, nutrition and smoking; and on psychosocial risk factors such as living alone and without social participation (Alzheimer Europe, 2008, p. 123ff.).

8

The Quest for Quality in Long-Term Care

Kai Leichsenring, Henk Nies and Roelf van der Veen

Introduction

Developing and ensuring the quality of its products and services has become an imperative task for any organisation's management. This understanding first developed in industrial settings, rather than in health and social services, where the traditional approach has been to consider quality assurance as an intrinsic part of professional ethics (Evers *et al.*, 1997). With the increasingly market-oriented provision of health and social services, however, different stakeholders have driven forward an agenda towards the development of methods to systematically describe and improve performance of personal services, first in the area of health care and, latterly, in social and long-term care (LTC) organisations (Blonski, 1999). Public purchasers of services started to question what kind of product they were funding ('value for money') and how to regulate market access for different types of providers. Providers, in turn, had to verify their performance in a more transparent way to satisfy accreditation guidelines and tender specifications in a progressively competitive environment. Service providers have also viewed quality management as a way to improve organisational effectiveness in the delivery of care. Further imperatives came from inspectorates, which were established to monitor the quality and safety of the care delivery process to assure a basic level of quality. Finally, service users have voiced changing expectations concerning the quality of care. In the context of LTC, where users' out-of-pocket payments represent a large portion of funding, users' needs and expectations have particularly gained momentum in terms of participation, co-decision-making and quality of life.

As LTC is an emerging field, it is – as in many other areas – a 'latecomer' in the introduction and rolling out of quality management and quality assurance mechanisms. A rather patchy pattern emerges across Europe, as the development of LTC differs across the various countries (Huber *et al.*, 2008; MISSOC Secretariat, 2009). Compared to, for instance health care settings, this is due to a lower level of professionalisation, lower education levels and well-known difficulties in the standardisation of disparate practices. These comprise a wide range of professional and 'informal' activities along the continuum from general health and nursing care responsibilities for the 'quality of care', to social care and even informal co-producers of care that contribute to the users' 'quality of life' as an holistic value and concept. It is a field in which the key product or service – LTC – is produced by multiple professionals and organisations. Due to these idiosyncrasies it is not by coincidence that, within the realm of LTC, it has initially been in the residential care sector where first steps were made to improve structural, process and outcome quality of LTC by means of quality management and respective criteria and indicators.

In this chapter, trends, recent policy initiatives and practice examples to promote quality management and quality assurance in LTC will be described and analysed. This synopsis is built on a literature review as well as on a European Overview Report (Nies *et al.*, 2010) based on ten National Reports from Austria, England, Finland, France, Germany, Italy, Slovenia, Spain, Sweden, and the Netherlands that were compiled within the framework of the INTERLINKS project (see http://interlinks.euro.centre.org/countries). First, some definitions prevalent in the quality discourse are clarified.

Quality management is defined as the cyclic process to assess and improve appropriate service delivery within any organisational structure (Bauer *et al.*, 2006). The term *quality assurance*, while being part of any quality management, is used for the purposes of this overview as the activity of third parties to ensure and certify defined quality criteria from an external perspective (Bauer *et al.*, 2006).

In the first section, the overview will demonstrate that quality management and assurance are activities that depend on a wide range of preconditions, some of which have already been realised at the systems level as well as at professional and organisational levels, but with some ongoing concerns about their adequacy regarding specific requirements in LTC. In the second section, some general trends and opportunities for quality management and assurance in LTC will be presented. In the third section, the specificities needed in the context of constructing integrated LTC systems will be highlighted and illustrated by innovative practice examples from selected European countries as well as from transnational initiatives. In the final part, future challenges for quality management and assurance will be discussed with a view to quality development across the spectrum of formal care providers and the role of public regulators.

Prerequisites for quality management and quality assurance in LTC

In general, quality may be defined pragmatically as the appropriate delivery of a mutually agreed service or product. To assess the quality of a service such as LTC, it would therefore be necessary to agree upon structural and procedural standards,[1] and upon expected outcomes within an acceptable range of costs/prices, between all stakeholders involved. This would not only include users, the different types of providers and professionals and the purchasers or funders, but also relatives, in some cases suppliers, researchers and the general public. It is this vast range of stakeholders that complicates the definition of quality in LTC, where frail older people often live together with professional or informal carers, where they might not be able to choose the kind of care they receive, and where many of them live in institutions with staff engaged in their most intimate spheres of living. Questions such as 'What is *appropriate delivery* and from whose perspective?', 'What can be *mutually agreed*?' or 'What is an *acceptable range of costs/prices* and for whom does it have to be acceptable?', 'What is dignity?' and 'What is autonomy in daily practice?' are difficult to answer due to the differentiated and often opposed interests of stakeholders that pursue their own specific goals. The quest for quality in LTC is therefore specifically interspersed with prerequisites, some of which will now be discussed.

A shared view

As pointed out by Nies *et al.* (Chapter 2), LTC has yet to fully develop a proper identity of its own. Providers and other stakeholders mainly rely on standards and practices from other areas such as health care which do not always meet the specific needs of LTC. A shared view on the LTC system, its goals, missions and the conditions of access as well as other rights and duties of stakeholders involved, would therefore be a first important step to defining LTC quality and quality management. Without any doubts, such a process is demanding and presumes relevant knowledge by a wide range of professional groups as well as by civil society, especially those whose voice may not usually be heard.

As a partial solution, for instance, in Finland, the Ministry of Social Affairs and Health developed the 'National Framework for High-Quality Services for Older People' (2008) with the Association of Finnish Local and Regional Authorities and other representatives to define quantitative and qualitative targets of the service structure. For instance, one target defined that 14 per cent of the population aged 75+ should receive regular home care, while only 3 per cent should live in nursing homes and 5–6 per cent in enhanced service housing. Based on such basic objectives, it is then possible to further develop strategies and improvement measures to reach these objectives and to monitor their achievement at all policy levels.

Other examples for a political dialogue to define common goals for LTC against which the quality of respective systems can be assessed would be the Dutch conferences on developing a shared vision on quality in health care, including LTC (1989–90, 1995, 2000; see Boot and Knapen, 2005) or the 'Big Care Debate' in the UK (Glasby, 2011).

Administrative entities, regulative frameworks and procedures for quality assurance (systems level)

For a regulator of LTC, it is a precondition to define services and facilities, their structural quality in terms of size, infrastructure and personnel as well as the quality of processes and expected outcomes. Usually this is done by legal regulations, tender documents and authorisation or accreditation guidelines. Furthermore, specific institutions and entities (inspectorates) have to be established to carry out these tasks as well as to verify and assess whether requirements have been met.

Legislation is certainly an important means to define general conditions for quality assurance. For quality assurance, governments have usually developed legal minimal quality standards, focusing mainly on structural characteristics of care provided such as room size, safety features, training and staffing levels, etc. For health care and social services, specific quality laws and acts with a national scope have been launched in the past decades to ensure equal treatment of patients and users and that their rights are upheld regarding their treatments (confidentiality, informed consent, etc.). In LTC, however, the fact that services and facilities are often regulated at regional or local levels has resulted in a wide range of regulations in countries such as Austria, Spain and Switzerland, but also in the UK, for instance with a view to different English and Scottish approaches. More centralised regulations can be observed in the Nordic countries and the Netherlands.

In Germany, the traditional federal governance structure has partly been overruled by a central framework to regulate quality requirements in LTC settings. The introduction of the German LTC Insurance (1994) gave rise to the establishment of the MDK (the Health Insurances' Medical Service) as a central body for needs assessment and quality assurance in LTC. This type of new institutional structures can also be observed in similar regulatory bodies that were newly established in England (Commission for Social Care Inspection, CSCI, now the Care Quality Commission, CQC), but also in France and Finland. Starting from the definition and verification (inspection) of minimum standards for structural quality criteria, most of these bodies have striven to define standards for the 'quality of care' derived from health care and respective professional expert standards. Specific legal regulations for LTC services that focus on users' quality of life are still scarce. In the Netherlands, in all sectors of health care including LTC, instruments have been developed to measure service users' experiences. These have been implemented in nationwide benchmark and users' information systems and are publicly reported on the Web (www.kiesbeter.nl; Delnoij *et al.*, 2006; Triemstra *et al.*, 2010; see also below).

There are several mechanisms at system level to ensure quality in LTC: accreditation and inspection, certification, benchmarking and quality indicators, standards and guidelines. *Authorisation or accreditation* is a voluntary or compulsory method to ensure quality *ex ante*, i.e. to regulate the 'market entry' of any service provider. Service requirements are defined by specific legal regulations and compliance is assessed by inspection. In many countries accreditation is a precondition for public funding and/or reimbursement. In this context it is necessary to set up procedures for controlling whether the required regulations have been fulfilled. This task has traditionally been accomplished by inspection units that, at best, visit provider organisations once a year to check mainly structural requirements. Inspection can be viewed as the a posteriori counterpart of authorisation, and as such it is usually carried out by state agencies. Certification is a voluntary method to certify the compliance with a specific quality management system, generally defined by the International Standards Organisation (ISO 9000; see www.iso.org) or the European Foundation for Quality Management (EFQM; www.efqm.org). This process is to regularly assess *ex post* whether an organisation complies with its defined quality standards, whereby certification is accomplished by a third party, usually an accredited certification agency. To qualify for certification, a care provider has to define and describe the organisational performance, spanning from the organisational structure, management of human and material resources and the content of services provided to diagnostic and therapeutic protocols, personnel qualifications and the correct performance of experimental activities.

Only recently, more specific indicators have been defined in terms of process quality, such as whether an internal quality management system is in place or whether expert standards are applied. In some countries like Switzerland, this has resulted in certification procedures by third parties that apply to all care homes.

Another form of general communication is public reporting of organisations' quality ratings, such as via the Internet. Though this may primarily be aimed at increasing information accessibility to service users, it is also used by professionals, for instance to choose an employer according to their quality rating. For instance, in Germany (MDK, 2008), the UK and the Netherlands, criteria and indicators allowing for ratings (stars, marks) that are publicly reported through the Internet have been introduced over the past decade (Mor *et al.*, 2012). In the context of New Public Management governance mechanisms (Allen *et al.*, 2011), public reporting of quality ratings is also considered to create a sense of competition among professionals and organisations to enhance quality performance. Indeed, there is some evidence from the acute health sector that publication of ratings or outcomes may improve the organisation's functioning but it is unsure whether it really affects users' choice (Hibbard *et al.*, 2005). Still, these practices are heavily debated in LTC as the evidence base as well as the role of users in assessment processes and their ability to choose are far from reaching a consensus among stakeholders.

Instruments for quality improvement (organisational level)

Partly as a response to ongoing changes in the national regulatory frameworks, and partly through a desire to improve performance, LTC providers have started to gain interest in quality management. The introduction of quality managers is therefore not only a response to legal regulations and inspections, but also a first step towards systematic efforts to improve quality and transparency. Today, many care homes and some home care providers have some kind of quality management in place, but for instance in Germany, not even 50 per cent of care homes have a certified quality management system (about 20 per cent in Austria). During this development, general quality management systems (ISO 9000ff. or EFQM) have been adapted to the specificities of the LTC sector in order to generate evidence for improvements of processes and structures. However, measures to enable authorities or mandated third parties to verify the implementation and to evaluate the effectiveness of the applied quality management systems are still scarce, in particular with respect to outcome indicators.

In order to quantify the level of quality assessed by any quality system it is useful to agree upon indicators for national or at least the regional level, but initiatives to improve the evidence and knowledge base of specific indicators have only recently started (BMG & BMFSFJ, 2011; European Centre *et al.*, 2010; Hammar *et al.*, 2011; Ródenas, 2011; www.interrai.org).

The introduction of a quality management system is an additional investment for providers of LTC services and facilities: it includes training of staff, consultancy, documentation, working time of staff involved, costs for third-party audits (certification) and improvement measures. Often however, only scanty information of dubious reliability is available to evaluate its potential benefits and cost-effectiveness. It is noteworthy that additional administrative costs and 'quality management bureaucracy' do not always result in improved outcomes (Groenewoud, 2008). But with a higher level of reflection and a systematic utilisation of tools such as the PDCA cycle (plan – do – check – act), it is likely that staff and other stakeholders can improve performance, if they are supported by management and enabled to implement improvement measures. In the LTC sector, where the level of education and professionalisation is still weak in most European countries, adequate training for quality management in LTC in particular will be necessary to realise its potential (Leichsenring, 2010). The role of professional organisations in developing specific norms or standards of good practice appears to be particularly influential. Compared to acute health care, protocols and standards must be fine-tuned in much more local and context-specific settings and – as they should focus on the improvement of users' quality of life (see Chapter 2) – it is not only 'protocols' that can be applied and controlled in the same way as in the acute sector. Flexibility, skilfulness and the autonomy of professionals are needed due to the various settings in which they are operating, where they encounter multiple challenges and the context-sensitive character of LTC. Moreover, there are a large variety of professionals with a range of qualifications as well as other carers (including informal carers) that need to be involved.

The methods and concepts for ensuring quality in LTC are still strongly influenced by the hospital sector, where management systems have been much more developed. Common quality management systems and supportive systems are being introduced in and across various countries (e.g. RAI, E-Qalin®, TQM), in some cases combining classical quality management instruments with organisational development and adequate learning and training methods. The board of directors of an organisation can thus assess whether management is 'in control' of quality. Self-assessments address aspects linked to clinical quality, outcomes for service users (including their experience) as well as to intra- (e.g. effectiveness, access, safety) and inter-organisational quality (e.g. care pathways, discharge protocols, multidisciplinary assessment methods). As mentioned before, evidence of validity and reliability of these instruments is often lacking.

The integration of health and social care services and cooperation between different professionals is emerging in several countries. New interventions, such as case or care management, discharge and integrated care programmes, multidisciplinary teamwork, joint care planning and regional networking are being developed (Nies and Berman, 2004; Alaszewski and Leichsenring, 2004). Indicators to measure their quality and to enhance impact, costs and efficiency across different settings for different patient groups are, however, not implemented as yet.

The tendency towards self-assessment has the potential to strengthen and empower organisations and professionals to enhance service delivery and innovation at the organisational level, including the exchange of good practice and mutual learning. However, all the while that results of self-assessments are perceived as just a requirement for funding or as a bureaucratic procedure that has to be completed to avoid a fine, organisations may be reluctant to report poor performance and to 'learn from failure'. Leadership issues in relation to collaboration, participation and delegation as well as an openness for transparent marketing strategies in increasingly competitive markets are therefore relevant factors that impact on decisions of LTC organisations to implement proper quality management.

Enabling professionals to manage and assess quality

The professionals' ability to carry out defined tasks with respect to quality management and quality assurance depends on their skills to use assessment systems, to communicate with and to involve relevant stakeholders, and to manage improvement projects. These skills are complementary to a range of other challenges of care delivery to older people with complex needs, with which professionals are currently confronted. This entails, for instance, new methods of needs assessment, coping with dementia care, electronic patient records and other IT applications, enhanced multidisciplinary working, the blurring of professional boundaries, case and care management or other activities that are not strictly within their traditional job and skills profile. Also, regulatory frameworks have defined increased standards concerning staff's skills and vocational education. These requirements lead towards more and often mandatory training, with professionals and employers being made responsible for appropriate and up-to-date education (e.g. Italy, Finland). This involves formalised degrees and diplomas with an emphasis on LTC (geriatric aides, health and care management, diplomas for care home managers, etc.) that have started to spread in some countries (e.g. Germany, Austria, the Netherlands). As higher qualification often implies higher wages, employers (and purchasers) are facing additional costs, including those of training itself. Incentives or disincentives thus depend, apart from ethical and professional codes of conduct, on the regulators that set requirements for the levels of education and training, and the purchasers that, however, often do not sufficiently consider costs for training and staffing levels in defining reimbursement rates.

In this context, skills for quality management are one additional area for which continued vocational education and skills development are needed, as managers and staff in LTC are usually not specially trained in quality management. At the same time, they are confronted with more and more legal regulations, standards to comply with, and a rather shaky basis for funding additional requirements (Leichsenring, 2010, p. 14). Examples from Finland (Hammar *et al.*, 2011) and other countries (Leichsenring, 2011b) show evidence that investment in training can trigger better outcomes and enhancement of defined quality indicators. Also the organisational model of Buurtzorg ('Care in the neighbourhood') in the Netherlands with its small, self-managing teams of a maximum of 12 professionals indicates a way forward. Apart from providing care in the neighbourhood by integrating with other local organisations and informal caregivers, the teams are also responsible for organising their further training according to the team's and the individual professionals' needs (Huijbers, 2011).

Strengthening the role of users in quality development

The general development towards a more consumer (user and informal carer) oriented way of thinking in the delivery of care and other services has also improved service users' and informal carers' roles in ensuring the quality of services, such as through service contracts, shared decision-making, enriched choice, opportunities to express their degree of satisfaction, access to information and even the involvement in quality assessment. Degrees to which service users are provided with such opportunities vary, however, within and between European countries.

In all countries there are mechanisms for including the service user in care planning and ways of reaching informed consent (stemming from the acute health care sector), sometimes supported by law or merely as guidance for good practice. The rationale behind these principles is that service users have the right to participate in and influence the planning and implementation of their treatment and services. This may be guaranteed even if service users are not able to make decisions about their care themselves any more (e.g. by means of user advocates, in England and the Netherlands). If the above principles are not followed, general mechanisms are complaints procedures that are stipulated in most countries.

An interesting initiative in this context stems from the realm of consumer protection and aims at more transparency and better information for users of residential facilities in Germany. The criteria, against which quality of life is measured and assessed in the voluntarily participating care homes, were developed by associations of care homes, representatives of health insurance funds and interest groups of seniors. Seniors are trained as volunteers to assess the 121 criteria during one day in the participating care homes (Stiehr, 2011d).

Choice and control of service users are also enhanced by cash benefits for older people with care needs living at home, including 'personal budgets' or 'direct payments' that allow services to be purchased individually (Huber *et al.*, 2008). This increase in purchasing power partly implies a shift of responsibilities also in relation to quality assurance (Leichsenring *et al.*, 2010). With the dearth of evidence on how well engaged the most vulnerable and frail groups are, it may however be anticipated that there is a need to empower and enable service users to choose their provider individually and to act accordingly when services are poor. It also needs a certain level of independence for service users (and their families) to seek appropriate services and in some cases even to act as an employer. This may not be possible for more frail and vulnerable service users whose choice is constrained by the extent of their care needs (see Chapter 3). Such arrangements would therefore call for specific quality assurance mechanisms that have not yet been sufficiently developed.

Finally, there are two additional instruments that serve as a precondition for the involvement of users in quality assurance: client satisfaction surveys and the previously mentioned public reporting of quality.

Client satisfaction surveys are used as a method for measuring quality and/or getting feedback from users. This can be arranged within an organisation as part of its voluntary or mandatory quality management system, but it can also be organised on a national level. One of the few examples of a national system has been implemented in the Netherlands, where the Institute for Research in Healthcare (NIVEL) and several other organisations have developed the 'Consumer Quality Index' (CQ-Index) as a standardised system for measuring, analysing and reporting customer (clients') experience in health care (NIVEL, 2009). The respective questionnaires provide insight into what clients find important and how they rate their experience with care. Surveys are carried out by accredited external agencies in accordance with the standards and guidelines of the *Centrum Klantervaring Zorg* (Centre for Customer Care). By law, each health care and LTC organisation has to arrange a survey once every two years. Results can be used by clients to choose a health insurer or health services, by client organisations representing the interests of their members, by insurers to assess the quality of services purchased, by managers and professionals who want to improve their performance, and by public authorities and monitoring agencies such as the Healthcare Inspectorate (IGZ) and the Dutch Care Authority as well as the Ministry of Health, Welfare and Sports. The CQ-Index is a substantial part of the 'Quality Framework for Responsible Care' (QFRC) as it provides a range of measurable indicators that show whether health and LTC organisations provide responsible care. The results of the QFRC are publicly reported for each organisation on a website (www.KiesBeter.nl).

The intention of presenting these results online is to provide transparency of data and to stimulate organisations to improve their performance. Public reporting of specific quality indicators in LTC settings has also become an important issue in Germany and England to enhance the users' information base for choosing between providers. Apart from methodological and ethical problems already mentioned above, this practice can only be a very first step towards better transparency and quality reporting for users as, for instance, many older people still do not have access to the Internet so that a significant number of service users remain excluded from this information.

Against the background of the general challenges to establish preconditions for quality management and quality assurance in LTC that have been outlined in this section, defining and positioning the role of users in assessing quality remains a specific challenge in LTC. At the policy level, users are typically represented by organisations that defend their interests and act as advocates to make sure concerns are heard. The relationship of such organisations to actual users may vary, as well as their degree of professionalisation and their integration into the policymaking process. Comparable data on users' opinions, experiences and expectations are rare, while the assessment of needs and outcomes of interventions are still disputed and vary from country to country. All this information, however, would contribute towards important preconditions and indicators for quality management and governance. Instruments and legislation to support ways for giving a voice to users' needs and opinions at a group level and an individual level are still at a very initial stage in most countries. Examples might be if representatives of users are invited to specific committees or administrative bodies to express their view, or if an ombudsman of users is appointed (see for instance Wiener Heimkommission, 2009).

Trends in quality management and quality assurance

Many individual aspects of the preconditions described above have been developed and implemented within emerging LTC systems across Europe. This includes both regulatory measures at the systems level (institution building, regulatory frameworks) and the organisational level (quality management systems), but also at professional and user levels. These policies and activities have gained momentum over the past 15 years, partly inspired by 'New Public Management' approaches and market-oriented governance (Evers *et al.*, 1997; Huber *et al.*, 2008), partly by the sheer growth of provider organisations, by innovations in other sectors (health care) and by a growing interest in quality assessment by users and other stakeholders.

Above all, LTC has become an acknowledged area of social welfare systems, which took place within the framework of European welfare regimes and respective health systems, calling for political choices that were influenced by national traditions and preconditions. Evolving from the current state of the art in the context of different national frameworks, decisions for future development have to be made at the national level in order to ensure quality, access and sustainability. Based on most recent developments, findings and decisions, some common trends on a number of governance issues concerning quality management and quality assurance can be identified.

Centralisation and decentralisation with ambiguous consequences for quality assurance in LTC

Governance of quality in LTC is organised at various administrative levels (national, regional, provincial or cantonal) but most often with mixed responsibilities at national, regional and local levels. For health care, quality assurance is a national responsibility all over Europe. For quality in long-term and social care, national frameworks exist in some countries (e.g. the Netherlands, Finland, Sweden), but more often it is regional and local authorities that hold an important position in defining quality requirements and (mainly structural) standards. However, a general tendency can be observed also in decentralised countries (e.g. Germany, Austria, Spain) to support and develop quality assurance in LTC from the federal level by means of quality frameworks and criteria that are supported by multiple stakeholders. At the same time and against the backdrop of market-oriented governance, individual performance criteria and definitions of quality, requirements are increasingly moved to the level of service contracts between local purchasers and provider organisations, which may be explained by the fact that funding and quality assurance are usually, but not always, linked. It has therefore generally become more difficult to disentangle the different roles of stakeholders, as various de- and recentralisation tendencies are coinciding with reforms in various areas of welfare (Kazepov, 2010; Leichsenring *et al.*, 2010).

For instance in England, it is local authorities that act as purchasers of services from the independent sector, while quality standards and quality assurance have been centralised by the introduction of specialised entities (CSCI, later CQC). Also, compared to health care, LTC inspection tends to be based on less developed and mutually agreed standards. Furthermore, as governance of health, LTC and/or social care may be positioned at different administrative levels, regulations tend to focus on one or the other type of service, rather than on integrated services or 'chains of services'. With different quality assurance systems operating alongside each other it is often difficult for users as well as for commissioners, insurers and inspectors to monitor across services to compare services and performance.

Centralised regulation may certainly capitalise on economies of scale in developing quality systems and instruments and on ensuring equal conditions across the country. With existing economic and cultural differences within states, local or regional sensitivity in relation to specific standards may be required, but several examples demonstrate that decentralisation can go together with a standardisation of quality assurance methods and instruments, rather than standards, even across national borders. These examples include the E-Qalin quality management system that is being applied in Austria, Germany, Luxembourg and Slovenia (Leichsenring, 2011) or interRAI which is used in more than 30 countries worldwide (www.interrai.org). In most circumstances, it would appear inefficient to develop quality management systems at a local level, but it would be necessary to support their implementation by regional and local authorities as there are indications that, if local financing is poor, the implementation of quality systems becomes a cumbersome task (Turk, 2009).

Complementing the rising importance of market mechanisms by quality assurance

The different national pathways and traditions in tackling LTC have resulted in a wide range of 'welfare mixes' (Evers and Wintersberger, 1990) with respect to the role of public, private for-profit and non-profit as well as households' contributions to care. With the overall domination of market mechanisms to govern LTC services over the past two decades, the number and proportion of for-profit providers has particularly grown in importance (Colombo *et al.*, 2011). This development has fuelled debates and policies to create equal access opportunities and regulations that apply likewise to all stakeholders within the quasi-market of LTC, whether they are public, private, for-profit or non-profit (see also Huber *et al.*, 2008).

Furthermore, debates on quality requirements and transparency in LTC have increased in importance due to the introduction of competitive (quasi-) markets. For instance, new private commercial providers have reached a market share in LTC markets of about 15 per cent in Sweden, of almost 80 per cent in England and of about 38 per cent in Germany (Leichsenring *et al.*, 2010).

With the appearance of this new set of stakeholders, questions about the potential impact of providers' legal status on quality has become an issue of concern (Scourfield, 2007).

Until now, there have been contradictory signs as to what extent quasi-markets have triggered improvements in the quality of LTC services. For instance, public tendering and contracting of services has often been based mainly on price, rather than on quality criteria (INSSP, 2010). Comparisons over time are not feasible, and comparisons between, for instance, for-profit and non-profit providers are difficult due to the lack of universally acknowledged indicators.[2] There are indications that the private sector is more service-oriented than public providers (Stolt *et al.*, 2010) but at the same time, scandals and bankruptcies, e.g. in England, Sweden and Austria, in the private for-profit sector have recently shown that monopolies of any kind may be detrimental to sustainable care provision, in particular if regulation is poorly developed (PSIRU, 2011; Scourfield, 2007).

In addition, contracting has generally remained confined to individual services rather than to a 'chain of services'. It would be an important indicator for progress in shaping emerging LTC systems if quality management and assurance mechanisms were developing across organisations, with respective indicators for structural, process and outcome at the interfaces between sectors. Indeed, more developed thinking in terms of 'care chains' and care coordination is needed to respond to users' real needs – 'care trusts', more joint training and financial incentives for coordination in care provision may help to bridge the divide between health and social care. This might also make it possible to address the 'quality–cost chasm', that is the definition of quality criteria by one agency and the decision on budgets and prices by another (Leichsenring *et al.*, 2010).

Finally, competition without overarching and cross-sectoral regulation seems to be detrimental to emerging LTC systems for two additional reasons: on the one hand, some authorities have produced new monopolies of (private) providers by tendering services for an exclusive geographical area; on the other hand, opening the market for a range of competing providers might result in reduced cooperation between these organisations, rather than integrated service provision. The idea that quality could play an important role in driving the choice of users has also failed to be completely tested, as users are faced with lack of information as well as with financial and mobility constraints that often prevent them choosing providers exclusively on the basis of any quality criteria, however they may be defined (Downey, 2011).

From inspection to self-assessment and certification of quality management by third parties

There is a general tendency towards transparency of LTC services and their performance with various objectives: to support users in choosing a service, to inform citizens what is being provided, to allow inspectors to control basic quality requirements, as well as to inform commissioners and insurers about what quality they pay for (public accountability). For (local) governments it is important to know whether they serve the population appropriately, while for managers and professionals who work in care providing organisations, it becomes increasingly important to know how they are performing and in what respect they may improve their work. These objectives are not always compatible, particularly if negative results have direct or indirect negative economic consequences for the organisation. If organisations are honestly intending to improve their services, they should feel safe to be transparent on both positive and negative outcomes in order to learn for improvement, rather than filling in tick-box questionnaires.

Traditionally, authorities have relied (and still rely) on inspections to assess quality in LTC. This is usually accomplished by one or several inspectors who visit the organisation once a year or even less frequently to check whether legally defined requirements have been met. This procedure has often been criticised as it is costly and might not be very effective in assessing the quality of care provided, if it is not backed up by a functioning internal quality management system (BMG and BMFSFJ, 2011). As a corollary, self-regulation and self-assessment are increasingly gaining ground in connection with a shift from merely assessing structure-oriented quality indicators to the implementation of more sophisticated quality management methods. Umbrella organisations of providers are playing an important role in pushing forward this development to ensure quality by their associates and to develop methods and criteria in negotiation with the regulator. This often results in further autonomy for service providers to choose their own quality management system (e.g. Switzerland, Germany, Austria).

Of course, supervision and inspection cannot be left to umbrella organisations of LTC providers but need to be carried out by third parties (in the context of classical quality management tools) or funding agencies such as insurance agencies or public authorities. In any case, with increasingly market-oriented governance mechanisms, traditional inspection and sanctions to assure compliance are bound to be complemented or even replaced by accreditation (*ex ante*) and certification of self-assessment by third parties (*ex post*).

From quality assurance to continuous improvement?

As an important part of the modernisation of LTC systems, a dialogue between all relevant stakeholders – including user organisations – about assessment methods and quality criteria has already started in some countries (Austria, Germany, the Netherlands). Based on mutually agreed criteria, public disclosure of data (see above) can be based on consensus to reduce the counterproductive effects of 'public humiliation' through the publication of indicators demonstrating bad performance. In any case, quality assurance cannot be reduced to mechanisms that assess quality *ex post*. It has to start at the very moment when (new) providers access the 'quasi-market' of LTC and pursue public funding or reimbursements from LTC insurance. With a view to quality assurance *ex ante*, all countries surveyed have stipulated defined minimum standards linked to authorisation or accreditation procedures.

The question remains, however, whether minimum standards (and respective reimbursement rates based on these standards) may motivate providers to implement a continuous improvement process of their service quality. As outlined before, existing standards still focus predominantly on structural or process quality, rather than on outcomes for the individual. A relatively recent trend is therefore to highlight more user-centred approaches to quality (Department of Health, 2006), and a focus on 'quality of life' indicators, rather than solely on quality of (clinical) care (European Centre, 2010). Whether respective initiatives will remain rhetoric (Downey, 2011) or lead to new kinds of measures and indicators, remains to be seen. In any case, one other important driver will be essential for providers and regulators to focus on with respect to quality development in LTC in the next few years, notably the lack of professionals who are willing to work in health and LTC, and this can already be observed (OECD, 2008).

Apart from the lack of mutually agreed indicators of quality in the chain of services in LTC, one of the main difficulties seems to be that legally inferred quality management is seldom underpinned by respective supporting or enabling measures. These include training of staff, organisational development, and funding of improvement projects to promote continuous improvement within and between organisations (for an exceptional example see: Minkman *et al.*, 2011).

Future challenges: towards integrated quality development in long-term care

Quality regulations still reflect the boundaries between health, social and LTC in most countries. A 'silo approach' is therefore still very much present. In this context it is not uncommon for quality in LTC to assume a subordinate role to health. Another consequence of the health and social care divide is often manifest within social care services and informal care arrangements, where quality issues are neglected and only referred to in the context of health reforms. Thus quality assurance becomes a reflection of the patchy landscape of funding and legislative systems, providers and other stakeholders. There is hardly any dialogue between quality assurance systems in health care, LTC and social care, let alone the 'private' and still rather unregulated sphere of informal care provided by families, migrant carers in domestic assistance or volunteers.

First initiatives and considerations to improve integration *and* the quality of coordination mechanisms can however be identified. These include the shared definition of goals across organisations, mutually agreed care planning and the implementation of procedures between services to meet objectives, evaluate results and implement changes. To support comprehensive quality management across LTC service providers, some key issues have been identified in the INTERLINKS Framework for LTC (http://interlinks. euro.centre.org) that will be described in the following, supported by innovative examples that have started to address respective challenges in their designs and their daily practices.

Promoting and facilitating quality mechanisms in relation to linkage, networking, coordination or integration of agencies and organisations

The complexity and the variety of how care and services are arranged at the client level calls for more consistency and the reconciliation of different professional cultures. Current practice from several countries reveals four promising options in this respect:

- integration at the area level
- integration in care pathways
- integration at the professional level
- integration of funding mechanisms.

Policy initiatives to *integrate services at the area level* are aimed at local strategies covering areas such as community planning, traffic and housing, cultural and recreational activities, education and participation, well-being and health, and regional networking of primary care units. Examples exist of quality indicators at a local level, related to timeliness of acute and social care delivery, assessments and care packages. As mentioned before, at the *professional level* several countries are introducing new professions and professional roles, such as case managers or care coordinators and/or multidisciplinary teams working across services. Also quality management has become an explicit job profile in the area of LTC. Further, financial incentives for stakeholders are introduced by selectively applying *funding mechanisms*. One approach is to provide money or vouchers to the users, the second is to enhance coordination by funding mechanisms to act as commissioners of integrated services.

Care pathways or clinical pathways that intend to integrate health and social care processes for specific user groups are based on guidelines and evidence. Care pathways appear to be primarily applicable at the interfaces of acute and LTC care. In this context, implementing quality criteria for integrated care within services can help achieve better outcomes for patients in terms of quicker and more appropriate care, which can be shown by enhanced satisfaction rates. Also carers and professionals express greater satisfaction, and at the same time cost-effectiveness may be increased significantly. One of the most integrated services in this respect is still linked to disease management programmes and according to how patient pathways are defined, such as in the Netherlands where stroke services show positive results when using specific quality criteria (see Box 8.1; Diermanse, 2011).

Box 8.1   Quality management of integrated stroke services (The Netherlands)

Patients with a stroke can occasionally stay too long in hospital, because there is no available place in the rehabilitation centre of a nursing home. Hospitals are not able to provide good care for a longer stay, because there are insufficient opportunities for therapy. This has a negative effect on the condition of the patient.

To develop new pathways, performance indicators and criteria for effective stroke service were set up by the Dutch Stroke Network in 2001. The indicators and criteria refer to the acute period, rehabilitation, chronic care and patient satisfaction of the care given. Two hospitals and a care home started with integrated stroke services with the performance criteria as their starting point.

The criteria for effective stroke services serve as a checklist and guideline for implementation. Integrated stroke care focuses on logistics and coordination of treatment between and within institutions, so that the patient gets appropriate care at the right time and in the right place. The pathway for cerebro-vascular accidents improved satisfaction among patients and carers. The implementation of stroke pathways and 'chain' monitoring has been facilitated by means of a 'stroke chain coordinator' who is responsible for networking, monitoring and evaluation.

Source: Diermanse (2011).

Ensuring preventive and rehabilitative structures and processes involving informal carers

The lack of political commitment to invest in adequate prevention and rehabilitation for older people in need of LTC calls for new 'bottom-up' strategies that seek to embed comprehensive quality standards within initiatives (structure, process and outcome measurements, including clinical, psychosocial, organisational and procedural aspects). For instance, the Geriatric Network and Geriatric Academy Brandenburg (Germany) successfully improves continuity and quality of care for older people at the regional level by increasing geriatric knowledge across professional and organisational boundaries, including prevention and rehabilitation (Box 8.2). Such examples need to be further scaled up and rolled out.

Box 8.2   Geriatric Network and Geriatric Academy Brandenburg (Germany)

GeriNet aims at innovative care approaches like for example rehabilitation at home ('Mobile Reha' – as a specific innovation). All activities are oriented at supporting the patient to be able to live at home. This initiative has the potential to overcome barriers to LTC integration, since medical as well as nursing, health and social services are addressed. The holistic approach to the needs of frail older people helps to adopt a preventive and rehabilitative attitude.

Care and care planning has to be patient-oriented. Family carers are addressed as 'external members' of the geriatric team, but also as having support needs of their own. Services and activities are mainly financed within the general financing system of the social health insurance and the LTC insurance. Rehabilitation at home has become refundable by the social health insurance. The 'Geriatric Academy' organises training for doctors and other health professionals as well as for informal carers.

Other activities include regular meetings of 'quality circles' on diverse topics. Joint discharge procedures have been developed and accepted in the region. Furthermore there are activities to inform the public, self-help groups and care recipients about geriatric topics in the widest sense.

Source: Kümpers (2011b).

Focusing on quality of structures, processes *and* results of LTC providers

As mentioned above, one of the difficulties in quality management and quality assurance consists of the lack of acknowledged criteria that go beyond the sphere of clinical indicators. In this latter area, interRAI is one of the internationally most accepted assessment instruments for LTC. It enables organisations and care units to reliably compare or even benchmark their own performance, resident structure and quality of care at a national, organisational and care unit level with figures from other organisations. The instrument can subsequently be used to improve the quality and productivity of LTC by comparing own results with other facilities or organisations (see Box 8.3). Further research is needed however to test the validity and reliability of individual indicators (see Box 8.4), in particular those that go beyond mere clinical aspects of LTC in care homes (European Centre, 2010) and those that address quality criteria across organisational settings.

Box 8.3   RAI benchmarking: An instrument for leadership and development (Finland)

The Resident Assessment Instrument (RAI) benchmarking is used in 26 divisions of a Finnish not-for-profit third sector social and health care organisation Folkhälsan, which provides LTC for older people in South and West Finland.

In Folhälsan sheltered houses, each resident is assessed using RAI at least twice yearly and when there is significant change in the resident's status. Furthermore, all staff are involved in using the RAI system as it requires multiprofessional cooperation. Folkhälsan has put a systematic strategy in place that includes senior management down to each individual employee.

RAI is a tool for planning both individual care (goal setting) and measuring quality and efficiency of care units. The quality of care is assessed by 27 clinical quality indicators that describe the prevalence and incidence of potential quality problems in care processes and outcomes. The RAI system establishes a web-based benchmarking database for facilities to compare quality and cost of care (to learn from best practice). The benchmarking databases are updated twice a year.

RAI facilitates the follow-up of large-scale quality targets and therefore is a concrete aid for management and leadership. RAI helps in drawing up training plans, familiarising new staff, enabling multidisciplinary cooperation, supporting staff development discussions, shaping strategic plans of action, and aiding in continuous follow-up of operations. RAI highlights the weaknesses and strengths of a unit and gives a clear picture of its case mix and staff requirements. Benchmarking enables participating organisations and units to compare their own performance, client structure and quality of care at care unit, organisational and national levels.

Source: Hammar *et al.* (2011).

Box 8.4   Outcome indicators for rating the quality of care provided by care homes for older people (Germany)

This project (2008–10) aimed to provide a scientifically grounded and practical set of tools through which outcomes related to quality of residential care of older people could be 'measured', rated and compared. In Germany standards of measuring care quality exist (and are legally anchored), but they are limited to structural and procedural quality indicators. Therefore, outcome quality criteria were developed that are suitable for external quality assessments and internal quality management of institutions, or comparison between institutions. In addition, there was a need to develop a system that reflected a more comprehensive concept of quality of care which included health and social aspects of quality of life and self-reliance of the residents.

In the course of this project, quality indicators and tools were developed in order to measure and rate quality of care in a reliable manner. Indicators were tested during a 10-month period in 46 care homes. Voluntary non-profit, private and public institutions participated. The resulting indicators were divided into three different categories: (a) health-related indicators in three areas: maintenance and support for independence; prevention of illness or disability; assistance with special needs; (b) indicators for identifying aspects of quality of life in two areas: housing and housekeeping assistance; daytime activities and social relations; (c) indicators to address the quality of care concerning the links and interfaces of services; they included, for example, measurement of informal carers' involvement (informal carers' rating of the institution), the use of shared documentation, or whether current and future care had been planned with the resident and his or her informal carers.

The results of the project have been widely accepted among experts as future standards for outcome measurement but national implementation is pending. The outcomes of the project may promote the development of internal quality in residential care and also provide an important basis for quality assessments by external agencies.

Source: Weritz-Hanf (2011).

Measuring and considering user satisfaction and the perspectives of relevant stakeholders involved

Quality as such is a multidimensional concept focusing on multiple aspects that can be assessed in terms of structural, process and results-oriented features (Donabedian, 1966). This level of complexity is particularly amplified in LTC by the fact that it has to be assessed through the perspectives of the various stakeholders involved – from users and their families to professionals, management and external stakeholders.

From a service user perspective, perceived quality of life and quality of care are usually measured by user satisfaction surveys, even if this has its limitations as described by Billings in this book (Chapter 14). Other ways to involve users (or their advocates) in quality assessment could be inspections by user organisations or older volunteers (see Stiehr, 2011d) or to have them participate in the internal quality management system to facilitate a dialogue between the various stakeholders' views (see Box 8.5).

Box 8.5   The E-Qalin Quality Management System (Austria, Germany, Luxembourg, Slovenia)

The E-Qalin quality management system is based on training of at least two E-Qalin process managers per care home and a self-assessment process in the organisation during which about 60 criteria in the area of 'structures and processes', and 25 foci in the area of 'results' are assessed from five different perspectives (residents, staff, leadership, social context, learning organisation). By involving all stakeholders in the self-assessment and the continuous improvement of quality, E-Qalin strives to strengthen the individual responsibility of staff and their ability to cooperate across professional and hierarchical boundaries.

The criteria guide staff, management and other stakeholders to describe and analyse the reality of the care home and to assess their appropriateness according to the classical quality management cycle plan – do – check – act, but with an emphasis on whether all relevant stakeholders are 'involved' in these steps. Individual criteria are assessed in professional groups and in a core group that is responsible for the consensual final rating of each criterion and the performance indicators (facilitated by a special software program). The final aim of this internal assessment from different perspectives, with an emphasis on the user perspective, is to develop improvement projects, to set priorities and to implement improvements as quickly as possible. E-Qalin also attempts to involve residents of care homes and their relatives in quality assessment and measures for improvement, e.g. in specific assessment workshops alongside 'classical' satisfaction surveys. An external audit by an accredited certification agency certifies results every three years.

The final result of an E-Qalin assessment gives the same weight to structures and processes (50 per cent) and to results (50 per cent). The latter are mostly based on quality of care indicators as well as on indicators derived from satisfaction surveys with residents, family members and staff to, again, reflect the different perspectives of stakeholders involved.

Source: Leichsenring (2011b).

Conclusions

This chapter has described and analysed the development of quality management in LTC showing a wide range of challenges that need to be addressed in a coordinated approach by policy, practice and research. To conclude, some key governance issues and considerations for embedding quality thinking in emerging LTC systems will be flagged up for further consideration. This debate is still widely influenced by a number of unsuccessful initial experiences and quickly changing regulatory frameworks. For instance, in the UK both the regulatory bodies and respective quality guidelines have been changed several times over the past ten years (Holdsworth and Billings, 2009); in France, ongoing debates about more medically oriented or more socially oriented guidelines have unsettled providers when seeking to invest in quality management (Naiditch and Com-Ruelle, 2009); and in Germany many providers have challenged the latest framework of quality indicators even through legal action (see also www.moratorium-pflegenoten.de).

What are the challenges in designing appropriate quality assurance mechanisms in LTC?

The quality assessment of structures, processes and results needs to match the idiosyncrasies of LTC and of the end users, older people in need of care and support. This implies further efforts to operationalise criteria and measures to assess quality of life not only in a methodologically sound manner, but also in a way that care workers and their managers are able to implement. As purchasing care services cannot be compared to everyday consumerism, it is also necessary to develop support mechanisms with defined quality criteria for structures, processes and results to address legal, ethical and economic issues that are linked to, for instance, employing a personal assistant or paying a family member for personal care (Chapter 10).

Leading on from this, it is notable that what appears to be lacking in all countries are systems to ensure and measure the quality of care by informal carers, due to privacy and other ethical codes of conduct that forbid intrusion into this personal care environment. Given that the bulk of LTC is provided by informal carers, a way forward would be to develop methods of improving the quality of life of all stakeholders involved by overcoming the 'formal–informal care divide' through greater partnership and cooperation through the lens of quality management. This becomes even more urgent in countries where informal carers are not sufficiently supported and/or where family care is supplemented or replaced by employing migrant care workers (Di Santo and Ceruzzi, 2010). Serious issues of quality care may become evident in these circumstances where forms of regulation are generally absent.

Quality assurance mechanisms, including legally defined standards that only focus on structures and processes do not guarantee 'good' quality as an outcome experienced by service users and their relatives. The need for instruments and indicators that operationalise quality of life as an outcome variable, though challenging to translate into an instrument for daily practice, has been addressed but only in a few cases. These concepts are facing the inconvenience that, on the one hand, there is a need for standardisation, while on the other, individualised approaches are required.

The cross-cultural aspects of such an endeavour are further complicating the task, particularly as all European countries are facing increasing cultural diversity – and this includes LTC for older people. International efforts will therefore have to be developed that consider the differences between individuals and incorporate enough flexibility to adjust to specific target groups and settings.

No quality management system will be able to meet all contrasting objectives and the requirements of all stakeholders involved. For instance, as outlined above, for improving care delivery at the workplace, it is an intrinsic part of professional ethics to report errors and to be able to openly discuss potential remedies. However, if an inspectorate can close down facilities in cases of poor performance, free and open reporting and (internal) discussion will be compromised. The same holds true for the introduction of performance-related payment systems. If poor performance is badly paid, it is likely that staff and management will not be motivated to report errors, mistakes and incidents.

What kind of quality assessment is needed to attract practitioners in LTC?

There is growing evidence that LTC can be considerably improved by involving and rewarding staff to work on improvements (Bode and Dobrowolski, 2009; De Prins and Henderickx, 2007; Frerichs *et al.*, 2003). Working on improvement by employing clearly defined and easily applicable indicators helps staff and management understand that their work is meaningful and valued. In an area where the workforce is becoming scarce and payment is often poor, work itself should be rewarding and motivating, as staff's quality of working conditions has an impact on the quality of care. Examples for such improvement programmes can be found in the LTC sector. From 2005 to 2011 in the Netherlands for example, nearly 900 teams in hundreds of organisations systematically worked on improvement. The result was quality gains of 30 to 50 per cent with regard to individual indicators such as pressure sores, medication errors, use of restraints and behavioural problems (Minkman *et al.*, 2011). This relatively cost-effective methodology could transfer to other settings to enable cross-national benefits.

What kind of incentives could promote quality management and improvement within and across organisations?

A singularly challenging governance question concerns appropriate incentives for organisations to strive towards excellence or optimum performance. Funding is a principle and certainly the most well-established mechanism to achieve this, particularly as defined quality standards are usually required to obtain public funding (authorisation, accreditation). A further development would be the introduction of 'pay for performance' which is a mechanism that appeals at first sight, especially to governments, and would imply that those who perform best get paid the most. Thus far, examples for pay for performance can be primarily retrieved from the acute health sector, with outcomes that are not unanimously positive (Mullen *et al.*, 2010; Cromwell *et al.*, 2011). Commentators argue that pay for performance can lead to calculating behaviour and has to be based on clear performance indicators that, as we have explored, are not easily at hand in the LTC sector and warrant further development.

While quality management for individual services and facilities is already a challenge in LTC, it is even more complex in settings where various multi-disciplinary teams of professionals work together, especially when they interact between different organisations and on the basis of distinct contracts. There are only a few quality systems that are applicable across systems and their empirical basis is not or poorly demonstrated. Nevertheless, the integration of services is at the core of LTC as an immoveable requirement. Therefore, quality management that links organisational structures (such as case management, multidisciplinary teams, single points of access, respite services, chains of care) need further attention from policymakers and health services researchers.

The crucial questions for quality management in LTC will remain whether it really leads to better outcomes, how it contributes to LTC organisations' success, under which conditions and for which target groups. These questions are still widely under-researched and give rise to often poorly reflected anecdotal evidence. For instance, many professionals have argued that quality management is yet another bureaucratic exercise that takes away time from working with users. This may be true if quality is understood only in terms of ticking boxes, but the imperative task of quality management should be to ensure and improve the quality of care and services which, in turn, should be at the heart of daily practice in each health and social care organisation. It might still take some time until this message is embedded in all LTC organisations and professional practice but, as this chapter has shown, some tools and methods to support them have already been conceived.

Notes

1. In a context of quality and quality management, the term 'standard' has a double meaning and thus often provokes misunderstandings: on the one hand, the term is used to describe management standards that provide requirements or give guidance on good management practice; on the other hand, in measuring quality, standards are predefined normative values to be achieved in order to judge quality as 'good'.

2. Based on a very restricted range of four (clinical) quality indicators, a meta-study revealed better results for non-profit care homes (Comondore *et al.*, 2009).

9
Making Sense of Differences – the Mixed Economy of Funding and Delivering Long-Term Care

Ricardo Rodrigues and Henk Nies

Introduction

On average citizens of European Union (EU) countries that reach the age of 65 can expect to live for another 19.1 years and at the EU level this group of people represent 16.6 per cent of the population (Eurostat). This testifies to the current ageing profile of the European population. Although gains in life expectancy are a positive outcome in itself, not all of these years will be spent in good health or without physical limitations, which means that the need for care may arise at some point. Ageing could thus increase the demand for care, since at the population level old-age is usually associated with poorer health conditions. But as far as long-term care (LTC) systems are concerned, ageing could also reduce the availability of carers of working age. As a consequence, the pool of informal carers could be diminished, unless the reduction is compensated by informal care provided by older people, and in turn shrink the labour force available to work in the formal care sector.

While in the past most of the needs related to LTC would possibly be addressed within the family, a number of societal changes, from increased labour market participation of women to changing living arrangements, may contribute to a change in this picture. For these reasons there are mounting concerns about social protection systems that have to devote more public resources to LTC on the one hand, and reduce public expenditure on the other, thus raising the urgent question of how to finance and deliver benefits and services in a sustainable way.

Related issues were one of the main topics addressed by INTERLINKS. Although the principal purpose of this project was not to gather and analyse statistical information such as on public expenditure indicators, it has nevertheless allowed for the compilation of a number of important quantitative and qualitative data on LTC funding and organisation. Building on this and other existing sources (Huber *et al.*, 2009; Rodrigues *et al.*, 2012) the purpose of this chapter is to discuss existing financing arrangements to deliver support for people in need of LTC, for a better understanding of differences between selected countries, namely Austria, Denmark, Finland, France, Germany, Greece, Italy, the Netherlands, Slovak Republic, Spain, Sweden, Switzerland and the United Kingdom.

The present chapter is organised as follows. First, some of the frameworks that have been put forward for comparing LTC will be discussed, and building on these, a series of dimensions for analysing differences in public resources devoted to LTC are proposed. Three of these dimensions will be used in the following sections to describe and compare the financing mix and delivery of LTC, as well as its governance structure. The final section discusses the results and concludes what can be learned from different experiences in Europe for designing finance and governance of LTC provision.

A framework for analysing differences in public resources devoted to long-term care

In the recent past a wide body of literature has proposed several frameworks for the comparative analysis of LTC systems, most notably in the wake of Esping-Andersen's (1999) reassessed analysis of welfare state regimes through the inclusion of the family as producer of welfare (e.g. Anttonen *et al.*, 2003; Bettio *et al.*, 2006; Jensen, 2008). Although various clusters of countries and also welfare regimes have been put forward, most of the typologies have developed around the degree of division of responsibilities between the state and the family – what is termed as different degrees of defamilialisation (Leitner, 2003) – with the market sometimes taken as being implicitly associated with the family (Saraceno, 2010). The advantage of this approach is that it allows for the inclusion of the family as the most important producer of care. However, as the financing and delivery of care are mixed, so is the governance of LTC a hybrid as well. Traditional classifications such as Bismarckian and Beveridge regimes or the well-known classification by Esping-Andersen (1990) of liberal, corporatist-statist, social-democratic and (later added) Mediterranean systems, do not do justice to the many idiosyncrasies of LTC systems in Europe.

We argue instead that the discussion around welfare regimes and the degree of defamilialisation is more likely to be enriched if included in a wider framework for the analysis of financing, delivery and governance of care services. For this purpose we use the framework put forward by Jens Alber (1995) and the concept of the 'welfare diamond' as presented by Marja Pijl (1994), following earlier work by Evers (1990). In the context of this chapter, financing of care pertains to public and private resources that are gathered to cover the costs arising from dependency and need for care, and delivery of care refers to the production of care by several stakeholders. Finally, the definition of governance follows that of Hodges *et al.*

(1996, p. 7) in accordance with its application to the INTERLINKS project (Allen *et al.*, 2011):

The procedures associated with the decision-making, performance and control of organisations, with providing structures to give overall direction to the organisation and to satisfy reasonable expectations of accountability to those outside it.

In his framework, Alber (1995) proposed four areas that should merit greater attention when analysing care services from a comparative perspective: the regulatory structure, the financing structure, the delivery structure and the degree of consumer power. The first of these dimensions can be related directly to issues of governance and the different levels and agencies of government responsible for LTC. The second pertains to the mix of funding, and although Alber (1995) originally applies the funding structure to different public agencies only, this could be broadened to include the public–private mix of funding. The delivery structure refers to the mixed economy of care provision (public, private for-profit and voluntary or non-for-profit providers) and highlights issues of horizontal coordination between different stakeholders. Finally, consumer power is linked to issues of agency of users – that is to what extent users have the final say in deciding the care they need and the care provider to supply it – and concomitantly to the existence of mechanisms that allow users to participate and influence decisions, i.e. to 'voice' and 'exit' (Hirschman, 1970).

Within the context of the then still nascent discussion concerning what would be denominated cash-for-care schemes, Pijl (1994) referred to the concept of the welfare diamond to characterise the relationship between different stakeholders involved in the provision of care (Figure 9.1).

The salience of the welfare diamond for LTC is that it highlights the different welfare mixes possible that characterise LTC systems in Europe. Furthermore, the diamond also incorporates the formal and informal dimensions of care service organisation. Quadrants 1 and 4 of the welfare diamond represented in Figure 9.1 would refer to the formal dimension (shaded area in Figure 9.1), while quadrants 2 and 3 to the informal one, albeit the second quadrant could actually include elements of both areas depending, for example, on the connection of third sector organisations with the state.

The reasons why these earlier frameworks remain pertinent for the comparative analysis of LTC have to do with a series of characteristics inherent to this sector.

Given the relational nature attached to care and its relatively unsophisticated profile, at least in comparison to health care, the family is able to be a much more important and viable co-producer of care (Anttonen *et al.*, 2003). This means that the family is, unlike in health care, still a

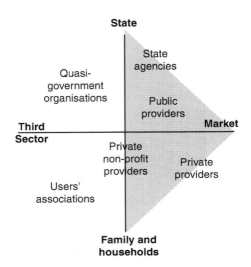

Figure 9.1   The welfare diamond in long-term care
Source: adapted from Pijl (1994, p. 4).

predominant provider of LTC. As such, informal carers could be seen as important stakeholders in the different welfare mixes as they are very much evident in care services (Jensen, 2008).

In a number of countries LTC has evolved from social assistance schemes, rather than from health care systems and therefore means-testing and out-of-pocket payments are present at varying degrees in all countries, contributing to quite a varied picture in terms of the financing mix of care. This again relates to Pijl's (1994) welfare diamond, mainly for analysing the financing of care.

Partially because of its roots in social assistance, the governance of LTC is much more decentralised and fragmented with regional or local levels of government playing a much greater role in financing or regulating the sector. Alber's (1995) concepts of regulatory structure and financing structure between different agencies and levels of government are directly related to this. The same can be said of the degree of consumer power, as the issues of using market-based mechanisms of governance and consumer choice mechanisms have gained increased importance in the context of recent reforms carried out in LTC systems.

For historical reasons the third sector has played an important role in the provision of care in many countries (e.g. Austria, Germany, the Netherlands). This is evident through the inclusion of the third sector as one of the vertices of the welfare diamond, as well as the importance attached to it within Alber's (1995) delivery structure.

Synthesising the above, it is possible to devise three broad dimensions along which differences in public resources devoted to LTC among European countries can be analysed. These are:

- Financing mix: the division of responsibilities for the financing of care between the state, the family and the market. Within the state it is also possible to distinguish the financing mix between different agencies or levels of government;
- Delivery structure: the extent to which the state and the family decide to 'make or buy', i.e. how is provision defined along the formal–informal divide and within the formal area where stakeholders play a role in the provision of care;
- Governance structure: under this heading fit the issues related to governance between different agencies and levels of governance, including regulation of quality, as well as the use of market-based mechanisms for the delivery of care (cf. Nies *et al.*, 2010).

The following analysis is based on these three dimensions described above. The claim made here is that the comparative analysis of LTC systems should not only focus on the financing of long-term care, but also on its delivery and governance, so as to fully capture the diversity of arrangements between the state, the family and the market.

The public–private mix of financing long-term care

Within their social protection systems, countries have followed different pathways to protect people in need of care from the potentially catastrophic financial consequences of this risk. For example, the Organisation for Economic Co-operation and Development (OECD) (2011) distinguishes between those that have established universal coverage (i.e. based on assessed need only) within a single integrated system that covers cost with care (Denmark, the Netherlands, Sweden); those that have means-tested systems where access to public funding is based not only on need but also on income and wealth (England); and those that have somewhat mixed systems combining elements of universalism with means-testing (e.g. social assistance) or that have parallel systems in place (Austria, Finland, Germany, Greece, Italy, Slovak Republic, Spain and Switzerland). These different arrangements can entail different levels of generosity and therefore imply dissimilar public–private mixes in the funding of LTC (Table 9.1).

The public expenditure devoted to financing LTC need is indeed quite heterogeneous among European countries, reflecting, albeit only partially, the differences in the proportion of older citizens that qualify for respective benefits.[1] Systems coined as universal tend to be more generous in principle and thus devote a bigger share of their GDP to publicly funded long-term

Table 9.1   Public expenditure on long-term care and share of beneficiaries in the old-age population for selected countries – most recent available data

	Public expenditure on LTC as a percentage of GDP	Beneficiaries of public benefits (1) as a percentage of the 65+
Sweden	3.6	15.2
Denmark	2.5	17.5
Netherlands	2.2	27.8
Finland	1.8	12.0
France	1.7	12.7
Italy	1.7	13.7
Austria	1.3	21.5
United Kingdom	1.2	17.5
Germany	1.0	11.4
Switzerland	0.9	19.3
Slovenia	0.9	6.7
Spain	0.6	6.9
Slovak Republic	0.5	6.7
Greece	N/A	N/A

Source: Rodrigues *et al.* (2012), calculations based on OECD Health Database, national sources and INTERLINKS National Reports (http://interlinks.euro.centre.org).
Note: (1) in kind and/or in cash.

care (e.g. Sweden, Denmark and the Netherlands). However, dissimilarities in public expenditure may reflect different eligibility criteria (definition of care needs) and the extent to which social protection covers these needs. Although it is difficult to establish direct comparisons, Austria and Germany provide benefits to people in need of a similar amount of care hours per month (50 and 45 hours respectively), but in practice the assessment proves much tighter in Germany, where the proportion of the older population covered is much less than in Austria. Countries such as the Netherlands and Denmark that rely more on the provision of formal care services have higher expenditure ratios than those that provide public benefits that aim to supplement or even support the use of informal care. The latter rely explicitly on the family for the delivery of care, such as the aforementioned examples of Austria, Germany, but also Italy.

Public expenditure ratios may also reflect dissimilarities in the relative generosity of LTC benefits, given unit costs of producing and delivering services. Here, generosity is used as a broader term that means how much of the individual care needs are covered by the public social protection system, as well as differences in the prevalence of people with LTC needs among the total population. Figure 9.2 depicts the relative generosity of each system on the vertical axis in terms of the breadth and depth of coverage of care needs (i.e. how much each beneficiary receives from the available pool of

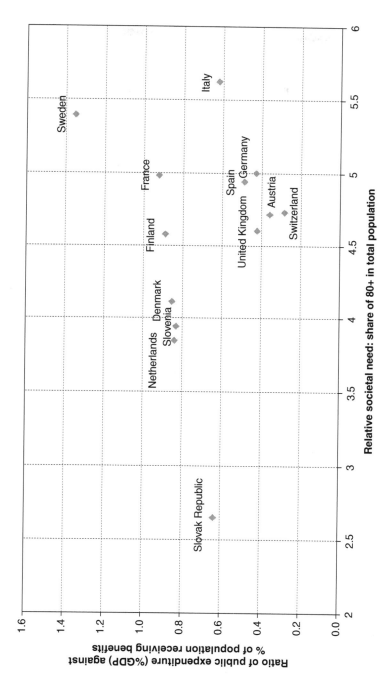

Figure 9.2 Individual and societal coverage of needs by long-term care systems (selected countries)

Sources: Reproduced with permission from Rodrigues *et al.* (2012), calculations based on OECD Health Database, national sources and INTERLINKS National Reports (http://interlinks.euro.centre.org).

resources – of GDP allocated to publicly fund LTC). The horizontal axis provides a measure of potential societal LTC needs, measured as the proportion of people aged 80 and older in the population (i.e. the proportion of those more likely to be in need of care). Figure 9.2 therefore provides further insights into the differences in public expenditure depicted in Table 9.1. For example, Spain, Germany and France all face similar relative societal needs as they have a similarly aged population. But the French LTC system provides a relative amount of benefits per beneficiary that is closer to Northern countries, such as the Netherlands and Denmark, thus accounting for its higher public expenditure ratio than Germany or Spain. France however also provides more generous benefits per beneficiary than Italy, and yet their public expenditure ratios and proportion of people aged 65 and older receiving benefits are roughly similar (Table 9.1). However, Italy has an older population so its population LTC needs are higher, and thus a similar proportion of older people receiving benefits actually means that a greater proportion of the population in Italy is receiving benefits than in France.

Slovenia provides an individual coverage of needs (i.e. ratio of public expenditure per beneficiary) that is similar to that of Denmark or the Netherlands, with the crucial difference being that access to these benefits is much more limited. One could expect means-tested systems to target more resources at those with greater care needs. Surprisingly though, the UK seems closer to the German example of providing benefits of relative lower amount to a wider proportion of its population – albeit with one crucial difference: unlike Germany, the means-tested nature of the UK system in practice means that personal resources must be drawn upon before public support can be accessed. Similar public expenditure levels may therefore translate into quite different approaches to the financing of LTC needs for individuals.

Given that most LTC systems have evolved from social assistance or 'poor laws' rather than as an extension of the health care system, co-payments from users and/or their families are required in all countries, even in those where eligibility is conditional on need only. There are however marked differences in the regulations governing these co-payments. Even though co-payments are present in all systems, they explicitly entail different views on family responsibility for financing care, on intergenerational solidarity and on the role of the state in supporting those in need of care. This is particularly the case with institutional care (Table 9.2).

Co-payments are usually set in relation to income, either as a fixed percentage or a variable share that increases with income, rather than a flat rate. Furthermore, co-payments are usually limited by a cap, set at a relatively low level in Sweden, for example, or a minimum income is left to the discretion of the beneficiaries for their personal use.[2] Where differences become apparent is on the contribution required from relatives (usually children) and on the more or less explicit requirement to spend down

Table 9.2 Rules governing co-payments in institutional care

	Income-related fee	Fee as % of income	Minimum income	Maximum fees	Payment by relatives	Assets
Denmark		Yes				No
Germany		Yes			Yes	Yes
Greece	N/A	N/A				N/A
Spain		Yes	10%			Yes
France		Yes	10%		No	Yes
Italy	Yes				Yes	
The Netherlands	Yes		Fixed amount	Yes		No
Austria		Yes	20%		Yes	Yes
Finland		Yes	15%		No	No
Slovak Republic	Yes		Fixed amount			
Sweden	Yes		Fixed amount	Yes	No	No
United Kingdom	Yes		Fixed amount			Yes

Sources: Adapted from Rodrigues *et al.* (2012), based on MISSOC database (date accessed 18 October 2011), ANCIEN and INTERLINKS National Reports, Leichsenring *et al.* (2010), Smith (2010), Rodrigues and Schmidt (2010).

Notes: 'Minimum income' refers to money that is left for the user to use at his or her discretion. 'Maximum payment' refers to caps imposed on the co-payment amounts.

convertible assets, such as own housing. LTC benefits usually only cover the 'care-related' component of institutional care, leaving board and lodging costs to be paid for by the individuals. As these can be substantial, or LTC benefits can be included in the calculation of user fees (e.g. Austria), co-payments may exceed the users' disposable income. In these cases, private contribution to the financing of care may be higher in countries such as Austria,[3] France, Italy, Germany, United Kingdom or Spain, where contributions from relatives may be required or assets need to be spent down before further public benefits (social assistance) can be granted. In mixed systems, co-payments can thus be a differentiating factor in terms of the public–private mix of financing.

The implications of all this in terms of the public–private mix of financing of care are not however entirely straightforward (see Table 9.3). Although private expenditure is indeed marginal in Denmark, with its universalistic LTC system, the public–private mix of financing care needs is much more skewed towards households in Germany, despite its LTC insurance, than for example in a means-tested system in the UK.

Contributions from different stakeholders, either the state or private households, to finance care only allow for a limited analysis of the relative

Table 9.3 Private expenditure on long-term care in selected countries – latest available year

Country	Private expenditure as % of total expenditure	Proportion of private expenditure for	
		Institutional care	Home care
Austria	11.1	81.6	18.4
Denmark	5.3	94.0	6.0
Finland	22.5	N/A	N/A
Germany	31.3	77.8	22.2
Slovenia	23.3	87.6	12.4
Switzerland	61.0	97.0	3.0
United Kingdom (a)	22.9	85.0	15.0

Source: Reproduced from Rodrigues and Schmidt (2010, Table 3).
Notes: (a) Data refer to England only.

burden placed on the state or users (and their families) to meet care needs. The fact that the family can ultimately substitute for formal care services, especially for lower degrees of care needs, means that families contribute to care not only through financial assistance, i.e. by participating in the mix of financing, but also through direct provision of care. Jensen (2008) has referred to this as the 'high degree of plasticity of social care'. Public resources transferred to individuals in need of care are themselves far from having a neutral impact on the 'make or buy' decision by households, meaning whether they provide it themselves or purchase it externally. For example, cash benefits act as a potential incentive to the direct provision of care by informal carers, if their level comes closer to wages that can be earned in the labour market (Jensen, 2008; Leitner, 2003; Pfau-Effinger, 2005; Saraceno, 2010).

The delivery of long-term care

The 'make or buy' decision concerning the provision of care not only per-tains to households, but also to the state. The provision of LTC can thus be characterised as being at the intersection between the family, the state, the market and the third sector.

The figures displayed in Table 9.4 provide an estimate on the delivery structure of LTC in selected countries in terms of its decomposition between formal (state, market or third sector) and informal care (e.g. family).[4] For most countries for which information is available, it is up to the family to provide the bulk of care for people with LTC needs (or alternatively there are unmet needs). It is only in the Netherlands that care services cover most people who reportedly had at least moderate activity limitations, while in

Table 9.4 Proportion of people (65+) with care needs receiving formal and informal care – most recent data (2006–10)

Country	Institutional care as % of 65+ with care needs	Home care as % of 65+ with care needs	Total formal care as % of 65+ with care needs	Informal care 'by default' as % of 65+ with care needs (a)	Intensity of informal care provided (median weekly hours) (b)
Netherlands	14.0	55.1	69.1	30.9	6.4
Denmark (d)	14.1	35.8	49.9	50.1	4.4
Sweden	18.6	29.8	48.4	51.6	3.8
Switzerland	15.8	31.1	47.0	53.0	5.7
Austria (c)	7.1	39.6	46.8	53.2	9.4
United Kingdom	13.6	25.3	34.7	65.3	N/A
France (c)	11.3	11.8	23.1	76.9	7.9
Finland	8.6	13.1	21.7	78.3	N/A
Germany	6.0	11.8	17.8	82.2	6.9
Slovenia	10.1	3.5	13.7	86.3	N/A
Spain	2.8	9.6	12.3	87.7	20.6
Italy	3.2	5.8	9.0	91.0	15.5
Slovak Republic	3.9	4.2	8.1	91.9	N/A
Greece	N/A	N/A	N/A	N/A	12.7

Source: Rodrigues *et al.* (2012), calculations based on OECD Health Database, national sources and INTERLINKS National Reports; EuroHex Database and Information System, Eurostat; figures for intensity of informal care provided are own calculations based on SHARE.

Notes: (a) Informal care 'by default' is merely the remaining proportion of 65+ not covered by services; (b) Figures refer to care provided to relatives outside the household; (c) Overestimation as figures on beneficiaries refer to the 60+; (d) Data for Denmark is not directly comparable and should be interpreted with caution; 'Population 65+ with care needs' refers to those with at least moderate self-reported activity limitations.

Denmark, Sweden, Switzerland and Austria close to 50 per cent of older people with self-reported care needs receive formal care services or cash benefits aimed at securing adequate care.

However, comparative research on informal care has highlighted the differences in the incidence and intensity of this type of care across European countries (Bolin *et al.*, 2008a) and has given credence to the hypothesis that formal care services at home could supplement rather than substitute for informal care (Motel-Klingebiel *et al.*, 2005). The intensity of the informal care 'by default' in Table 9.4 can thus be quite different and the coverage of formal care services does not rule out the possibility of combining services and informal care. In the case of Austria and Germany, where users may opt for cash to pay informal carers, this is indeed very likely to be the case (Eichler and Pfau-Effinger, 2009). The provision of hours of informal care (right-hand column in Table 9.4) provides additional insights about the division of labour between the family and formal care providers in the delivery of care. The provision of informal care resembles much more a full-time occupation for family members in countries that have devoted only a limited share of their public resources to LTC (e.g. Spain and Greece). In these cases, the family is not just the financer but also provider of care to its members with care needs, which is actually an idiosyncrasy of LTC in all countries, even where the state is an important financer of LTC. For instance, if LTC benefits take the form of cash benefits that, even if not earmarked to be spent on care, can be used to pay for informal care (e.g. Austria, Germany and Italy), it may be stated that the actual provision of care has generally been 'outsourced' to the family and from the family to migrant carers (see Chapter 10).[5]

This supplementary nature of formal and informal care, particularly in countries where users are given the possibility to choose between cash and in-kind services, also becomes evident when observing the figures for employment in the care sector (Figure 9.3). The ranking of countries does not completely follow that presented in Table 9.1, highlighting the clear

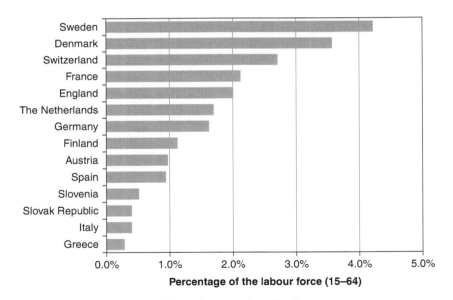

Figure 9.3 Percentage of the labour force employed in the care sector

Sources: Simonazzi (2009); SOTKAnet; BMASK; Statistics Denmark; Statistisches Bundesamt; SALAR; Statistics Slovenia; OECD Health Data; BSF/OFAS; UN Population Division, World Population Prospects; UK Office for National Statistics.

Notes: Own estimates for Austria. Figures for Switzerland include care for people of working age.

option for formalisation of care in Sweden and Denmark, as opposed to Italy and Greece, seen as extreme cases. It also becomes apparent that the relatively high number of older people with care needs receiving state support (right-hand column of Table 9.1) is not really matched by the employment figures in the care sector, especially in Austria. In comparison to Sweden and Denmark, public policies in Austria thus favour a delivery structure that still relies heavily on the family. Similarly, the Netherlands have moved towards making their universal public LTC provision less 'carer-blind' by introducing the notion of 'customary care'. This relates to the amount of informal care that users already receive, or that family members are expected to provide, which is considered for benefit eligibility purposes. This in effect restricts the eligibility for those people in need of care with family members.

The concept of welfare mix central to both Alber's (1995) and Pijl's (1994) framework of analysis of social services pertains not only to the division of responsibilities between the state and the family but also between the state and the market and the third sector – the mixed economy of care provision (Forder *et al.*, 1996). In the majority of the selected countries there has been an increased reliance on market mechanisms in the governance of LTC (Nies *et al.*, 2010). This has taken the form of choice mechanisms (e.g. vouchers or cash benefits), where public funding is attached to the actual use of care by people in need and opening of provision characterised by public monopolies to private providers (e.g. the aforementioned example of Sweden) or new entrants (e.g. Germany). However, the extent to which the market mechanisms have shaped the currently observed mixed economy of care provision is hard to pinpoint due to lack of data on the evolution of the relative importance of different types of provider (Table 9.5).

In the case of Austria, France and Germany, for example, third sector organisations have traditionally played an important role in the provision of care services and this carries on to the present day (Bode *et al.*, 2011). The increased purchasing power of users created by the introduction of cash benefits and vouchers only partly translated into increased supply of care by formal providers. The deregulated nature of cash benefits in Austria and Italy may have actually hampered the development of formal care services (Simonazzi, 2009), while the more regulated French 'Allocation Personnalisée d'Autonomie' fuelled an increased supply of care by 'formalised' informal carers (Le Bihan, 2012).

Private providers have certainly expanded after the introduction of the purchaser–provider split in England and Sweden and the LTC Insurance in Germany (Knapp *et al.*, 2001). However, in Sweden the market share of private providers remains limited even though the 'privatisation' of care provision has been one of the main driving forces of changes introduced in LTC since the 1990s. First this took the form of the purchaser–provider split, and more recently, with the adoption of 'consumer choice' through

Table 9.5   The mixed economy of formal care provision

Country	Public providers		Private non-profit providers		Private for-profit providers	
	Residential	Home care	Residential	Home care	Residential	Home care
Austria	55%	8%	24%	91%	21%	1%
United Kingdom	7%	14%	13%	11%	80%	74%
Finland	56%	93%			44%	7%
France	23%	15%	55%	65%	22%	20%
Germany	5%	2%	55%	37%	40%	62%
Italy[1]		30%		50%		20%
Netherlands[1]		0%		80%		20%
Slovak Republic[1]		75%		23%		2%
Spain[1] [2]		23%		24%		53%
Sweden	75%	N/A	10%	N/A	15%	16%
Switzerland[1]		30%		30%		40%

Source: Allen *et al.* (2011), Leichsenring *et al.* (2010).

Notes: (1) Only aggregate data for residential and home care available. (2) No clear distinction can be made between private providers in Spain – in this table, non-profit providers include all those with a formal contract with the Autonomous Communities; private for-profit providers include those with an authorisation only – all costs have to be covered by the individual resident.

the use of vouchers that allow for users to choose their provider. This is an option that policymakers hope will become widespread and contribute to the empowerment of users (Szebehely and Trydegård, 2012). In Spain, the market has to some extent compensated for the limited state support (either through public services or cash benefits) that till recently characterised the area of LTC.

It is clear that even when LTC is to a great extent publicly financed, it will still be provided mostly by the family. Similarly, private (for- or non-profit) providers have not always filled the void of insufficient public support of provision of care or emerged in response to increased user choice and purchasing power brought by the introduction of cash benefits or vouchers. The mixed economy of formal care delivery has thus been determined very much by path-dependency and national contexts.

The mixed governance structure

Governance by public authorities primarily relates to defining what and how services should be publicly funded and to what extent; how public resources are raised (e.g. taxes, insurance, personal payments) and how public accountability is regulated; how eligibility is circumscribed (which citizens have a right to receive publicly funded services); and how quality of the services is ensured (see ESN, 2010; Colombo *et al.*, 2011). These tasks in fact represent the often cited triad of governmental responsibilities: sustainability, access and quality.

Since LTC is closely associated with health and social care, its governance structure is related to these systems. However, while governance of LTC and social care is much more decentralised than health care, responsibility for a well-functioning system of health care is often laid down in nations' constitutions, which is not the case for LTC. Situated at the interface between health and social care, governance of LTC is characterised by horizontal dispersion of powers.

As the INTERLINKS National Reports show, the countries have laid down their various governance responsibilities at different levels of government: national, regional or at the municipality level. Some countries have a long-standing decentralised governance structure, of which Switzerland is probably the most well-known example. Typically with LTC, governments at various levels are responsible for planning, policy development, regulation, implementation of policies and management of the eligibility assessment and benefit provision. Even when the division between the purchasing and provider function has been established, local governments may well be the main providers as well as regulators of LTC, as shown in the previous section. Adding to the horizontal dimension of governance stated above, there is also a vertical distribution of functions and responsibilities among different layers of government (Figure 9.4).

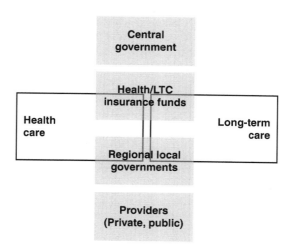

Figure 9.4 The mixed horizontal and vertical governance of LTC

Issues of coordination at the horizontal level also arise because LTC and health care do not always reside under the same ministry or department, even if they are at the same governance level. Moreover, as LTC serves people with chronic multiple conditions, their housing and income needs are often part of the problem and need to be taken into consideration. Also at the horizontal level, different government structures are responsible for service provision to older people in need of LTC. Additional stakeholders such as health care insurance funds, which play an important role in health care and LTC in countries like Austria, Germany and the Netherlands, may pose different problems of planning and coordination. These funds are responsible for their insured members, who are not necessarily geographically bound, whereas LTC usually is.

Health care services are usually organised at a higher level of governance than LTC. This distinction is typical for most European countries, with the exception of Austria, Germany, France and the Netherlands. In Bismarckian and other systems where health care and LTC care are parts of an insurance system, health care insurers usually carry governance responsibilities with respect to funding, eligibility and quality control.

Governance responsibilities within LTC are often dispersed within local, regional and national levels of accountability. National frameworks exist for LTC and social care, expressing what is desired or required at a regional or local level (Austria, England, Finland, France, Germany, Italy, Sweden). In some of these countries it is a local or regional responsibility to integrate national recommendations according to their idiosyncrasies (England, Finland, France, Italy). Local care provision is in some countries monitored at the national level, providing feedback on performance to local or regional authorities (England, Finland, Sweden, the Netherlands) (Nies *et al.*, 2010). Also responsibilities for raising taxes or insurance fees differ across countries, with the Scandinavian countries laying down responsibilities largely at the local level. Responsibilities for monitoring and ensuring quality of care are positioned at a local or regional level in countries such as Austria, Finland, France, Spain and Switzerland. The development of quality systems, standards and indicators is in some member states organised at the national level but implemented at a local level (England, Finland, France, Spain, Sweden) (Nies *et al.*, 2010).

These examples demonstrate that within countries and systems various aspects of governance are laid down and executed at various levels of government. In practice this means that the collaboration between health care professionals with LTC and social care professionals can be difficult because of non-compatible funding, quality systems and different eligibility criteria (see Box 9.1). Furthermore, integrated planning of social services may be difficult to achieve. One classical example is hospital discharge, which can be hampered by these mechanisms and – in worse cases – by shrugging off responsibilities.

Box 9.1 An attempt to bridge governance divides: Guidelines for Dementia Care (Sweden)

In 2010, the first Swedish Guidelines regarding dementia care were published by the National Board of Health and Welfare (NBHW). The development of the Guidelines goes hand in hand with the considerably increased focus on dementia care in Sweden. The Guidelines present evidence-based and evaluated treatments and methods within health and social care regarding people with dementia and informal carers. The Guidelines are in the first instance directed to the municipalities (e.g. social care, home help, etc.) and primary care, managed by the county councils. The purpose of the Guidelines is to provide support to decision-makers within municipalities, county councils and regions, so they can govern the different organisations according to systematic priorities. Other aspects of the Guidelines address doctors, nurses and other health and social care professionals on a managerial level. By using the Guidelines it is possible to overcome barriers between different organisational bodies, because the Guidelines put the patient in focus.

Source: Emilsson (2011).

While there is a wide variety of stakeholders with different roles in LTC, these same stakeholders also operate within different modes of governance. Among the many different modes of governance, Allen *et al.* (2011) typify the most prominent ones (see also Table 9.6):

- *Public/Social programming*: public regulation of supply by planning services and personnel based on population size and estimated need. Although public and social planning as a whole approach tends to have been replaced by market mechanisms, aspects of this approach remain in some countries.
- *Market mechanisms and New Public Management (NPM)*: mechanisms based on competitive tendering, contracting and competition within quasi-markets of LTC provision. These mechanisms are evident in the majority of countries in relation to LTC.
- *Consumerism* as a subset of market mechanisms: focuses on maximising user (and carer) choice, giving them a greater role in decision-making about their care. To be seen in greater use of cash, vouchers, personal budgets and direct payments, often used to support home care and personal assistants and to choose from various services and benefits on offer.

Public/social programming seems to have become less prominent in the wake of reforms inspired by New Public Management that introduced competition, purchaser–provider split and other quasi-market arrangements for the

Table 9.6 Overview of main governance mechanisms used

Country	Main governance mechanism used
Austria	User-oriented through the LTC cash benefits, combined with subsidies to providers (fixed prices). Limited social programming (e.g. building quasi-public care homes).
Denmark	Social programming with direct provision, block grants, activity-related subsidy, a local basic contribution and a local activity-related contribution. Minimal user-oriented (cash/voucher).
Finland	Public/social planning with highly decentralised organisation of care; local (municipality level), funders, organisers and providers of services. Increasing role of private for-profit service providers. Low user-orientation (cash/voucher), while service vouchers in LTC are becoming more acceptable/widely used.
France	Previously more social programming orientation (in-kind services) with recent strong trend for a consumer orientation (through vouchers and fiscal incentives and for provider competition).
Germany	User-oriented through the LTCI cash benefits for service users (Care Attendance Allowance). Little user orientation in formal care. It is distributed as limited in-kind services. Limited social planning of LTC services. Municipalities have almost no impact on LTC steering. Regional LTCI Boards supervise the care delivery with little ambitions for a comprehensive planning of nursing and social care.
Greece	Informal care in LTC is the rule and the private sector, including migrant care workers, predominates in both home and residential carer provision for financially weak groups without family support. Social planning includes subsidies, partial coverage of costs, direct provision in residential care homes.
Netherlands	National framework of a regulated market with quasi-competing providers. Decentralisation of steering and commissioning power (regional budgets) to regional LTC offices (acceptance of service providers to operate on the market), and of social support to local authorities; tendering of domestic care services. National needs assessment scheme. Significant number of cash benefits/individual budgets.
Slovakia	User-oriented mechanism, with cash benefits, *ex ante* needs assessment and subsidies for informal carers.

Switzerland	Between a commissioning and a programming approach, depending on the canton. In the French-speaking cantons, social planning is prominent. In German-speaking cantons, there is a more commissioning approach involving managed competition between service providers.
Sweden	With the growth in outsourcing by means of public tendering by municipalities, the long-standing monopoly of public provision has been eroded as new private for-profit providers have succeeded in winning tenders against provider agencies within municipalities. Public procurement had until recently been based on a rationale that guaranteed a local monopoly (replacing the public monopoly) to providers who won the tender, denying users any choice among providers. Nowadays, municipalities may choose to introduce the 'user choice' model by giving vouchers to users who may then choose between different accredited providers.
United Kingdom (England)	Older people's services are commissioned by local authority adult social services (social) and Primary Care Trusts and GPs (soon to be Clinical Commissioning Groups), LTC services are commissioned from a mixed economy of acute hospital trusts/social enterprises, mental health trusts, GPs and various community care providers (private, public and third sector). Within this commissioning approach there is a strong commitment to user-orientated mechanisms of personal budgets throughout adult social care and personal health budgets.

Source: http://interlinks.euro.centre.org; Allen *et al.* (2011, Table 4.1).

provision of care. More recently, a strong consumerism focus has emerged in LTC, partially as a 'natural' development of market-based approaches to the delivery of care and partially as an attempt to both empower users and to bring about changes in the standardised care delivered by market mechanisms (cf. Kremer, 2006).

The outcomes are not entirely clear for, in some cases, the introduction of these novel modes of governance has been linked to the pursuit of concomitant policy goals such as increased employment in the care sector (Da Roit and Le Bihan, 2010). It is clear nevertheless that the move away from public/social planning has led to the emergence of new actors or the revamped prominence of older ones in third sector organisations in some countries, as discussed above. Private and non-profit providers are increasingly contracted to deliver public services, replacing a more trust-based system. Balancing responsibility and activity between government and non-government sectors is viewed as central to the successful governance of LTC in Europe (Wittenberg *et al.*, 2002, p. 227).

While the reliance on market mechanisms seems to emerge as a common development in the governance of LTC, it is harder to pinpoint other trends in terms of reforms in the governance of LTC. The INTERLINKS reports have shown diverging paths in the governance of LTC:

- *De- or Recentralisation (Rescaling) of responsibilities*: involves both shifting of administrative, financial and regulatory mechanisms to other government layers;
- *Proactive government initiatives to promote coordination/integration*: where government initiatives focus on gaps and interfaces in LTC and on funding and financial flexibility between health and social care organisations (Nies *et al.*, 2010), and/or on the development of more integrated care pathways;
- *'Muddling-through' and relying on informal carer*: places care responsibilities firstly with informal carers, with consequences such as 'illegal' migrant care. These countries often do not have a coherent policy.

Decentralisation has often been advocated on the grounds of increasing health and LTC integration at the patient level, encouraging innovation, achieving greater cost-efficiency and bringing LTC systems closer to the users. However, this risks bringing about marked geographical differences and difficulties in the recognition of competences (see Chapter 8 for examples of regulations on quality), or inequities in accessing care – a prominent issue in countries with a strong universalist tradition in the provision of LTC, such as Sweden (Trydegård and Thorslund, 2010).

In sum, governance of LTC is a matter that concerns different levels of government and stakeholders applying their own modes of operation. The resulting complex governance structure of LTC may cause problems for older people, which is one of the reasons why integrative mechanisms such as single points of access and case management take a prominent place in well-functioning LTC systems.

Conclusions

LTC can still be characterised as an emerging area of welfare policy which is in a transitional state in Europe. For some countries the current levels of public expenditure seem unsustainably low as they are based on a heavy reliance on the family for the satisfaction of care needs. At the other extreme, among those which have more mature LTC systems, the concern is that current levels of benefit generosity and systems coverage may be fiscally unsustainable in the future, even if public expenditure on long-term care is dwarfed by other social expenditure such as health and old-age pensions. This may explain the newly found importance and visibility attached to the support for informal care, as a way to keep public expenditure under control.

In all their diversity and complexity, what this chapter has tried to argue is that making sense of differences in public expenditure on LTC across Europe also involves understanding the differences in the delivery and governance of LTC. The issues surrounding both the delivery and governance of LTC remain two relatively unexplored avenues of discussion and research.

As most systems currently stand, governance of LTC is characterised by multiple stakeholders, division of responsibilities in the regulation, delivery and funding of care and multiple governance mechanisms. Regardless of the origins of such complex arrangements, the fact remains that they pose challenges for users navigating the system, as there is not always a single point of access. This may compound inequities in the use of services and create possibilities for cost-shifting among stakeholders disguised as efficiency gains. One of the main conclusions of INTERLINKS is the potential losses, both in terms of users' well-being and tax payers' money, incurred at the interfaces between services and different providers along the continuum of care.

The most recent and relatively widespread reform in terms of the governance of LTC has been the introduction of market-based mechanisms. Despite its touted advantages in terms of increased efficiency and responsiveness to users' needs and preferences, the impact of market mechanisms in LTC has not yet been fully understood. It has brought a new focus on important issues like user agency and user-centred care. But it has been superimposed on existing systems, sometimes with the accompanying necessary changes in terms of funding and definition of quality, thus resembling more a piecemeal reform rather than a comprehensive one. It has appeared to have changed incentives, although it is not clear if it has hampered integration of care by eroding trust-based relations between providers, or if the high-powered incentives of competition have helped to bring down barriers for joint working.

As far as the delivery of care is concerned, it is not just about the traditional, albeit still pertinent, division between formal and informal care provision and the different profiles of informal care. The introduction of market-based mechanisms has created different mixed economies of care in the provision of care. These different mixed economies of care denote also different degrees of 'marketisation' of LTC – an issue that remains far from consensual at a European level.

Notes

1. The broader definition 'user of long-term care benefits' is used here so as to encompass not only those receiving in-kind services financed by the state, but also beneficiaries of cash benefits (attendance allowances) that may or may not be used to pay for informal care, depending on the national regulations. Wherever possible, beneficiaries of care allowances (benefits paid to carers) were excluded so as to avoid double counting. In some cases, however, double counting could not be avoided as, for instance, individual beneficiaries might qualify and/or benefit concomitantly from two or even several services.

2. For instance, in Austria, residents of care homes are charged their pension and their LTC allowance to contribute to total costs, but 20 per cent of their pension and a flat rate of about €45 per month from their LTC allowance have to remain at their personal discretion.

3. At a governance level, the Austrian LTC allowance also serves as a shifting mechanism of federal funding to the users that then use it to pay for services owned or managed by the regions.

4. Based on EU-SILC (Eurostat), it is possible to calculate the prevalence of people aged 65 and older who have at least some degree of limitation to their activity and thus take the resulting 'population with care needs' as the target group of LTC systems. Against this it is possible to estimate the proportion of the population with care needs that receives formal care services or cash benefits that might be used to secure access to adequate care (e.g. Austria and Germany).

5. This movement may be seen as part of a wider picture with far-reaching implications in terms of gender, whereby women's labour market participation in the receiving countries is sustained by outsourcing care tasks to, again mostly female, migrant care workers (see Chapter 10).

10
The 'Care Gap' and Migrant Carers

Rastislav Bednárik, Patrizia Di Santo and Kai Leichsenring

Introduction: The common issue

The demographic transformation due to population ageing, changes in household structures and roles, together with more demanding patterns of care, such as dementia, and an inadequate provision of long-term care (LTC), have significantly influenced the steadily increasing demand of families for alternative care resources, namely for assistance and support within the household by third parties. This has resulted in the emergence and spread of a new phenomenon by which families establish care arrangements with private care workers, who are mainly female migrants from neighbouring East European countries (Bettio *et al.*, 2006) or from Asia, Africa and South America (Cox, 2006; Lutz, 2008). Private care workers actually fill the gap between family care that is no longer able to sustain 'traditional' caring tasks for older parents, and the lack of public services: in fact the EUROFAMCARE study on family carers already found several years ago that, in 17 out of 23 countries in Europe, families reported relying on private migrant care workers at least from time to time (Mestheneos and Triantafillou, 2005).

Over the past decade, the phenomenon has become widespread in Southern European countries, with Italy as a leading country with an estimated minimum of at least 700,000 migrant care workers (Da Roit and Castegnaro, 2004; Pasquinelli and Rusmini, 2008). This development has partly been fuelled by the Italian welfare system and its high prevalence of cash allowances, rather than service provision. In Italy, more than 13 per cent of all families with at least one family member in need of care are hiring private care workers (Ufficio Studi Confartigianato, 2011, while in Greece the proportion of families hiring migrant care workers is around 7 per cent. It must be noted that these 'official data' are underestimated because of the widespread phenomenon of undeclared work in this market. In reality, there is a distinction between several types of 'migrant care workers':

- Migrant care workers providing household chores and care in a broader sense in private households with legal immigration documentation (also temporary) and a legally declared employment contract; some may also have relevant training in health or social care (Piperno, 2009);
- Undocumented migrant care workers providing assistance for a defined period in private households with neither regular immigration documents nor an employment contract or training (FRA, 2011);
- Migrant care workers residing in the receiving country and with employment in the formal care sector based on relevant education and training which is widespread in Western and Northern Europe, where migrant workers are employed by care homes or by home care organisations (see for the UK: Hussein, 2011; Spencer *et al.*, 2010).

This chapter is based on results from INTERLINKS research (Triantafillou *et al.*, 2010; Di Santo and Ceruzzi, 2010; see also Chapter 1) and deals with the specific characteristics of migrant carers at the interface between informal and formal care in emerging LTC systems. In the first part, the phenomenon will be described and analysed from the perspective of receiving countries with a focus on Italy, Germany and Austria. This selection was made for pragmatic reasons as the phenomenon of private care provided by migrant carers has developed more prominently in these three countries, triggering legal and political action at various levels. However, some facts and figures from other European Union (EU) countries have been taken on board, where appropriate. The second part is dedicated to the situation from the perspective of a sending country, the Slovak Republic. This focus was chosen as Slovakia was the only 'sending country' represented in the INTERLINKS consortium, but nevertheless provides some important messages from this sole perspective that have relevance to other contexts. Finally, some practice examples will be presented to show how both private care workers' and families' rights can be protected to improve the quality of care at the interfaces between formal and informal care.

Migrant care work in European receiving countries

It is generally difficult to retrieve data about private migrant care workers in receiving countries as most of them still work in undeclared conditions, and because the phenomenon is relatively underestimated or even neglected. However, Triantafillou *et al.* (2010) and Di Santo and Ceruzzi (2010) have made inroads by describing and analysing the phenomenon in some selected European countries to shed light on its extent from a quantitative perspective, and examine the consequences for formal and informal care.

In Greece, the past 10 to 15 years have seen a substantial expansion in the numbers of migrants who undertake caring work in the homes of older people with long-term care needs. Although the phenomenon is widespread, it is seldom revealed, because private carers are often undeclared migrants. Data from the largest Social Security Fund in Greece (IKA) report that the mean number of employees who work in private households in a legal way, i.e. registered domestic workers (not all of whom are providing care), amounts to about 20,000. The vast majority of these (85 per cent approximately) are migrants (Di Santo and Ceruzzi, 2010). While the majority of female immigrants in Greece are undocumented, data on those with a work permit show that 46.5 per cent were employed in private households in 2006 (Kapsalis, 2009).

In Germany, the topic of migrant care workers in private households is relevant and has been the subject of political debate. A special regulation had originally been established in 2002 to facilitate the procurement of domestic help in households with people in need of care. By overhauling the recruitment barriers for citizens from third countries, this regulation enabled citizens from the former EU accession countries of Poland, Slovenia, the Slovak Republic, the Czech Republic and Hungary to legally work in Germany as a domestic help for a maximum of three years. This regulation was integrated into the German Immigration Act, which entered into force in 2005, when the number of countries for recruitment was also extended to include Romania and Bulgaria. However, families continue to illegally employ migrant domestic help or care workers with estimated figures of up to 200,000 households.

Migrant care workers are numerous in Spain and constitute around 30 per cent of the total paid workforce in the area of care. They are mostly migrants from Spanish-speaking countries in South America. Most of the domestic Spanish care workers are employed by private companies or cooperatives (97 per cent), in contrast to migrants who are mainly employed by private households (75 per cent). In 2006, more than 223,000 migrant workers in domestic care were involved in domestic tasks. Moreover, Social Security data established that, by December 2008, 56 per cent of the domestic care service staff were foreign born (Di Santo and Ceruzzi, 2010).

In Austria, about 5–10 per cent of older people in need of care are cared for by so-called '24-hour assistants', that is migrant carers from neighbouring Central and Eastern European countries. Due to the geographic closeness, this type of care is usually organised by means of intermediate agencies in Austria or in the sending countries, with two carers per family each covering 14 days per month. In most cases they are formally self-employed on the basis of an amended legal regulation for 'domestic workers'. People with care needs who are employing migrant carers receive an additional subsidy to cover costs for social insurance contributions. Also private non-profit organisations that have traditionally provided formal home care have started to offer

'24-hour assistance', either by employing migrant carers or by providing paid support coupled with administrative duties so that people with care needs do not have to bother with the bureaucracy surrounding employment (Rodrigues and Schmidt, 2010; Prochazkova and Schmid, 2009).

In France the phenomenon exists, though to a lesser extent, but there are no data on migrant care workers due to the French tradition of avoiding 'ethnic statistics'. This means that, even in cases where immigrants are legally entitled to stay and work in France, it is difficult to find data in routine statistics concerning their employment status (Naiditch, 2009).

The situation in Italy

In Italy, the changes in the demographic structure and in family roles, combined with a weak welfare system and support schemes that offer mainly cash benefits rather than services, have triggered an explosion in the demand for private assistants from families. Female labour market participation and the rising mobility of the labour force have created a reality in which cultural heritage and socio-demographic structures are not compatible any more. With 14 per cent of the Italian population above the age of 75 and a rising proportion of older people in need of full or part-time assistance (6.7 per cent in 2010 and an estimated 7.9 per cent in 2020, according to Del Favero, 2011), the bulk of care is still provided by informal family carers, but increasingly by private care assistants who step in to compensate for the lack of formal services. The latter are provided by public, private non-profit and commercial providers but the network is rather weak. At the national level, in 2008, so-called 'Integrated Health and Social Home Care' provided by public services only reached 4.9 per cent of the older population with an average of 24 hours of services provided per year to the user; residential care facilities only cover about 3 per cent of older people in need, while 9.5 per cent of all to an attendance allowance (indennità di accompagnamento)' are entitled to cash benefits (NNA, 2009). These are entitled to *indennità di accompagnamento* and paid as a single lump sum (about €550 per month in 2012) to those who have been assessed by a medical commission as being in need of care. The number of older people in need of care who are assisted by private workers has significantly increased to approximately 700,000 (Del Favero, 2011) and outnumbers by far those assisted at home by public services, estimated at 313,000 (Burgio *et al.*, 2010; see also Iref-Acli, 2007; van Hooren, 2010).

As in other European countries, policies in Italy also aim to keep older people in need of care at home for as long as possible (Mauri and Pozzi, 2005). At the same time, the facilitation of female labour market participation is a specific priority of labour market policies (Italian Government, Equal Opportunities Department, 2007). However, both goals are contradictory all the time the offer of formal care services remains at a very low level. As a 'solution' invented by civil society, private care workers, who are usually migrant women, have taken over to cover the needs for LTC at sustainable costs. In practice, these costs are between €850 for non-registered care workers and up to €1500 per month for registered care workers. This amount can usually be covered by the income of an older person's individual pension and the attendance allowance. As public services are not able to provide the same amount of care at such tariffs, the demand for care work undertaken by private workers is directly linked to the need of older people and their families.

While in the past it had been mainly middle class families who hired domestic assistants, over the last decade private care at home has become an issue involving all social classes. Many older people with small pensions are supported by contributions from their sons and daughters who often play an important role as coordinators of the private assistants, including the management of the contract and organising cover for absences due to sickness or holidays.

The National Social Security Agency (INPS) calculated that, between the years 2000 and 2007, the number of foreigners working in the field of family assistance had increased by 173 per cent (Iref-Acli, 2007). When public policies became aware of the phenomenon, an initial reaction was, in 2002, to facilitate the 'regularisation' of migrant care workers. Regularisation is defined 'as any state procedure by which third country nationals who are illegally residing, or who are otherwise in breach of national immigration rules, in their current country of residence are granted a legal status' (Baldwin-Edwards and Kraler, 2009, p. 7). Employers and/or employees who decided to 'regularise' their relationship were exempt from fines and received a regular residence and working permit. Such permits were offered to migrant workers in all sectors, but almost fifty per cent (about 330,000) of those who registered reported a job in the field of domestic work and personal assistance. Thus, between 2002 and 2009 at least 900,000 private assistants (working for about 1.5 per cent of the Italian population) were 'regularised': 300,000 through the Regularisation Act of 2002, another 300,000 between 2003 and 2008 in the context of defined 'Immigration Flow Decrees' and finally 300,000 with another 'Regularisation Act' in 2009 (Pasquinelli and Rusmini, 2010).

This last Act failed to achieve the desired outcome: according to the Italian Home Office, 700,000 applications had been expected but by 30 September 2009, only approximately 295,000 had been received. These were mainly Ukrainian (42,000), Moroccan (38,000), Moldovan (29,000) and Chinese (22,000) workers (Pasquinelli and Rusmini, 2010). This means that many employees had decided to keep working illegally. This could be due to the fact that it would seem more advantageous for a migrant care worker to receive a gross wage in the short term, and not pay taxes and pension contributions that might never produce a return. It must be remembered that most migrant care workers do not stay in Italy until they are of pension age. Furthermore, many undocumented migrants, i.e. those without actual residence or authorisation, did not trust the waiver regulation, which promised exemption from punishment in the context of 'regularisations', and preferred not to get in touch with the authorities. As a result, a large amount of migrant care work in Italy is still carried out 'illegally' – with continuing and impressive growth rates. The overall amount of private assistants has grown over the past ten years, according to Censis (2008), as one of the few occupational sectors with an increasing trend even during a period of economic crisis. In 2009 the official number of private domestic assistants in Italian households reached 1,538,000 (data from the 'Regularisation Act').

The profile of migrant and domestic care workers in Italy

Most of the regularly registered private assistants are of foreign origin; only 22.3 per cent are Italian, even though the economic crisis is currently making this sector of care and household work increasingly attractive for Italians, in particular for women between 45 and 55 years of age who have become unemployed. Based on social security data the number of Italians in domestic employment has risen from 34,000 in 2005 to 59,000 in 2009, with a particularly sharp increase between 2007 and 2008 (Sarti, 2010).

Women make up 87 per cent of all foreign care workers and 96 per cent among Italian private assistants. Care and domestic workers in Italian households count for more than a fifth (21.9 per cent) of all people enrolled in the national social security register (INPS) (Di Santo and Ceruzzi, 2010).

Most migrant care workers could or would be able to work in other economic sectors since they often have a university degree, for instance 70 per cent of Moldovan women working as private assistants in Italy are graduates (Di Santo and Ceruzzi, 2010).

Young, 'illegal' and segregated are the characteristics of this 'new' type of caregiver (Rossi, 2004). The latest trend is that hourly jobs seem to currently replace the 'classical' cohabitation with the person in need of care (Caritas, 2008). This seems to be a sensible choice also in terms of social inclusion, as being a *badante* (Italian for assistant or caregiver) is not conceived as a lifelong occupation. Indeed, for those migrants who want to stay in Italy, care work is restricted to a limited period during which they try to construct an economic and social stability in order to find a regular job in another sector or generate an economic basis for returning back home.

Migrant care workers usually go abroad to seek a better future for themselves and their families: Italian salaries are still significantly higher than those in their countries of origin. The average salary of a private assistant is about €1350 per month; without a formal employment contract it is about €850–1000 per month if the private assistant lives with the older person 24 hours a day. Decreasing amounts of time trigger a gradually lower salaries are commensurate with hours worked, e.g. private assistants who work only 4–5 days per week for four hours per day will receive a salary that might be roughly 50 per cent of the above amounts. Working as a live-in private assistant includes free meals and accommodation, with opportunities to save money to send to families in the home country.

Migrant care workers have to deal with all kinds of tasks in the household of an older person with care needs: personal care and hygiene, cooking, housekeeping, washing clothes and shopping. They even provide health care, including injections and treatments, which adds yet another legal problem to the often already precarious situation in which private assistants are working. Those with the relevant professional training as registered nurses sometimes even work night shifts in hospitals in addition to their daily work in a private household. Both workloads and the type of activities required from a private assistant tend to increase in relation to the age and the level of care needs of the older person.

If the supply of migrants decreased, the provision of LTC for older people in Italy would face serious problems, since it is estimated that the economic value of migrant care work currently produces a public saving of about €45 billion. In fact, a place for a single older person in a residential care facility would cost on average about €26,000 yearly for public subsidies and a further €18,000 per year for the older person and their family. However, as only a small proportion of private assistants are formally employed, the state is losing more than €800 million yearly on taxes and contributions for these activities (NNA, 2009).

Improving the situation of migrant care workers in receiving countries

Considering these figures, the relationship between migrant care workers and the formal service system has become a crucial issue, not only in Italy. While scant attention has been given to this issue by policymakers and social and health care services over a long period of time, more recently many local authorities have started to introduce measures aimed at ensuring and improving quality of care in households that employ migrant carers, for example by matching demand and supply, introducing training initiatives, providing economic support, establishing a register of migrant care workers, and tutoring (Rossi, 2004; Sgritta, 2009).

However, neither coordination in governmental policies nor uniformity at local or regional levels have been achieved so far. The scant attention by national and regional legislation, with the exception of the Friuli Venezia Giulia, Sardegna and Emilia-Romagna regions, is complemented by a lack of guidance for professionals in formal care services on how to cope with the presence of migrant care workers in private households. For instance, the so-called 'Integrated Health and Social Home Care service' is striving to identify tools to coordinate social and health care services, while suitable ways to coordinate formal and informal services remain of little interest.

Still, the presence of migrant care workers has also triggered largely unanswered concerns about the quality of services provided in private households. While in previous years the approach of public authorities towards family assistance and migrant carers was mainly about 'regularisation' and contributions to cover the costs (van Hooren, 2010), some local authorities in Italy have started to ensure and improve the quality of private assistance by means of formal training, setting up a register of trained private assistants, providing information and counselling, tutoring, etc. (see Box 10.1; see also Di Santo, 2011b).

Box 10.1   ROSA – A network to promote formal employment of migrant care workers (Italy)

This project, supported by the Italian Department of Equal Opportunities and Rights, is hosted by the Puglia region and promoted by a wide range of stakeholders, including the main national trade unions. In response to the large number of migrant care workers in this region, ROSA aims to improve the integration of local home care networks, families and appropriate working conditions for migrant care workers, in particular to encourage the regulation of undeclared work in home care. To improve the gap between supply and demand in home care by including migrant care workers as a key resource, the following activities are offered by ROSA:

- training programmes to develop the skills of migrant care workers,
- involvement of local stakeholders to create a system of services that allows women to improve the balance of work and family care (information on available resources, information and consultation);
- incentives for families to establish regular contracts with migrant care workers, in particular through the introduction of a care voucher scheme and a related information campaign.

In a context of scarce public services and a welfare system that tends to place most responsibilities for LTC on families, migrant care workers have become a key support mechanism for older people with LTC needs. The amendment of legal, educational and organisational questions linked to this phenomenon, namely the lack of employment contracts, moonlighting and illegal work as well as issues of quality and safety, is therefore high on the agenda in countries like Italy. ROSA is based on a stable collaboration between different governance levels and various local organisations and therefore succeeds in integrating public and private resources to enhance the continuity of care at home for older people and, at the same time, to support migrant care workers and families.

Source: Di Santo, 2011a.

Also in Austria, the political debate about migrant care workers has triggered some initiatives in an attempt to regulate access, to create legal relations between employers and employees, and to better integrate private assistants in the LTC system (see below, Box 10.2).

Such initiatives are only a first step on a very long pathway of reforms that will have to be carried out to deal with the following issues in particular (Rossi, 2004).

Information

Social help desks can support families with information provision, guidance on local services able to meet their needs, and requests about the whole range of social and health care services for people in need of care, how to employ private family assistants, and the referral to agencies providing support in relation to contracts and subsidies. Additional information needs to be provided to migrant care workers who are about to start working in a family. In Italy, for instance, there are two professional associations to support domestic workers, private care workers (Italians and immigrants) and families who are about to employ a private worker. These associations offer information on procedures to be followed in order to employ a private assistant (for personal and home care), and how to manage a regular contract with the private assistant already working in the household, e.g. with respect to social security contributions and taxes (Di Santo, 2011b; see also www.qualificare.info/home.php).

Similarly, in Austria a range of private non-profit and commercial organisations have taken up the role as brokers and/or employers of migrant care workers. For instance, most of the larger non-profit organisations providing home care have added '24-hour assistance' to their portfolio of services. This includes brokerage of selected and trained migrant care workers (mainly from Slovakia), administrative support and information about subsidies, as well as organising cover in case of the migrant care worker's illness or other formal care service absence to complement the 24-hour assistance (www.hilfswerk.at).

Recruitment of migrant care workers

Matching demand and supply should be organised as an integrated service that supports both the family and the assistant. In Italy, for instance, the responsibility to deal with these issues lies with different institutions. The municipalities are responsible for social care issues while employment is a matter for the provinces, resulting in two types of help desk that do not necessarily match labour market supply with older people's demands. As a result, specialised agencies, but also cooperatives that are self-governed by Italian and migrant private care assistants, have developed to provide support in matching this supply and demand, and to 'regularise' migrant care workers in terms of visa regulations and work permits (see, for instance, www.servizibadantimantova.it).

Promoting formal employment (incentives to legal contracts)

Public subsidies to support families in paying the caregiver's social security contributions are an important means to combat moonlighting and illegal work (Box 10.2). In order to incentivise the establishment of formal employment contracts, respective schemes need to be underpinned by intensive information and counselling activities ('one-stop shop'), the coverage of social security contributions without income thresholds, and quality assurance by developing a partnership with formal health and social care services (Prochazkova *et al.*, 2008).

Box 10.2   Public support of 24-hour assistance in Austria

In order to regulate the 'grey' market of migrant carers providing 24-hour assistance in private households, the Austrian government adopted the so-called Domestic Assistance Act ('Hausbetreuungsgesetz') in 2007. The objective of this regulation was to acknowledge the need for 24-hour assistance by amending labour market, social insurance and professional regulations. Furthermore, means-tested subsidies were foreseen for people in need of LTC employing migrant carers and/or other informal carers, including family members. The regulation stipulates two possibilities to legalise the relationship with a 24-hour carer: she or he may either be directly employed or register as a self-employed 24-hour assistant (BMSK, 2009). People directly employing assistants are entitled to monthly subsidies of between €550 and €1100, or €275 to €550 for self-employed assistants as a contribution to cover additional expenses for social insurance contributions. These subsidies are means-tested, i.e. those entitled to it may not earn more than €2500 per month (excluding LTC allowance, family allowance and similar benefits). This threshold is increased by €400 for each household member entitled to maintenance, and by €600 for each member with a disability. Assets of the person in need of care are not considered, but the following preconditions have to be fulfilled:

- Need for 24-hour assistance, i.e. at least an entitlement to level 3 of the Austrian LTC allowance;
- There has to be a defined relationship with someone for assistance (with the person in need of care, with his or her family or with a non-profit provider).

The subsidy is only given if the 24-hour assistants have completed a training course covering the theoretical aspects of care work – usually according to the training guidelines for home helps – or if they have proof of already having successfully carried out assistance over a period of more than 6 months, or if they already have a specific qualification in the area of health or social care. In the meantime, most non-profit organisations providing home care now act as brokers for 24-hour assistants, as they are able to offer administrative support (subsidies, social security contributions, etc.), cover for illness or holidays, and training. Clients (families) are charged an enrolment fee of about €1000, a monthly contribution of about €200 as well as a lump-sum payment for the 24-hour assistant of €1800 monthly.

By the end of 2010, about 30,000 had registered as self-employed 24-hour assistants while the number of directly employed 'personal assistants' had remained below 500. About 10,000 families are receiving subsidies for social security contributions.

Source: BMSK, 2010; see also Prochazkova *et al.*, 2008.

Training for migrant care workers

The amount of care work performed by migrant care workers in households with older people needing care is usually underestimated and, as it is undertaken in an informal setting, perceived as less professional. It is generally assumed that the skills needed to carry out these activities are easily acquired ('learning by doing') or else perceived as a 'natural skill', evident particularly in women who have always been a reference point for caring work. In reality, specific skills are required, including social skills, knowledge about older people and technical expertise in order to provide quality care. The professional profile of a private family assistant is still not officially recognised in most countries. In Italy, some regions have started to define and formalise such a profile, by certifying professional skills or creating a professional qualification (first level). Training modules of varying durations are now available in certain areas in Italy. A programme focusing particularly on migrant carers was launched in 2011/12 by the Ministry of Employment and Social Policies to integrate information, brokerage and social care training (Conclave *et al.*, 2012).

Ensuring the quality of services

Quality assurance in the context of private assistance can only work if the older person, the caregiver and the family share responsibility in cooperating with the local formal care network which includes the general practitioner, hospitals, care homes, day care centres or home care service providers. First steps towards these types of networking have been successfully implemented in some regions in Italy by means of home tutoring, discharge management and the arrangement of cover for absence (Rossi, 2004). Some elaboration will be provided regarding these three areas.

Home tutoring is a professional service, usually offered by a home care nurse or a social care worker on request of the family or the older person with care needs to provide support when a migrant carer initially starts work, and if difficult situations occur later on. The tutor helps the migrant carer to improve skills in planning, arranging and managing daily tasks. The formal care system is thus working in partnership with the migrant carer and the families, rather than being perceived as controlling (see Box 10.3).

Box 10.3   Care at home by integrating formal and informal care – Casa Amica Notte e Giorno (Italy)

The example was originally a pilot project run by the Consortium CidiS-Orbassano in the province of Turin (Piedmont region, Northern Italy) with co-financing by the Foundation of the San Paolo Bank. The consortium consists of six municipalities responsible for social services delivery. One of their services is to provide older people in need of LTC with vouchers to directly purchase home care services, including registered migrant care workers with legal documents, home care services provided by private registered organisations, and respite services (e.g. during weekends) provided by residential facilities. According to an individual needs assessment the vouchers can amount to a minimum of €700 and a maximum of €7000 per person per year and can help to partially cover service costs. Cidis-Orbassano used project funds to carry out a major service reorganisation aimed at strengthening all services for older people by: supporting older people, migrant care workers and informal carers (e.g. cover for absence, home tutoring); creating services aimed at matching supply and demand for private care; mixing public and private services to provide a selection of different services that can be offered; providing specific training courses for migrant care workers.

Source: Di Santo (2011b).

Discharge management is common practice in some hospitals and particularly in wards with a high proportion of older people (e.g. heart surgery and orthopaedics). Respective procedures stipulate that the family as well as the private assistant must be contacted before the patient's discharge, so that all carers have adequate information about any new treatments to be followed at home, dietary advice, sleep patterns, mobility, as well as any changes to activities of daily living that need to be introduced during the convalescence period.

Arranging cover for absence is a service much appreciated by families, as it is important to rely on alternative care provision in cases of absence due to sickness or holiday. Currently, families cover these periods of leave during their own holidays, taking time off work, or hiring other assistants as replacements. Some services allow these families to get temporary access to home care services, a day care centre or respite care. So dealing with the problem of cover can turn out to be an opportunity for the families to get to know public services and to experience new forms of care, but they have to pay a fee as a contribution for the service received.

Slovakia as a sending country of migrant care workers in LTC

The necessity to distinguish between receiving and sending countries of care workers in the current situation of emerging LTC systems in the EU is not so much the result of different levels and numbers of older people requiring care across Europe, rather an expression of the current labour market situation. Sending countries are marked by demographic changes as well, with an increasing number and proportion of older people in need of care. However, if citizens from 'sending' countries are attracted by 'receiving' countries, this is due to the deficient labour market and labour conditions in their home countries, in particular concerning the wage differential between Eastern and Western, but even Southern Europe. Moreover, in many sending countries the demand is weak, so that a large proportion of households in countries with low GDP per capita do not have the purchasing power to hire caregivers and pay a market price. Care for older people in need of care in the sending country is partly provided by public social services, with only a small proportion of people entitled to them, but it is primarily provided by informal carers who, in turn, are partly supported by some public benefits (care allowances). The main burden of caring for older people thus remains with close family members, who are mainly women (daughters or daughters-in-law) and often themselves at an older age.

Temporary migration of female Slovakians to care for older people in Western and Southern countries of the EU usually takes the form of commuting to the host country for a period of two weeks per month or for a period of several months over some years. This activity has been perceived as an opportunity to earn better wages and contribute to the household income. In general, providing care abroad is not a permanent migration, but a separation from family lasting several weeks or months with the prospect of a visit home to Slovakia for a few days several times a year. The established model of care for older people in the host country is to provide care for several months or years for the same client. Slovak carers are usually women aged 30–50 years, providing household chores, meals and various forms of personal care for their clients – and they return home at infrequent intervals. This model certainly has an impact, where emotional ties and relationships with families are severely disrupted.

Nevertheless there are several push and pull factors that encourage Slovak middle-aged women to perform care work abroad. Among the push factors there is primarily the poor economic situation of households caused by low domestic wages and high unemployment. Households living in regions of eastern and southern Slovakia in particular have to cope with high levels of unemployment and significantly lower income in their region.

Table 10.1 provides statistics for unemployment, average monthly wages and the number of migrant workers for Slovakia and its Southern (Nitra and Banska Bystrica) and Eastern (Presov and Kosice) regions, highlighting the difficult situation on the labour market in Slovakia as whole and even more in individual regions. Hence, both unemployment and low wages are pushing the workforce towards searching for work abroad. According to data from the Statistical Office in 2010, the total average number of workers in Slovakia was 2,318,000 people. At the same time the number of unemployed was 389,000 and the unemployment rate was 14.4 per cent. It is necessary to add that the high unemployment rate (over 10 per cent) has persisted in Slovakia since the transition to the market economy (1990) – the exception was in 2008 when it fell slightly below 10 per cent – while wages have remained far below the average level in the EU. The number of Slovaks registered as working abroad increased after Slovakia joined the EU on 1 May 2004, reaching a peak in 2007 with 177,200 people and then decreased again since the global crisis to 126,700 in 2010. Given these general conditions and push factors it is therefore surprising that only about 5–7 per cent of the labour force are working abroad, which may be partly explained by the lack of language and professional skills, but also by improving wage conditions in Slovakia

Table 10.1   Push factors for migration from Slovakia and number of Slovak workforce abroad (2010)

	Slovakia	Nitra region	Banska Bystrica	Presov	Kosice
Unemployment rate (total per cent)	14.4	15.4	18.6	18.6	18.3
Female unemployment rate (per cent)	14.6	18.2	18.5	20.6	15.7
Average net wage per month (in €)	659.0	592.0	579.0	548.0	637.0
Number of workforce abroad (in 1000)	126.7	28.2	10.4	32.0	16.1
– of which women	37.3	9.8	5.0	5.5	6.9

Source: Statistical Office of the Slovak Republic, http://portal.statistics.sk

(see below) and reduced pull factors from Western European countries due to the economic crisis.

Notwithstanding the extent of migration (for health care professionals see Benušova *et al.*, 2011), the situation in the health and social care sector of the Slovak labour market has remained relatively stable, with 138,000 people employed in 2000, slightly over 137,000 in 2005 and 140,000 in 2010. However, the number of vacancies in this sector is low: in 2010, there were only 1079 job vacancies all over Slovakia. As for professionals in social care, their number is currently at about 11,000 (Bednárik, 2010). Some of them are employed in social services, including care and nursing homes, mostly as public employees of municipalities, but others are also employed by private providers. The activities of all professional carers are almost entirely funded by public sources. In addition, about 56,000 informal family carers are receiving public care allowances (Social Situation Report, 2011).

In this context it is interesting to compare these numbers with the estimated number of people in need of care in Slovakia. The country's population is relatively young, but with an accelerating trend of demographic ageing. At the end of 2010 (Statistical Office, 2011) there were 671,000 people above the age of 65 (12.4 per cent of the population) and 151,000 people above the age of 80 (2.8 per cent). Only 22,000 older people live in care homes, with another 15,000 waiting for placement. According to EHIS (2009), in Slovakia there are 40–50 per cent of people aged 70–74 years who declare a bad health status and 20–33 per cent of people aged 70–74 years who declare major limitations in daily activities. It can thus be estimated that, in the current situation, about 180,000 older people in Slovakia have a potential need of LTC, of which not even 50 per cent are supported by one or the other kind of public support. Forecasts however predict a rapid growth in the number of older people, in particular those of 80 years and above, resulting in an increase of up to 250,000 older people in need of care by 2020.

The dynamics of growth in the demand for LTC in Slovakia is accompanied by the development of social services which is still hampered by insufficient public funding and private purchasing power. What is developing, however, is education and training in this area. Prior to 1990 the social care profession in Slovakia had been given a relatively low value: social care had been carried out mostly by untrained women, while treatment was provided by registered nurses. The system of nursing schools was and is relatively stable, with a steady interest among young girls in the profession. For example, in 2010, 7200 full-time students and 2800 part-time students had studied at a secondary medical school in part-time courses, with about 5100 students specialising as Medical Assistants and 1200 as Sanitarists (National Health Information Centre, 2011).

The labour market in Slovakia is not able to absorb many of these graduates – job opportunities in the health and social care sector have remained stable at a very low level over the past ten years. Another push factor for people trained in social care to look for work abroad is the huge wage differential both within Slovakia and in relation to wages in neighbouring EU Member States. For a long time the social care sector has reported lower wage levels than the national average wage. For example, in 2010, the average gross monthly wage in health and social care was €698, while the national average wage was €769 (Statistical Office, 2011). At the same time average monthly wages in social services only reached €531 (Bednárik, 2010).

It is therefore no surprise that many workers, and in particular social care professionals, continue to leave the country to work in Austria, Germany, the Netherlands, the UK and Switzerland, as well as Italy. Brokerage of work abroad for carers is arranged through several agencies with sites in every region of Slovakia. These agencies specialise in specific countries (e.g. Austria) or groups of countries (Germany, the Netherlands, UK), but care staff brokerage for Slovak carers is also arranged by foreign personnel agencies, in particular from the Czech Republic.

This array of agencies shows that providing care services in households abroad has become an attractive business both for intermediaries and for private assistants from Slovakia. Indeed, these agencies are not only arranging work for professionals, but also for job seekers without any prior experience or specialised training in social care. The list of countries to which Slovak job seekers are migrating (Austria, Germany, the Netherlands, UK) gives an idea of the fact that, apart from professional skills, at least a basic knowledge of the language of the receiving country and a minimum of social skills are important. Many arrangements therefore also include language courses free of charge as well as a reimbursement of travel costs for more distant countries. In addition to wages, live-in migrant carers are usually offered board and lodging by the host family. Some brokering agencies also arrange health insurance (for the Netherlands), or a small business licence, as in Austria (see Box 10.4).

Box 10.4   Key features of Slovak migrant carers working in Austria

Framework conditions and skills required for care work abroad:

- at least a basic knowledge of language in receiving country
- training or experience in care work with references are welcomed (optional)
- age 25–55 years
- rotation (14 days per month) for a minimum period of 2–3 months within one family
- net monthly wage €1150–1500
- board and lodging covered by the client

Job characteristics:

* household chores (cooking, shopping, washing, cleaning)
* depending on the health status and mobility of client, help with dressing and undressing, personal care and hygiene, taking for walks and keeping the client company (reading, conversation with client, etc.)

Services provided by brokering agencies that have to be licensed by the Slovak Labour Office:

* placement of job seekers
* support of migrant carers abroad (health insurance, arrangement of small business licence)
* cooperation with employers and partners abroad
* transport and/or reimbursement of travel costs
* provision of certificates and references.

Source: www.pracavonku.sk

Although the Slovak labour market offers only a limited number of vacancies for job seekers in the field of social care, the long-term care sector in general is growing in terms of trained home helpers and related training agencies. As receiving countries are increasingly requiring at least a minimum standard of training, various agencies have specialised in providing private courses to obtain a certificate as a home help. For instance, the R & D Agency offers courses in 47 cities throughout Slovakia with 230 hours of training. To obtain a certificate as a home help, 13 days of theoretical lessons (e.g. basic hygiene, physiotherapy, client orientation, safety) and 10 days of practical training in a care home have to be successfully completed to obtain basic knowledge in social care. Many participants are only able to pay the fee for such a course from their first wage earned abroad (€180–300). As there is also a rising request for language courses (German, English, Italian and French), the phenomenon of migrant care workers contributes to a growing labour market in Slovakia for language teachers, trainers and human resource managers in brokering agencies.

In the current situation, having citizens working abroad in domestic care still seems advantageous for Slovakia as a sending country: individual families benefit from their increased household income and the national economy takes advantage of the higher purchasing power. Still, little is known about the individual and social consequences of this practice from a mid- and long-term perspective: What impacts will the absence of mothers and grandmothers have on children? Will migrant work lead to a shortage of labour within Slovakia over the next ten years? Will extended opportunities for migrant care work hamper domestic development of formal and informal LTC services? The lack of carers for people with care needs in Slovakia has not yet been explored sufficiently well, but there is some initial evidence of an ongoing 'domino effect' with migrant carers from eastern neighbouring countries providing care in Slovakia (for Hungary see Széman, 2012 and Turai, 2011). With the rising number of older people in need of care, however, pressure on public policies to ensure support by formal services and for informal carers will certainly grow over the next decade. As mentioned above, the main driver for migrant care workers to offer labour abroad has been the huge wage differential between the sending and the receiving countries – this difference has, however, steadily reduced over the past decade. For instance, while the average wage in Slovakia had been around €400 by the year 2000, it has grown to about €800 in 2010 (though only to €660 in the health care sector). It is likely, therefore, that over the next ten years the rising number of older people with care needs who are able to pay market prices for care may activate the labour market for care in Slovakia. It has to be seen, however, whether the standard of living in Slovakia will continue to rise at the same pace and whether the large share of women returning with improved caring and language skills will trigger further development in the domestic care market. This would certainly leave the receiving countries with a serious care gap.

Conclusions: Migrant care for ageing societies as a low-cost solution with an expiry date?

This chapter has highlighted the challenges and potential remedies to improve the situation of migrant care workers and older people with care needs as well as their families who are often forced to resort to this type of support. This development has represented one of the main social innovations in long-term care over the past 15 years. However, as it was an unintended effect of public policies it created a wide range of legal, social and economic inconveniences as well as a challenge for 'society's moral framework' (Weicht, 2010) raising issues such as equal opportunities, fair working conditions across Europe, 'ethical recruitment' (WHO, 2010) and general questions regarding policies concerned with the mobility of health (and social) care professionals in Europe (Wismar *et al.*, 2011).

Information, support for matching supply and demand, the regularisation of the employer–employee relationship, and training and quality assurance by means of enhanced coordination between formal and informal care are therefore certainly necessary to create a basis for including private care work by migrant workers in the LTC system. In fact, modern LTC systems can neither rely on families' private initiatives alone nor on their willingness to take care of older people with long-term care needs. However, there is also the possibility that the supply of migrant care workers might dry up. In a mid-term perspective, the work of migrant carers might also change its identity, characteristics and function (Pasquinelli and Rusmini, 2009). With regard to family care, the increasing labour market participation of women and more equal gender relations have contributed to a changing role of family carers as coordinators and organisers of care, rather than providers of hands-on care. So similar developments might occur with regard to migrant care workers. Indeed, the general labour shortage in health and care sectors across Europe (and beyond) has already created fierce competition between receiving countries to attract professionals (and marginally trained migrant workers) in a global context (Yeates, 2009), and international organisations have started to acknowledge the challenges of migrant care arrangements (FRA, 2011).

As long as migrant care arrangements remain a functional equivalent to family-based love and intimate relationships, rather than paid work and employment, public and private responsibilities will remain blurred, thus also contributing to a further under-regulation, lack of transparency and integration with the formal care system. In a context of limited service provision and economic restraints, resorting to crudely market-based solutions while sustaining the ideal of family-based care (Lyon and Glucksmann, 2008) may explain why policymakers in some European countries tend to ignore or rather neglect this phenomenon. Denial is certainly a short-sighted strategy. Policymakers will therefore have to seriously start careful considerations for investing in services and new solutions to guarantee continuous care for frail older people and the construction of an integrated system of LTC.

11
Volunteering in Long-Term Care for Older People: The Potential for Social Innovation

Kvetoslava Repková, Karin Stiehr and Barbara Weigl

Introduction

Volunteering in long-term care (LTC) has long been associated with care delivery in Europe. In countries with a developed civil society such as the Netherlands, the UK, Sweden, Denmark, Finland and Germany, intensive ongoing dialogue concerning LTC policies and practice reveals a high proportion of volunteer activities and their estimated economic value. In post-communist countries, however, such as Slovakia, Bulgaria and Poland, after decades of an overall state monopoly, central regulation and provision of social welfare, volunteering is only recently being 'rediscovered' and established in the public consciousness and within policies. This is particularly so for the area of LTC, where new functions and roles for volunteering activities are being discussed with a view to integrating volunteers in the network of LTC services. Similar developments can also be observed in Mediterranean countries such as Greece or Italy (GHK, 2010).

At present, despite the different starting points, all European countries are facing common challenges arising from changes in the overall understanding of care and tasks that are attributed to the wider provision of LTC. According to Triantafillou *et al.* (2010) these common challenges embrace demographic changes and the expanding need for care; changes in social and family structures; gender aspects of care; user empowerment and privatisation; economic crises including growing income inequalities and respective consequences for care arrangements; and expectations for the increasing labour market participation of women and older workers who are currently considered as a traditional source of care. In addition, the development of volunteering is often linked with the inability of public budgets to cater for the increasing LTC needs of older people (Allen *et al.*, 2011) by guaranteeing and improving the quality of services (Nies *et al.*, 2010). Other challenges come from the changing nature of voluntary commitment and a mismatch between the needs of voluntary organisations and the aspirations of the new generations of volunteers, which also affects LTC.

All these challenges call for changes towards developing an integrated strategy that leads to empowerment, enabling citizens to become co-creators of innovative social relationships and models of collaboration in health care and environmental services, ultimately making a valuable contribution towards achieving greater value for money in public services (Barroso, 2011; European Union, 2010). In this context, the search for effective solutions in LTC by boosting voluntary contributions is currently seen as a potential area for social innovation.

Promoting volunteering in LTC was a prominent issue and high on the agenda at the European Union (EU) level as well as in individual EU Member States during the European Year for Combating Poverty and Social Exclusion in 2010, the European Year of Voluntary Activities Promoting Active Citizenship in 2011 and the European Year for Active Ageing and Solidarity between Generations in 2012. The culmination of these activities was that almost all key authorities agreed upon the importance of voluntarism for sustainable socio-economic development on a national as well as on an international level. According to the final report on 'Volunteering in the European Union' (GHK, 2010) the importance of volunteering has long been acknowledged by the EU. At the same time, however, there is a lack of a systematic and structured EU approach towards volunteering as a very diversified phenomenon, with almost every European country having developed its own understanding, reflecting the differences in history, attitudes and perceptions. Volunteering thus covers an enormous diversity of activities, such as provision of education and services, health and social care, sporting and cultural events, mutual aid and self-help, advocacy, human rights campaigns, electoral assistance, and poverty eradication. While LTC is yet another area of diversity and complexity, squaring LTC with the different understandings and approaches to volunteering, its nature and extent across Europe is certainly a huge challenge.

This chapter will try to identify some key concepts of volunteering and show their relevance for the development of LTC systems in Europe. Alongside the presentation of some terminological considerations, new patterns of volunteering related to its human resource management will also be briefly described. In the second part of this chapter, volunteering in LTC in Europe will be scrutinised from an empirical perspective. Political concepts will be contrasted with organisational perspectives as well as the motives and benefits for both volunteers and older people with care needs. Finally, conclusions will be drawn from the findings which are based on INTERLINKS examples as well as on other sources.

Theoretical background

Terminological considerations

There is no common, formal and institutionally recognised definition of volunteering in Europe. Various societies have different expectations from volunteering which also depend upon the type of national welfare system. The result is a condition in which '*what is considered volunteering in one country is not necessarily considered as such in another*' (FUTURAGE, 2011, p. 45). The diversity refers particularly to LTC. LTC services have been traditionally understood as an internal 'family affair', so that 'helping' and 'providing assistance' has often been culturally conceived as a normal part of life in the community. Volunteering is therefore struggling to become identified as a distinct and easily defined area of activity (ILO, 2011, p. 12).

This lack of a common definition of volunteering contributes to the significant differences in findings on its extent in individual countries. Within Europe, it is obvious that there is a need to share a common definition and develop methods of unifying relevant data sources in this field, in all the sectors concerned with volunteering. If these factors were aligned across Europe, it would be possible to better understand and compare the scope of volunteering and the value it has for people who perform it, as well as for the overall welfare of society (FUTURAGE, 2011).

However, definitions from some key actors in this field provide an initial starting point for how volunteering can be articulated:

- The Association for Voluntary Organisations (AVSO) distinguishes between 'volunteering' which can be occasional or regular, part-time or full-time, and 'voluntary service' which refers to specific, full-time project-based voluntary activities that are carried out on a continuous basis for a limited period of time.
- The European Volunteer Centre defines volunteering as an activity that can occur in different settings, either informally or formally within the structures of non-profit organisations. Its nature can vary from part-time to full-time and from one day to many years in several different fields (GHK, 2010).
- The definition used by the International Labour Organisation (ILO) in its 'Manual on the Measurement of Volunteer Work' is of particular interest as it shows an implicit acknowledgement of volunteering by an international organisation that is mainly dealing with employment and paid work: '*Volunteer work is unpaid, non-compulsory work; that is, time individuals give without pay to activities performed either through an organisation or directly for others outside their own household*' (ILO, 2011, p. 14). The ILO definition covers both formal and informal voluntary engagements and is considered important for the area of LTC, where a significant part of voluntary work is performed beyond the realm of any formal or professional rules and schemes.
- The 'Final Report on Volunteering in the European Union' applied a more comprehensive approach and highlighted some institutional, legal, economic and cultural strategies to overcome the challenges to volunteering. The report stresses the need to develop strategies that engage

volunteers, as well as to secure professionalisation of the voluntary sector, a legal and regulatory framework, and sustainable funding. At the same time, the report highlights the need to overcome the lack of recognition experienced by volunteers, and to develop a clear strategy to overcome a political landscape that is divided regarding how volunteers are perceived. EU authorities, Member States and organisations engaging volunteers were identified as the main actors to respond to these defined challenges and opportunities (GHK, 2010).

'New' types of volunteering

Frič and Pospíšilová (2010) concur with the opinion of other authors (e.g. Dekker and Jalman, 2003; Hustinx and Lammertyn, 2003; 2004) in that new patterns of modernisation also bring about new patterns of volunteering, from the traditional collective pattern to a new reflexive pattern of volunteering. The traditional collective style is based on the membership of volunteers in a specific association, with relatively high loyalty to the organisations and a low level of turnover. A new generation of so-called reflexive volunteers is not tied to membership in a specific organisation any more and shows little organisation loyalty. Their motivations for voluntary engagement are based upon a diverse range of activities; they choose their voluntary work by themselves and manage it with respect to their own objectives, personal integrity, personal development and own career. In this context, quality management becomes an increasingly important aspect of effective volunteering, particularly in terms of maintaining a balance between the individual interests of volunteers and the interests of organisations engaging them to carry out voluntary work.

Even so, the authors of the report on volunteering in Europe (GHK, 2010) consider professionalism of the voluntary sector as one of the main challenges and opportunities for the development of voluntarism as such. Professionalism in this context means strengthening the human resource management in organisations that engage volunteers. This means improving their ability to balance the requirements for a qualified voluntary performance, and enhancing the motivation of volunteers to work in a non-paid regime. Human resource management must therefore become aware of the specific status of volunteers in comparison to the professional and paid employees in organisations and their mutual complementarity in relation to fulfilling the requirements of people with different types of social care needs. Volunteering is described as being a distinctive sector of the society which, despite certain boundaries and limited possibilities, cannot be replaced by an alternative (Frič *et al.*, 2010). On the other hand, its linkage to other sectors is crucial, such as with regard to the formal work carried out by paid employees.

A concept which is explicitly based on the personal interests of volunteers in terms of future benefits is that of 'Time Banking'. In German 'Senior

Cooperatives', which have existed since the early 1990s, older people exchange support services between each other. 'Senior Cooperatives' are a variation of 'Local Exchange Trading Systems', open to members of all ages, which apply non-monetary systems to compensate for voluntary work (Pacione, 1997). Those who help other members with household chores or other tasks in daily life can choose to be paid in cash or, as an alternative, 'save' the number of worked hours for receiving the same amount of support which may be needed in future.[1] The specific aim of Senior Cooperatives is to offer the members the opportunity to stay in their home environment until the end of their lives and restrict care home confinements to when a very high level of care is needed. The Dutch pension fund for workers in the health care sector is developing a similar system at present.

Volunteering in the life course perspective

In order to enable societies to face the challenges arising from the demographic changes in postmodern society, consideration of the life course perspective is crucial (GHK, 2010; FUTURAGE, 2011). The life course perspective means being aware of the motivations for volunteering, assessing a person's capacity to undertake voluntary work at different stages of their life (Tang, 2006), and identifying and adopting specific strategies for supporting different age groups of volunteers to do voluntary work (FUTURAGE, 2011). Tang (2006) confirmed that, to ensure continuity of volunteering in different stages of the volunteer's life, the provision of required resources must be guaranteed throughout their career and at the relevant times.

- The 'Report on Volunteering in Europe' (GHK, 2010) also emphasises the life course perspective and recommends supporting volunteering among senior citizens and young people. It highlights the increased engagement of older people in the voluntary sector, which is considered an indicator for the vitality of the sector and the capability of society to respond to challenges of ageing. According to Broese van Groenou and van Tilburg (2010), there are three key social changes that may facilitate volunteering of older people: (1) Expanded educational opportunities among the younger generation of seniors, since work as a volunteer in older age is positively associated with educational engagement; (2) Trends moving away from individualisation, resulting in the increasing importance of friendships beyond the family and involvement in social networks; and (3) Increased employment in later life. As the authors stressed, the increased societal opportunities for education, employment and expanding personal networks may positively affect volunteering rates of the younger generation. Their conclusions correspond with the findings of the Special Eurobarometer 'Active Ageing', confirming a higher prevalence of volunteering among older (40+) and employed people (Active, 2012).

It has been suggested that there are multiple benefits from older people taking part in voluntary activity. Their civic engagement is considered one of the possible ways to remain active, to make a contribution to society, and to profit from opportunities offered by engagement. To ensure this 'chain' of benefits is realised it is necessary for younger and middle aged people to support voluntary campaigns as it is suggested that people with voluntary experience from an earlier age will be more likely to revisit this role in their later lives (Broese van Groenou and van Tilburg, 2010; FUTURAGE, 2011).

The life course perspective is used not only as a response to challenges of population ageing, but also in relation to the family course perspective, when the chances for engagement in volunteering can be adversely influenced by commitments of people in the area of informal caregiving. In order to support volunteering in LTC, establishing a type of 'social support chain' has been suggested. Within the chain, informal caregivers are supported by voluntary organisations in their caring commitments, and this helps to overcome any social isolation derived from being solely within the family environment. It also encourages informal carers to become volunteers once their caring period has ended (FUTURAGE, 2011).

Volunteering in LTC for older people

There is a relatively large amount of research evidence pointing to a high prevalence of voluntary activities in health and social services (e.g. Salamon and Sokolowski, 2001; Tang, 2006; GHK, 2010). Conversely, there are also high expectations within societies that older people will become active in caring for dependent older and disabled people (Active, 2012). Hence this section will address both these perspectives and present findings from INTERLINKS research and other sources.

Contextual factors of volunteering in LTC

While professional approaches to LTC have improved over the past decades, there is increasing awareness of volunteer work. Two factors may play a significant role in this context. These are, first, ongoing changes in the socio-demographic structure with increased physical distance between parents and their adult children, more older people with no children, older people living on their own, and the tendency for older women, who traditionally were available to care for their parents once their children left, to take up jobs. Second, funding bodies for formal LTC face increasing cost pressures and a rising number of older people in need of care. Though volunteers, who usually work free of charge, may not be able to compensate for professional care in the strictest sense, their commitment ensures a special quality, based on a personal relationship with the person in need of care.

However, volunteers constitute a separate and additional category of stakeholders in between family carers and professionals. Like family members, they are lay workers irrespective of the fact that they may be trained to a certain extent or receive some financial compensation for their efforts and personal expenditures. Nevertheless, volunteers are embedded in a non-familial context, either organised by a professional care provider from the third sector or, to a growing extent, the private market, or they organise themselves in a formal or informal association. Their activities vary from providing practical support to older people in everyday life to advocacy roles, but they also focus on the support and counselling of family carers, either as an external assistant or in the form of mutual self-help activities. Their assignment to the informal care sector is essentially due to the fact that the variety of their organisational contexts, educational backgrounds, objectives, and work methods is not in line with definitions in the realm of formal care.

Based on the 'Welfare Diamond' (Evers *et al.*, 1994), which classifies care provision into four sectors of public, market (private businesses), voluntary (NGOs, non-profit associations and cooperatives, etc.), and household (family, friends, neighbours) services, Triantafillou *et al.* (2010) summarise the orientations of different stakeholders in LTC according to the categories of formal and informal care. Extending these categories, volunteering blurs boundaries and, by challenging mainstream definitions, shows a profile of its own (Table 11.1).

From the perspective of the increasing financial pressure on systems of formal LTC, any kinds of measures to support the position of family and

Table 11.1 Volunteering between formal and family care

	Political responsibility orientation	Stakeholder orientation	Employment orientation	User orientation
Formal care	Care as a responsibility of the welfare state	Services delivered by various types of organisations and considered as a source of income for professional carers	Care as an activity belonging to the labour market	Older people and informal carers as marketable clients
Family care	Care as an individual or family responsibility	Services provided freely by family members or other private individuals	Care as an activity not entitled to working regulations	Older people considered as belonging to the private sphere
Volunteering in care	Care as a responsibility of the social community	Services provided freely out of personal interest and altruistic reasons	Care as an activity based on voluntary agreements	Older people considered as community members

Source: based on Triantafillou *et al.* (2010).

social networks appear to be in the public interest in all states throughout Europe. This holds a *'high risk that policy makers will continue to view informal care as a relatively free and accessible resource – as long as it is not publicly acknowledged – that will ultimately fill the gaps and substitute for the deficiencies of publicly provided service delivery'* (Triantafillou *et al.*, 2010).

Focusing on the aspect of a specific quality of voluntary work, however, another feature emerges. Volunteers are neither bound by employment contracts nor private obligations, but the voluntary work as such must provide benefits for those who are performing them. Active citizenship has long been recognised as a form of grassroot democracy and includes all organisations in the civil society, which are actively and democratically involved in defining and tackling the problems of their communities and improving their quality of life. Nevertheless, to maintain the readiness of citizens to invest time and efforts in serving the public good, their motives must be met and the framework conditions must be satisfactory. Where a culture of volunteering has been developed, governments and associations have recognised that contributions by volunteers may be highly desirable but are not available for free.

Volunteering and LTC policies

From a European comparative perspective, the occurrence and recognition of voluntary work is extraordinarily diverse. Key distinctions between Member States comprise, among others, the regulatory framework for volunteering (GHK, 2010). In some Member States a legal framework specially related to volunteering exists or is in the process of being developed, e.g. Belgium, Bulgaria, Czech Republic, Hungary, Italy, Latvia, Poland, Portugal, Romania, Slovakia and Spain. In other Member States, volunteering is regulated by or implicitly within other existing general laws, as in Austria, Denmark, Estonia, Finland, France, Germany, Greece, Ireland, Lithuania, the Netherlands, Sweden and the UK. Another diversity refers to the estimated economic value of voluntary work in European countries amounting from less than 0.1 per cent of the GDP for example in Slovakia, Poland or Greece to more than 3 per cent in Austria, the Netherlands and Sweden (GHK, 2010). A Dutch estimate on the financial benefits resulting from 450,000 volunteers working in care for dependent people, based on one hour per week, 40 weeks per year and an hourly rate of €14, results in €252 million which are generated per year by volunteers in LTC in the Netherlands (Scholten, 2011).

A good source which allows the direct comparison between the position of volunteering in LTC in the EU Member States is the 'National Reports on Strategies for Social Protection and Social Inclusion' in the framework of the Open Method of Coordination of the European Commission.[2] LTC is – besides poverty, social exclusion, pensions and health – one of the areas under study. The indicators against which the developments in LTC are assessed are quality, access and financial sustainability, all of which might be of relevance to volunteering. The occurrence of voluntary work can thus be considered an indicator for its position in the national strategies to cope with the challenges in LTC provision.

Volunteering as a part of civil society represents a bottom-up democracy approach, which is especially thriving when developed in an historical process. Thus active citizenship combined with fostering hands-on work is the predominant concept in the UK. In the planning of health and social care, the UK government engaged with a large number of citizens, including older people; this process was documented in the White paper 'Our health, Our Care, Our Say' (2006). In 'The Future Care of Older People in Scotland' (2006), a vision was set up in which older people are seen as both citizens and the biggest providers of support to other older people, and consequently having both rights and responsibilities (UK National Report, 2008–2010). In the Netherlands, the country with the longest active citizenship culture besides the UK, mainly pragmatic aims are set. It is evident that the number of volunteers and informal carers is still very high but might come under pressure in the future. Therefore specific policy programmes have been developed to support municipalities in their effort to maintain and even increase the number of volunteers (National strategy report, 2008).

Promotional measures by state institutions can accelerate the development of a culture of volunteering, and some countries with less distinct traditions are trying to catch up. In this sense, the German report refers to a reform through the Long-Term Care Further Development Law of 2008, in which funds for volunteering and self-help activities were raised from €20 million to €50 million per year. They serve to strengthen the local networks of volunteers and self-help group members and to provide them with training. A certain rhetorical part can be assumed, since the term 'volunteering' is strictly avoided; instead in a paragraph on *'generation-spanning civil commitment'* it is pointed out that the *'committed involvement of citizens'* was reinforced (National Strategy Report, 2008–10, p. 102). A focus on more practical project outcomes is set out in the report from Austria, where a pool of formal and informal carers was created to reduce the strain on family carers of people with dementia, and an internet forum for the mutual exchange of family carers was set up (National strategy report on social protection and inclusion, Austria 2008).

Beyond any historical, cultural or political contexts, financial constraints of public bodies are another strong driver to encourage voluntary work. Representing the Mediterranean focus on the social care responsibility of the family, the Greek government is clearly committed to the concept of volunteering: There is an explicit aim to supplement national resources by offering services at a low cost by volunteer organisations (National strategy report, 2008–10). Furthermore, the cooperation with NGOs for the protection of vulnerable population groups is enhanced in drawing and assessing policies, and the organisation 'Volunteers' Society' is promoted to help

NGOs to capitalise on the organised volunteers' 'contribution to qualitative social care services' (National strategy report, 2008–10). But the interest in volunteering issues also emerges in Nordic countries, where the state had traditionally assumed responsibility for welfare issues. At the municipality level in Sweden, voluntary organisations such as the church and pensioner associations have long been engaged in volunteering activities, but are meanwhile regarded more and more as a substitute for the underfunded public sector – in parallel to the increasing caring responsibilities of relatives (Ljunggren and Emilsson, 2010). In Finland, too, volunteering is not an unusual phenomenon. About two-thirds of Finnish municipalities have voluntarily appointed a council of older people to give them a voice in issues of senior policies (Leinonen, 2006).

In Slovakia as a former socialist country the involvement of all relevant stakeholders, including civil society, in the production of social welfare has only slowly developed over the past two decades (National strategy report, Slovakia 2008). However, volunteering as such is far from being rolled out. Until 1989 the whole NGO sector was incorporated into state structures, and volunteering was deprived of its meaning of being an expression of active citizenship. The crucial change came in the early nineties when the right of people to associate freely for publicly beneficial purposes was legally restored. But among the general public, many prejudices about volunteering still persist. Consequently volunteering still has a poor status, and information on opportunities is still lacking (Nikodemová, 2009). Nevertheless, an Act on volunteering being prepared by the Ministry of Internal Affairs of the Slovak Republic in close cooperation with the National Volunteering Centre and other civic society organisations came into force in December 2011 (Act, 2011). LTC volunteering persists in the Act as part of a general concept of volunteering in Slovakia.

With a view to the very different traditions, cultures and governmental promotion measures, the occurrence of volunteering in LTC in particular societies is strongly connected to the image of volunteering in general. A certain tendency, however, can be shown by grouping countries according to different levels without striving for exact results: a comparatively high level with more than 30 per cent of the population aged 15+ is active as volunteers in Austria, Denmark, Finland, Germany, the Netherlands, Sweden and the UK. Between 20 and 30 per cent of adults are volunteering in France. The involvement of adults in volunteering activities appears to be at a relatively low level, with less than 20 per cent in Greece, Italy, Portugal, Slovakia and Spain.[3] The most commonly reported activities were administrative and supporting tasks; helping and working directly with people; preparing and supporting voluntary activities; managerial and coordination tasks; campaigning and lobbying; and the organisation of events. Helping and working directly with people referred to a wide variety of tasks and many target groups, depending on the volunteers or voluntary organisation. It involved care of individuals with severe health problems, such as visiting older people, supporting vulnerable groups in society or activities within the community (GKH, 2010).[4]

Research on volunteering and volunteers in LTC is a specific domain which requires a high priority in governmental strategies to ensure that funding is provided for it. Therefore relevant surveys are very scarce and, due to different methodologies as reported earlier, findings cannot be compared directly.

However, elaborating on the above statistics, Dutch estimates suggest that overall 450,000 people, mainly women, take on unpaid care tasks on behalf of organisations for dependent people with whom they had – at the beginning – no personal relationship. One hundred and fifty thousand volunteers are active in residential care for old and disabled people where they form between a quarter and a third of the total workforce. Two hundred and fifty thousand volunteers engage in domiciliary care totalling 4.5 million hours per month (Scholten, 2011).

In England, too, only volunteering within organisations providing LTC was measured, not taking into account other social networks, community or religious groups in this sector. This was considered one of the reasons for the result that volunteers constitute only one per cent of the total LTC workforce in England. A large proportion of employers indicated that their workforce does not include volunteers. But in organisations in which at least one volunteer was active, the contribution was considerable; in some places their number amounted to more than a quarter of the paid staff. Small organisations, regardless of the private, public or third sector to which they belong, were more likely to actively involve volunteers than large organisations. In terms of their socio-demographic characteristics, volunteers were older, more gender-balanced and less ethnically diverse than the paid staff in LTC (Hussein, 2011).

In Germany, the total number of volunteers in LTC is not known but some surveys exist with results on their special characteristics. Between 80 and 90 per cent of volunteers working in care homes are women, and the majority are aged 65 years or older. A large number of the volunteers are already experienced in caring for older people, be it privately (approx. 40 per cent) or professionally (approx. 20 per cent) (Engels and Pfeuffer, 2007; Freiwilligen-Agentur Halle-Saalkreis, 2006a; Klie *et al.*, 2005).

In Slovakia, no overview about volunteering in LTC provided in home care settings currently exists. There are only data on the number of volunteers in residential care by non-public providers of social services, which make up about 30 per cent of the total staff working in these services.[5] Approximately one-third of them are women (Repková, 2011a). Volunteer centres provide assistance to LTC organisations (e.g. in senior homes, care homes) by embedded volunteering programmes.

Volunteering in LTC organisations

With special regard to LTC volunteering in organisations, Hussein alludes to literature suggesting that where professional staff are present, the volunteer role often augments the professional one, especially in providing companionship and information, but also transport and other day-to-day assistance such as shopping (Hussein, 2011). In contrast to that, the German National Network for Civil Society *(Bundesnetzwerk Bürgerschaftliches Engagement)* in their policy paper on volunteering in LTC injects a more political role by underlining that the most important task of volunteering citizens is to advocate for older people and those in need of care, representing them in dealing with patronising authorities in organisations and associations. The main goal is to maintain or improve the quality of life for people in need of care and their relatives. This is the determining factor for the quality of professional care as well as caregiving volunteers (BBE, 2011).

These contrasting positions comply with the ambiguous position of volunteering which can be considered supplementary to formal care, but also based on personal relationships to the older person in need of care and his or her family, and advocacy functions alongside or counter to the professional system. A further issue relates to the fact that voluntary work is not restricted to external help but can be performed just as well by the members of the original target groups, dependant older people and family carers themselves (FUTURAGE, 2011).

When it comes to the pervasiveness of voluntary work, it can be found in all areas and contexts of LTC: residential care, day care, domiciliary care and community care. It embraces individual support as well as group activities, and it may be directed through legal and organisational settings for formal and informal care. Activities of volunteers happen at local, regional, national and international level.

Table 11.2 below seeks to differentiate between approaches (volunteering organised by external institutions or associations vs self-organised activities) and the main target groups of voluntary work in LTC.

Although care tasks in the strictest sense of the word are frequently mentioned in the framework of volunteering (Hussein, 2011), their overlapping with functions of trained professionals is precarious. Usually official rules and regulations restrict volunteering activities to 'soft' tasks such as spending time together and providing practical support.

Various practice examples illustrate volunteering activities across the countries. According to research findings in German residential care facilities, in more than 60 per cent of care homes volunteers performed visits, helped out in festivities and were engaged in palliative care. Also, in more than a third of the residential care facilities, shopping and escorting services were performed, and volunteers contributed to religious services, ran cafeterias and organised charity sales. More seldom, with only 10 per cent of respondents, volunteering activity was evident in quality assurance, administration or nursing (Klie *et al.*, 2005). Other voluntary work took place in outpatient care similar to the examples below from Greece and France (Boxes 11.1 and 11.2).

Table 11.2   Examples of volunteering activities in LTC

	Target group: Older people in need of care	Target group: Family carers
Externally organised activities	• Befriending • Reading correspondence, newspapers etc. • Shopping support • Household chores • Transporting • Escorting • Support in financial matters • Organisation of events • Cafeteria service in care homes • Library service in care homes • Hospice service	• Respite care • Information • Training • Advice • Guidance
Self-organised activities	• Work on boards or committees • Representation of interests • Quality assurance • Organisation of events • Peer support	• Work on boards or committees • Campaigning • Advocacy • Exchange of experiences

Box 11.1   Help-at-Home services (Greece)

These services, run by the municipalities, involve multidisciplinary teams of social workers, nurses and family assistants. They evaluate the client's situation and needs, provide advice and social support and maintain contact with competent agencies, making referrals where needed. Help-at-Home services have developed strong volunteer networks which support and extend the work of the professional care service providers.

Source: Kagialaris (2011a).

Box 11.2   Equinoxe (France)

The alarm system Equinoxe is based on the voluntary support of people who are nominated by the older person and living nearby. After an emergency call they are sent to assess the situation and support the older person if medical assistance is required. The call centre staff will continue monitoring until the issue is sorted out.

Source: Naiditch (2011b).

- Voluntary work by older people with actual care needs is often neglected but also possible. It is legally binding in Germany that care home residents are represented in a self-organised advisory board with a right to a say in management matters. The board members are in charge of taking residents' complaints and recommendations and negotiating them with the director or relevant staff. Further, they support the integration of new residents, must be heard when changes in contracts or prices occur, are involved in the planning of events and all aspects of accommodation and meals, as well as the quality of social and health care. The advisory board can make use of the support of external volunteers; in cases where an advisory board cannot be established by the residents themselves, an external 'advocate' will be nominated by the care home supervisory authority.
- In the Netherlands, the needs of family carers were addressed by targeted projects. As a consequence volunteers were equipped with skills to perform respite care in complex care situations, and trained to support beside the family carer as a companion. In addition, female immigrant family carers who have limited access to formal care and help were matched with a 'buddy', a volunteer who gave social and emotional support. Finally, a project was directed at care volunteers themselves who needed more coaching than the organisation could offer and were subsequently supported by voluntary coaches (Elfering and Scherpenzeel, 2009a–d).
- When family carers organise themselves into an advocacy group, they act as volunteers in their own affairs. EUROCARERS is a transnational association which aims to represent the interests of informal carers and give them a voice. It acts as an interface by disseminating information about existing types of support for informal carers and by directing recommendations for improvement of their working conditions towards political decision-makers. National spin-offs, mainly run through voluntary work, are found in Germany (Stiehr, 2011a), Scotland, Finland and the Netherlands.

In order to provide and improve opportunities for volunteering, some functions in the promotion of voluntary work are particularly well performed by independent bodies, such as municipalities or a consortium of different welfare associations. Promotional functions embrace, for example:

- Information about volunteering opportunities combined with support in selecting the right placement according to the desires and needs of interested people. This task is performed by Volunteer Agencies or Volunteer Centres in many European countries;
- Provision of accident and third-party liability insurance for volunteering, which has become a standard in all German regions within the last ten years;

- Investigation of training needs among volunteers in a town or region, be it general knowledge such as legal and tax regulations for voluntary associations or specific expertise, like in palliative care, and the offer of appropriate courses and seminars. Since 2003, a programme for the training of volunteers is for example funded by the region of Hessen;
- Design of measures to recognise and honour voluntary efforts for the common welfare, often practised in the form of festivities on the International Volunteer Day on 5 December each year;
- Stimulation of further development and innovation in the field of volunteering, be it through conferences, consultations or research – for example Scholten (2011) worked with key actors to develop and highlight a number of relevant themes such as professional profiles for volunteer coordinators, diversity issues among volunteers in LTC, limitations for voluntary work and cooperation between paid staff and volunteers;
- Publication of handbooks, brochures and flyers with examples of good practice on how to facilitate voluntary work in organisations. In the realm of care, for example, the Volunteer Agency Halle-Saalkreis issued handbooks for local coordination offices as well as for care facilities (Freiwiligen-Agentur Halle-Saalkreis, 2006b and c) and MOVISIE, the National Centre for Social Development, issued good practice examples (Elfering and Scherpenzeel, 2009a–d).

Recruiting, coordinating, guiding, supervising and acknowledging volunteers and their efforts are tasks that must be performed at an organisational level. With a view to residential care and respective working conditions, elements of good volunteering were derived from research findings and included, among others, regular information; open communication; having a defined contact person; continuous guidance; regular meetings for the exchange of experiences; training opportunities; participation in the residents' advisory board; acknowledgement of volunteering; and supervision (Engels and Pfeuffer, 2007). Self-organised initiatives also benefit from good framework conditions.

As shown by the above-mentioned examples, voluntary work in LTC has various dimensions. It can be organised by an association for older people in inpatient and outpatient care, but also to offer support to family carers. In addition to this, voluntary work is self-organised by older people in need of care as well as by informal carers. Nevertheless, a complex set of framework conditions is needed to promote and enhance the work of volunteers and ensure specific quality.

The relationship between volunteers and older people in need of care

This specific aspect of LTC volunteering must also be considered in the context of the central position adopted within the INTERLINKS framework, namely the importance of seeing the user and informal carer at the centre

(Billings *et al.*, 2011). Therefore, in keeping with the person-centred approach of INTERLINKS, the impact of voluntary activities on older people with care needs and their families, as well as the effect of this kind of voluntary work on the volunteers themselves, can be essential (Box 11.3).

Box 11.3 'Network careCompany' (Germany)

Voluntary 'care companions' connect with families, listen to needs and inform them about opportunities for respite care, acknowledge efforts and point out strengths, create networks for mutual exchange and support, and they guarantee confidentiality in what has been shared. Upon completion of 120 hours of training, care companions are organised in local groups and commit themselves to taking part in follow-up training sessions and assessment meetings, organised by a national umbrella organisation. Other volunteers, so-called 'project initiators' who have undertaken 160 training hours, are responsible for establishing and running local initiatives in pairs. The local initiatives are independent and responsible for their own funding but with back-up from care, education and welfare associations as well as local and regional authorities.

Source: Stiehr (2011b).

It is clear that there are benefits to service users resulting from the personal relationship between volunteers and 'clients', based on a human touch and an altruistic attitude in which time is a gift with no financial interest involved. Furthermore, there is no top-down relationship in volunteering, but knowledge is more equally divided, and mutual exchange can lead to fresh perspectives. This can result in a general improvement of the older person in his or her state. Finally, family carers feel supported and feel more included in a community (for impacts on family carers see also Stiehr, 2011b).

On the other hand, volunteering offers opportunities for volunteers, such as doing something meaningful, as well as broadening their horizons and strengthening their networks. After their professional career, active citizenship offers older people new roles and purposes in life. Research on German volunteers in LTC revealed that the broadening of knowledge and experiences, and contacts and relationships with others were reported as essential motives (Engels and Pfeuffer, 2007; Klie *et al.*, 2005; Freiwilligen-Agentur Halle-Saalkreis, 2006a). For Czech volunteers it is known that motives of civic duty ('feeling that it is important to help others') play a much greater role for those working in health and social services than for those in the area of sports, culture and leisure where more hedonistic motives ('that they enjoy it') dominate (Křížová and Tošner, 2011; Vávrová and Polepilová, 2011).

In a study at the University of Wales, a systematic review of 87 articles was conducted to ascertain the health effects of volunteering on health service users (including LTC) and on individual volunteers. Effects of volunteer activities on service users included increased self-esteem, disease management and acceptance, longer survival times of hospice patients, improved cognitive and physical health function, increased levels of physical activities, reduced depression and anxiety, less need for hospital and outpatient treatment, and life satisfaction. However, a number of studies failed to demonstrate statistically significant health effects on service users, and none of the quantitative studies found negative effects (University of Wales, 2008).

The reviewed studies also showed that volunteering had a healthy effect on the volunteers themselves. Improvements included positive effects on self-rated health, depression, mortality, activities of daily living, quality of life and life satisfaction. The only negative aspect was mentioned in one study which discovered lower caregiver satisfaction among volunteers than paid employees (University of Wales, 2008).

Further insights into the relationships between older people in need of care and volunteers were given by Hartikainen, who focused on two of the most drastic effects of a move from one's own home to residential care: 1) a feeling of 'homelessness' combined with lack of alternatives after giving up the environment in which autonomy and identity was experienced over lifetime, and 2) the bodily closeness in care which is associated with movement into other people's personal territory. With a view to the first point, volunteers open and create social arenas for older people by talking and singing together, joint celebrations or readings of books and articles at the bedside. These situations involve activities, experiences, humour and playfulness that introduce variation to daily routines in residential care. As regards the limitations of the private and intimate sphere, the importance of everyday routines, such as communal meals and rituals, create new elements that imply security. With stimuli in daily life, such as handicrafts in groups, volunteers help to pass time but also give older people the experience of meaningful activity and bring back positive memories (Hartikainen, 2009).

Again, voluntary work is not restricted to the fit and healthy, but an option for dependent older people, too. It is often assumed that residents in LTC facilities are unable or do not desire to help others. But research in the US provided evidence in the 1990s that residents of care homes are indeed willing and able to engage in activities designed to help others, namely mentoring skills for English as a Second Language students. The comparison of self-rated health, depression and life satisfaction after three months supported the hypothesis that residents in LTC facilities who participate in a volunteer activity report higher levels of well-being than those residents who receive customary care (Yuen *et al.*, 2008).

In summary, voluntary work provides various benefits for both the volunteers and older people in need of care in relation to their health status and general well-being. In order to thrive, these benefits need a number of framework conditions. Voluntary work is happening at the local level, which

Box 11.4   Social work with older people – Frequent home visits for socially isolated older people with lower levels of care needs (Germany)

This project was implemented in two boroughs of Bremen initiated by the county government. The main target group was older isolated people with insufficient social support from their families or neighbours. Activities followed the principles: visit – advise – assist – and befriend. The main actors were volunteers accompanied by professionals. The uncomplicated access that was facilitated by the volunteers allowed for contact with the older person on an equal level rather than using a sometimes paternalistic professional approach. This contact enabled the older person to live independently at home for longer with more options for participation in social activities, greater empowerment and increased life satisfaction.

Source: Weigl (2011).

means that national strategies may create a generally positive environment for volunteering but need to be implemented in the regions, municipalities and voluntary associations (Weigl, 2011; see Box 11.4). Basically, this refers to addressing and placing interested volunteers, guiding them during their work and acknowledging their achievements. Since voluntary associations can be competitors in recruiting volunteers, some functions are best allotted to local or regional authorities. Others of course must happen within the organisations to ensure the best possible working conditions. Since volunteers are not bound by contracts, their work satisfaction is essential to prevent them from terminating their activities.

Discussion and conclusions

This chapter reflected upon the latest developments in terminology related to volunteering and voluntary activities in Europe, mainly in relation to volunteering in LTC. As the highly diversified nature and shape of volunteering in different European countries complicate the adoption of any general definition that would be accepted and shared by all stakeholders, the main challenges and opportunities related to volunteering as a life course phenomenon were presented, in particular with a perspective on volunteering as a way forward for active and healthy ageing.

The analysed tension zones could be identified at different levels: (1) Volunteering in LTC is conceptually placed in between formal and informal care, though with more proximity to the latter; (2) In policymaking the approaches of civil society and active ageing on one side and financial pressures on the other may collide; (3) Organisations are attracted by the prospect of unpaid work but often not prepared to adequately respond to the specific expectations and needs of volunteers. These tensions on the level of political and organisational planning, however, do not seem to impact structurally on the individual relations between volunteers and older people in need of care, which are generally characterised by mutual benefits. Some conclusions can be derived from these findings.

First of all, volunteering, as an ambivalent and multifaceted concept in itself, represents yet another building block in the construction of LTC systems in Europe. The current situation is complicated by the fact that systems of LTC as such are under construction, and 'volunteering' is increasing the complexity of key issues to be addressed (http://interlinks.euro.centre. org). The practice examples demonstrate that it involves both formal and informal voluntary activities of people of different ages to assist older people and their family members in meeting their LTC needs. It also includes issues on various forms of organisation and management approaches (such as education or training to perform voluntary work), helping to ensure the acquisition of necessary competences, skills and relationships with all relevant stakeholders in LTC.

Second, despite this prominent position, and apart from empirical facts and data that were presented in this chapter, there is a dearth of information on volunteering in general and on volunteering in LTC in particular. Still, these issues have increased in importance and awareness; even the International Labour Organisation (ILO) suggests adding a 'volunteer supplement' to national labour force surveys on a periodical basis.[6] These data and various additional surveys will be needed for a solid and evidence-based policy and practice in this field.

Third, the increasing acknowledgement of voluntary work can also be observed in LTC provision across Europe. Volunteers are contributing extensively to the functioning of LTC services, both in terms of quality and quantity. For older people and their family carers, volunteers contribute to improving the quality of life and the feeling of social integration, with a considerable impact on health effects. For formal provider organisations of LTC, volunteers are not likely to replace paid staff, but they provide supplements to basic care work such as conversation, convivial activities, etc. Given their intrinsic motivation and commitment that is not based on contractual obligations, they provide services that are time-consuming and would not be financially feasible if supplied by paid staff. Though a cordial relationship may exist between older people in need of care and volunteers, they are emotionally less involved than relatives and thus are potentially able to provide effective support in situations of conflict or emotional stress. Instead of financial gain, volunteers experience personal benefits that seem to be on a parallel with the experiences of their clients.

Formal care systems have therefore started to recognise the positive effects and benefits that are achieved by volunteers. Future developments for securing this resource will necessitate a greater desire on the part of formal systems to cooperate within an equal partnership. This includes mutual respect and a willingness to learn from each other, necessary for the development of a care system which integrates all forms of informal and formal care. In most European countries, volunteering in the field of care and social participation is beginning to professionalise, embedded in a broader debate about 'quality management' in volunteering. Volunteer management becomes an important instrument to support volunteer organisations in successfully implementing conditions appreciating volunteer work as an important social value. The benefits created by voluntary work for both the beneficiaries and providers of these efforts are far from being free of charge and require secured means for personnel and infrastructure. The investments made in this area have two important objectives: they enable volunteer organisations to engage in the well-being of older people in need of care; and contribute to a wider social cohesion within society.

Notes

1. An example of non-monetary compensation: one member does shopping for another member and gets in return the watering of his or her plants during holidays. The more time that is invested in support for others, the more time can be claimed for one's own support at any future date.
2. http://ec.europa.eu/social/main.jsp?catId=757&langId=en
3. The 'Report on volunteering in EU member states' (GHK, 2010) also offers information about sectors in which individual volunteers in 24 European countries were actively included in sport and exercise; social, welfare and health activities; religious organisations; culture; recreation and leisure; education, training and research.
4. Existing data on the number and percentage of volunteers hardly serve as a valid basis for transnational comparisons due to different methodologies and definitions used for volunteering. The sources for some countries include estimates and sample sizes that are either very small or restricted to specific associations or sectors. For instance, according to data referring to the situation in Slovakia, recent research assessed the volunteering rate at 27.5 per cent in formal and 47 per cent in informal volunteering (Brozmanová Gregorová *et al.*, 2012).
5. Data on volunteers in public residential care has only been available via national statistics since 2012.
6. Including the question: 'Did you provide any unpaid assistance to people outside of your household, such as older people, children, the poor, or disaster victims, prepare and serve food, or transport people or goods?' (ILO, 2010, p. 24).

12
The Role of Information Technology in Long-Term Care for Older People

Jenny Billings, Stephanie Carretero, Georgios Kagialaris, Tasos Mastroyiannakis and Satu Meriläinen-Porras

Introduction

The role that information technology (IT) has in assisting older people to maintain independence in their homes is increasing in many European countries (European Commission, 2010a). Advocates argue that it permits a more person-centred approach, assists carers in helping to reduce the burden, and can support older people with a wide variety of long-term conditions, not only in coping with physical conditions such as diabetes, but also conditions such as dementia. Evidence is accumulating about the value of technologies such as telecare, telemedicine and telehealth and how their use is becoming more widespread (European Commission, 2010b). However, IT solutions do not come without considerations in their application and effectiveness.

The intention of this chapter is to provide some insights into the role of IT in long-term care (LTC). First, it will provide an overview of the European context with regard to the use of IT and associated political drivers, and then assess potential benefits of and challenges to current IT solutions for independent living and care. In addition, it will conduct an analysis of key issues within the INTERLINKS Framework for LTC with respect to those practice examples applying IT, in order to compare and contrast approaches. The importance of the user perspective (both staff as well as older people and their carers) is key to the success of technology in health care, therefore the role of users in IT applications regarding their design and implementation within LTC will be critically assessed. Finally, the chapter will provide a prognosis for outlook of the place of IT in future LTC provision.

First, and for the purposes of clarity, the terms used to describe the different sorts of technologies associated with health care need to be defined:

- *E-Health* is the overarching term for the range of tools based on information and communication technologies used to assist and enhance the prevention, diagnosis, treatment, monitoring and management of health and lifestyle (from http://ec.europa.eu/health-eu/care_for_me/ e-health/index_en.htm)

- *Telecare* refers to the provision of social care from a distance supported by telecommunications (Empirica, WRC and European Commission, 2010).
- *Telemedicine* is defined as the delivery of health care services, where distance is a critical factor, by all health care professionals using IT for the exchange of valid information for diagnosis, treatment and prevention of disease and injuries, research and evaluation, and for the continuing education of health care providers, all in the interests of advancing the health of individuals and their communities (WHO, 1998). Telemedicine involves secure transmission of medical data and information, such as biological/physiological measurements, alerts, images, audio, video, or any other type of data needed for prevention, diagnosis, treatment and follow-up monitoring of patients (WP European Commission, 2009). Telemedicine covers telehealth and other disciplines relevant in LTC such as telemonitoring.
- *Telehealth* applications are concerned with service delivery from a health care provider to a citizen, among health professionals, or among citizens and family members (Stroetmann *et al.*, 2011). Telemedicine and telehealth are similar and in many documents appear as the same concept, but the former is restricted to service delivery by physicians only, and the latter signifying services provided by health professionals in general, including nurses, pharmacists and others (WHO, 2010a). In this framework of LTC needs, home telehealth refers to the range of support needed, typically including not just clinical (medical) monitoring and intervention, but also a broader range of homecare support that more traditionally fall within the scope of social/homecare services (Empirica, WRC and European Commission, 2010).
- *Telemonitoring* designs systems and services using devices to remotely collect/send vital signs to a monitoring station for interpretation. Telemonitoring is the remote exchange of physiological data between a patient at home and medical staff at a hospital to assist in diagnosis and monitoring. This could include support for people with lung function problems, and diabetes for example. It includes among other things a home unit to measure and monitor temperature, blood pressure and other vital signs for clinical review at a remote location (for example, a hospital site) using phone lines or wireless technology (COCIR, 2011).
- *Smart home technology* is a collective term for IT as used in houses, where the various components are communicating via a local network. The technology can be used to monitor, warn and carry out functions according to selected criteria. Smart home technology also makes automatic communication with the surroundings possible,

via the Internet, ordinary home telephones or mobile phones. Smart home technology gives a totally different flexibility and functionality than does conventional installations and environmental control systems, due to how programming, the integration and the units react to messages submitted through the network. The lighting systems may, for example, be controlled automatically. Good physical access is a prerequisite for the optimal utilisation of the technology (Laberg *et al.*, 2005). Smart homes are applied to assure the increased safety and independence of older people. They are composed of sensors, actuators, controllers, a central unit, networks and an interface (Laberg *et al.*, 2005).

- *Ambient Assisted Living (AAL)* is a jointly funded European Union (EU) programme, the objective of which is to enhance the quality of life of older people and strengthen the industrial base in Europe through the use of IT. The programme has been motivated by demographic changes and ageing in Europe, which imply not only challenges but also opportunities for citizens and social and health care systems, as well as industry and the European market. The objectives of ambient assisted living are understood as:
  - to extend the time people can live in their preferred environment by increasing their autonomy, self-confidence and mobility;
  - to support and maintain the health and functional capability of the elderly individuals;
  - to promote a better and healthier lifestyle for individuals at risk,
  - to enhance the security, to prevent social isolation and to support the multifunctional network around the individual;
  - to support carers, families and care organisations;
  - to increase the efficiency and productivity of resources in ageing societies (aal-europe.eu).
- *Gerontechnology* is an interdisciplinary academic and professional field combining gerontology and technology. It refers to the research and development of techniques and technological products, based on knowledge of the ageing processes, for the benefit of a preferred living and working environment and adapted medical care for older people. Gerontechnology concerns matching technological applications to health, housing, mobility, communication, leisure and work of older people. Research outcomes form the basis for designers, builders, engineers, manufacturers, and those in the health professions to provide an optimum living environment for the widest range of ages (Graafmans and Brouwers, 1989).
- *Assistive technology or adaptive technology (AT)* is any product or service designed to enable independence for disabled and older people (King's Fund, 2001). It is an umbrella term for any device or system that allows individuals to perform tasks they would otherwise be unable to do

or increases the ease and safety with which tasks can be performed (WHO, 2004a). Examples of assistive technology include the 'curb cut' (or dropped curbs at street crossings), standing frames, text telephones, accessible keyboards, large print, Braille, and speech recognition software (Pew and Van Hemel, 2004).

The demographic and political context of IT implementation in Europe

The implementation of technological solutions across Europe has increased considerably in the past few years, according to the Eurostat Information Society Statistics (2011). It is of note that the percentage of households who have Internet access at home has increased from 49 per cent in 2006 to 70 per cent in 2010. In this period, the percentage of people in Europe using the Internet for seeking health-related information (injury, disease, nutrition, improving health) also increased from 19 per cent in 2006 to 34 per cent in 2010.

In addition to this, there has also been an increase in Europeans using the Internet for interacting with public authorities in the last decade, whether for obtaining information or downloading, completing and returning official forms. The percentage of Europeans aged 65 to 74 that used the Internet for such interaction increased from 10 per cent in 2008 to 15 per cent in 2010. This percentage increases to almost half of that population group in Denmark, Norway and Iceland. This could be due to the effectiveness of the implementation of E-administration services in these countries, the ease of access to the sites for this population group, and a cultural preference to deal with public authorities through this means. In fact, since 2003, these three countries have led the European rankings of population groups between 65 and 74 year olds who access the Internet once a week, reaching 52 per cent of this group in Denmark, 58 per cent in Iceland and 59 per cent in Norway in 2010, while the average across the EU reached only 25 per cent.

The political and strategic importance of developing new health technologies has been targeted for years by the European Commission (Carretero *et al.*, 2012a). E-Health, including IT for independent living and care, is considered highly innovative, with a demand that is accompanied by an existing technological and industrial base in Europe. The European Commission has long urged for public policies that promote growth and development of IT solutions for older people (European Commission, 2007). As can be seen by the policy imperatives, the potential contribution of positive initiatives in the field of telemedicine, telecare and Ambient Assisted Living (AAL) to reduce the pressure of population ageing on health and social services is being emphasised.

Recently, the Digital Agenda (European Commission, 2010b) has stressed the importance of new technologies to ensure sustainable health care and a life of dignity and independence, and has highlighted the need to improve the technological competence of older people. Especially in relation to telemedicine (telehealth and telemonitoring), activities have been planned to develop technologies for daily assistance that take into consideration the environment and technological capacities of older people. Specifically, Action 78 of the European Digital Agenda aims to strengthen the AAL programme to improve the quality of life of older people and augment the industrial base in Europe through the use of IT.

The potential benefits of IT solutions for independent living and care

There is a growing belief that new communication technologies can contribute to the improvement of older people's quality of life and health, to offset the reduction in resources required for an ageing population, and to help maintain an active life at work, leading to more sustainable health and social care systems and a life with dignity and independence for seniors (European Commission, 2010a). In particular, new technologies may contribute to these objectives through social media and other Internet applications not only to maintain and create social bonds, support daily activities, home alarm mechanisms and risk management, including medication reminders, but also to support telecare and telemedicine for the delivery of health care in the home, or portable monitoring systems for the management of chronic diseases (European Commission, 2010a).

Several technologies are available for independent living and care, mainly telecare, home telehealth and smart homes. Telecare and home telehealth are the more developed technologies and have been deployed in the market. Although the quality of the evidence based on the benefits and effectiveness of telecare and telehealth consists mainly of the cost-effectiveness of applications and cost savings (Barlow *et al.*, 2007; Bensik *et al.*, 2006; WHO, 2010b), other reviews and studies are emerging that reveal a broader range of benefits of these IT developments.

In general, findings indicate that the most effective telecare interventions appear to be automated vital signs monitoring (for reducing health service use) and telephone follow-up by nurses (for improving clinical indicators and reducing health service use) (Barlow *et al.*, 2007). Telehealth, telecare and telemonitoring would appear to be effective interventions in the areas of diabetes, mental health, heart failure and cardiac disease (Bensik *et al.*, 2006). With regard to other medical specialties such as psychiatry and neurology, Hersch *et al.* (2006) point out that interventions are particularly effective when verbal interactions are a key component of the patient assessment. Treatments administered in these specialties via telemedicine appear to achieve comparability with face-to-face care.

A specific area of effectiveness concerns the ability of IT interventions to reduce hospital admissions and the length of stay for patients. For example, telemonitoring programmes have been found more efficacious than home nursing visits in decreasing readmissions of patients with heart failure (Benatar *et al.*, 2003; Dar *et al.*, 2009). In addition, Cleland *et al.* (2005) found that heart failure patients receiving either telemonitoring or nurse telephone support at home had a similar number of hospital admissions and mortality, but those receiving telemonitoring had the mean duration of their hospital admissions reduced by six days. Reductions in hospital admissions due to congestive heart failure have also been reported in 27 per cent to 40 percent of cases and the length of stay for those patients has also been reduced (Polisena *et al.*, 2010).

Interestingly it has been suggested by Paré *et al.* (2010) that the beneficial effects on the health status of patients using telemonitoring systems to deal with chronic diseases such as diabetes, asthma, heart failure and hypertension are mostly observed in patients suffering from serious exacerbations of ill health, those interested in playing an active role in the management of their illness, and those interested in using health technological devices. Thus there appears to be a relationship between self-care motivation and reported beneficial effects.

While it would appear that most positive effects have been found within medical specialties, some applications of IT have also been reported as beneficial to the broader sustainability of social and health systems and use of resources. For example, using telephone and text reminders improve the non-attendance rates of patients to consultations and their average cost is estimated at only €0.50 per reminder (Hasvold and Wootton, 2011). Equally the use of store-and-forward telemedicine, an asynchronous form of telemedicine that consists of collecting medical data, storing it and transmitting it to the health professionals for later assessment, can reduce travel costs for both patients and health care professionals (Wootton *et al.*, 2011).

In addition, some programmes are coming to light that have provided reliable evidence that IT applications can improve the delivery and efficiency of health and social care systems (European Commission, 2010a) and can act as good practice for potential interlinking between health and social care among Member States.

An example is the ACTION Living with Dementia support programme in Sweden, a project that assists carers using telematics interventions to meet older people's needs (Magnusson *et al.*, 2005). This project began in 1997 and now exists in five municipalities in Sweden. It includes an information and educational programme which contains different computer exercises from the Lexia cognitive training programme, and a videophone system to maintain contact with health and social care staff and other families in a similar situation to their own. Cost analyses reveal that there is an approximate saving for municipalities of about €10,745 per family per year.

The potential challenges to IT applications

Despite these advances across Europe, the implementation of specific technological solutions associated with health care is facing several challenges. First, there is a lack of 'digital' training required by older people; according to Eurostat (2011), in 2010, 49 per cent of men and 61 per cent of women aged 55 to 74 years in Europe have either never used the Internet or not for more than a year. Second, equipment, interfaces and services designed to promote healthy and active ageing must be tailored to the specific needs of older people due to loss of functional and sensory ability, an aspect discussed more fully in a later section. Third, despite the emergence of knowledge, there is still a lack of information and robust studies on the effectiveness and efficiency of these technologies, and this creates a barrier to respective public and private investment to implement solutions at a wider scale. For example, there is scepticism regarding the ability of IT solutions to produce tangible benefits, and there are some areas that have failed to demonstrate effectiveness, despite their innovative potential in addressing health and well-being of older people and their carers.

Regarding evidence for the use of smart home technologies, for example, Martin *et al.* (2009) conducted a systematic review to explore their effectiveness as an intervention for people with physical disability, dementia or learning disability living at home. The search included randomised controlled trials, quasi-experimental studies, controlled before and after studies and interrupted time series analyses, as well as measures of service satisfaction, device satisfaction and health care professional attitudes or satisfaction. In fact no studies were found that met the inclusion criteria.

In addition, DelliFraine and Dansky's (2008) review aimed to determine the effect of home telehealth on clinical care outcomes. Studies selected for the review covered a broad range of patient diagnoses (mostly diabetes, heart disease and psychiatric conditions) and telehealth technologies (including web-based interventions, video and data monitors alone or in combination, and telephone interventions). It concluded that telehealth was an effective clinical intervention in many settings with different patient groups. Twenty-nine studies were included in the review of which 24 were randomised (n=1706 participants). However, beyond providing a subgroup analysis of randomised trials, the potential impact of study quality on the findings of the review was not considered. Since the review statistically combined a group of clinically and methodologically heterogeneous studies, the conclusion is necessarily broad. Given those limitations, this conclusion cannot be considered entirely reliable.

In summary, this section has highlighted the political imperatives that are assisting the drive forward in IT advances in Europe, and has indicted that IT applications have potential health and cost benefits in helping to maintain and improve independent living and care for older people and their carers, particularly when focused on medical specialties. It has also indicated that there are deficits in the evidence and challenges for both the end users (older people and carers) and for achieving the desired policy imperatives.

But with the focus primarily on disease groups, what appears to be fundamentally lacking in this review is a greater understanding of how and where IT solutions can benefit LTC structures and interfaces, particularly with respect to health and social care integration, and the links between formal and informal care. The next section therefore will critically assess the contribution of the examples accumulated through the INTERLINKS Framework for LTC, examining their ability to augment the knowledge base regarding the potential benefits of IT solutions.

Analysis of selected practice examples

Within the INTERLINKS Framework for LTC, there are a number of key issues addressing the role of IT in linking health and social care services, as well as formal and informal care. With this in mind, a range of practice examples have been identified to illustrate these key issues. In the following, a sample of eight examples from five countries (Greece, Finland, France, Slovak Republic and Spain) will be described and analysed transversally in relation to their approach, impact, sustainability and transferability. The analysis will also address the elements and key issues of LTC that may be supported by IT solutions and the nature and grade of their efficiency.

It must be noted that more examples illustrating the INTERLINKS Framework were found that referred to and used IT applications, suggesting that use across Europe is more widespread than is intimated here. However the eight examples were selected as they provided sufficient elaboration on the key issues and permitted a transversal analysis to be conducted, giving initial insight into the nature of activity in this area.

From the description of the applied practices found in Europe concerning IT and LTC within the INTERLINKS examples therefore, four key issues were identified as common fields of how IT solutions can be and have been implemented:

- IT solutions in ambient assisted living and smart housing;
- IT solutions in LTC management and administration;
- IT applications at the interfaces between health and social care professionals;
- IT applications at the interfaces between formal and informal care.

IT practice or implementation from the eight selected examples may cover more than one of these key areas, as they demonstrate a number of important cross-cutting features of technological benefits.

IT solutions in ambient assisted living and smart housing

There are two examples in this category: the E-Health unit from Greece (Mastroyiannakis, 2011b) and the AMICA telemonitoring system for chronic obstructive pulmonary disease (COPD) patients from Spain (Cordero, 2011). It must be noted that these examples are partially inter-linked. While the Greek E-Health unit has been running for over ten years and was granted the European E-inclusion award in 2008 by the European Commission, AMICA started in 2009, initially in Spain with seven partners, including Greece. Greek partners therefore were able to bring experience from their situation into the collaborative partnership, so enhancing the opportunity to embed learning into different environments (see Boxes 12.1 and 12.2):

Box 12.1   IT solutions in LTC management: The E-Health Unit (Greece)

This E-Health Unit was established in March 1999 and is situated in the Sotiria hospital in Athens, an 800-bed general public teaching hospital. The programme is funded by EU research programmes and the hospital budget. It aims to maintain older patients with multiple morbidities at home through effective IT supervision and monitoring, to avoid crises, reduce hospital visits and admissions, and to provide support and educate family carers. The unit has created a specially trained multidisciplinary rehabilitation and home care team of physicians and other health allied personnel. The E-Health Unit provides a) a two-way interactive video and voice communication to allow the patient to meet 'face to face' with a nurse or doctor; b) transmission of vital signs, logged into the patient's records; c) continuous monitoring of bio-signals through wearable systems; d) home hospitalisation and early discharge; e) home visits by nurses with diagnostic devices; f) individual multiprofessional care plans.

Outcomes include improvement of patient condition – increased autonomy and satisfaction, reductions in readmission rates and length of stay, and cost reductions. These effects have been sustained over a number of years.

Source: Mastroyiannakis (2011b).

Box 12.2   IT solutions in LTC management: AMICA (Spain)

Chronic obstructive pulmonary disease (COPD) is one of the leading causes of morbidity and mortality worldwide. COPD causes high levels of disability and care needs and it has a significant epidemiological, clinical, social and economic impact. New technologies enable the development of tools and systems to achieve a better management of chronic diseases. To this effect, the Spanish University of Cadiz is promoting a research project to construct and validate a telemedicine system for COPD patients. The project has been funded by the European Ambient Assisted Living Joint Programme. The AMICA solution aims to provide home care for older people affected by COPD. The technological support will allow for self-management of the disease, medical monitoring as well as the enhancement of patients' social interaction. The potential impact is to predict and prevent exacerbation, to reduce hospitalisation and health care costs, and to improve patients' and informal caregivers' quality of life.

Source: Cordero (2011).

In both examples, telemedicine equipment is used to encourage patients with COPD to manage their health at home with continuous monitoring or frequent proactive examination from health care professionals. There is similarity of approach towards the early detection of exacerbations of a common disease (COPD) by using IT. Outcome data from these two examples are available from Greece and Spain, but unavailable from the other five AMICA partners. There appear to be common benefits for the recipients, which are increased self-determination, increased satisfaction by both informal carers and the cared-for older people, cost-effectiveness, better quality of life and reduced hospital visits, admissions and length of hospitalisation.

Although the data from some countries are missing and the chance to fully compare across a number of contexts is not possible in this instance, AMICA provides a good opportunity to test transferability of interventions between countries in a limited fashion. From the aspect of sustainability, AMICA has a total budget for a period of three years and has been co-funded by the European Ambient Assisted Living Joint Programme, by each participant's national funds and by the seven project partners. Funding for the E-Health Unit in the Sotiria hospital has been partly incorporated into the hospital budget and is partly supported by research funds. The degree to which services can continue will depend upon a continuation of these funding streams, which is unpredictable and dependent upon national priorities. However, diverting partial funding through hospital budgets as in Greece would appear to create a more secure foundation for sustainability.

The role of IT in LTC management and administration

Ways in which IT solutions can facilitate, organise, support and improve management of patients and professional administration will be examined and four examples are listed in this area. The first two have already been mentioned: E-Health unit and AMICA from Greece and Spain (see previous section). Again, there are some common areas across these two examples. Due to the specialised software, the ability to keep electronic files, clinical data, laboratory results and interviews for all patients included in the programme not only reduces cost but facilitates the work of professionals enormously. For example, physicians responsible for the patients, as well as expert physicians, nurses, physiotherapists or other health care professionals, are able to communicate and share information, as well as generate and implement a common plan for the benefit of the person being monitored.

The ADaL example from Finland (Box 12.3) demonstrates the application of similar kinds of sensors and devices in the more controlled environment of a residential care home. The application of IT solutions in the ADaL programme was able to facilitate the management of care for people within LTC by ensuring constant monitoring and providing proactive alarms in stressful or potentially hazardous situations for the residents, during movements and transitions. By increasing clients' independence

Box 12.3   ADaL: Activating daily life programme with technology solutions (Finland)

The 'Activating daily life (ADaL) with technology solutions' pilot project was undertaken in the Kustaankartano Centre for Older People in Finland, a nursing home with 600 residents. The project was originally developed and implemented in four intervention wards from 2006 to 2008. Because the development in these wards was so successful, the programme spread to other wards in the centre.

The pilot hypothesised that the ADaL programme, coupled with technology solutions, could improve self-empowerment and enhance the quality of life of residents with LTC needs. The main aim of the ADaL programme was transform revise nursing patterns in nursing home wards with the help of technological solutions. The other objectives included improving patient safety, reducing falls, injuries, restraints and bedrails, and enhancing quality of life.

The programme consisted of a detailed multidisciplinary mapping of the residents' preferences, individual weekly exercises and the use of movement sensors to alert the nurses to those at risk of falling when trying to stand up from the bed or chair. The results showed that the ADaL programme was able to increase residents' independence without increasing staff costs. Their quality of life was increased: there was a reduction in the use of bedrails as well as bedbound residents because of technology solutions.

Source: Hammar and Finne-Soveri (2011).

and quality of life, together with employing technological equipment without increasing the number of ward staff, it was considered to be a successful investment. The cost was €2.7 million and was funded by Helsinki City from 2006 to 2010 based on outcomes of reductions in new fractures, the use of hypnotics and anxiolytics, and a reduced number of bedsores that needed significant treatment.

Regarding administration, the Smart Call example from Slovakia (Box 12.4) differs in that it responds to the key area of IT applications at the interfaces between health and social care administrations. Smart call creates a connection among clients who call for help with a press of a button, and an acute intervention is coordinated by the call centre, but it also brings services together at an administrative level. The Finnish home care example (Tepponen and Hammar, 2011b) is kept simpler through an all-in-one platform. It achieved a better management of home care provision through a digital model that increased the productivity of home care services, primary and secondary health care at the administrative level. However this is not one of its main purposes.

Box 12.4   IT applications at the interfaces between health and social care professionals: The Smart Call Technology (Slovakia)

Smart Call was developed in Nitra. It is an IT solution for the care of frail older people and coordinates health and social care providers in an emergency. Smart Call consists of monitoring and signalling technologies that promote independent living and safety for older people, especially for those living alone. The Assistive Distress Call Centre (ADCC) receives emergency signals for urgent help and coordinates the provision of suitable and fast care by health and social services. This coordination avoids duplication of needs assessment by different bodies and makes data and information available to the emergency staff. Currently 26 older people have installed Smart Call at their homes and according to the data provided by the ADCC, 4596 alarm calls were received from 2006 to 2010.

Source: Brichtova (2011b).

IT applications at the interfaces between health and social care professionals

This category includes examples from Greece, Slovakia (see above for descriptions of these examples), Finland and France.

In the Greek example (Mastroyiannakis, 2011b), the E-unit's specially trained multidisciplinary rehabilitation and home care team have established effective IT supervision and monitoring to avoid crises, reduce hospital visits and admissions of chronically ill older people. The unit uses IT to collect, store and analyse data collected by the patient or home care nurses into a common shared file, available to the multidisciplinary team that follows up the patient. Thus, in a joint meeting, shared data provides the input to lead decisions between health and social care workers to draw up a common care plan for the client. Although care is medically orientated, E-Health is considered a good example as of effective multidisciplinary team work.

Smart Call (Brichtova, 2011b) creates a connection with the press of a button for clients who need immediate help, and an acute intervention is coordinated by the call centre. IT facilitates action at the interface, through the communication and coordination of health and social care professionals including the fire brigade and police. The sense of safety and the cooperation between these professionals that was achieved by this IT solution has reduced hospitalisation and length of stay, enhanced preventive interventions, reduced admission to residential care and eased informal carers from the anxiety of a sudden unfortunate event happening to their family member. By using different kinds of equipment and planning, Smart Call can be likened to the Finnish home care example (Tepponen and Hammar, 2011b, see Box 12.5).

Box 12.5   IT solutions and new health technology facilitating integration in home care (Finland)

This IT solution was developed by the South Karelia District of Social and Health Services (Eksote) and achieved a better management of home care provision through a digital model that increased productivity of home care services, primary and secondary health care. The model consists of the use of web cameras, broadband and video phones to share information among the different health and social care services and provides rapid response to patients in need of care. In the project 185 patients participated, 30 of them during the three years of its implementation, one GP, two full-time nurses and several part-time home care workers, as well as two part-time IT engineers and several IT workers. Some of the collaborations tested in the project are still being used in daily care. Additionally, the evaluation of the project using the Resident Assessment Instrument to assess the needs of older people and their quality of life found that 27 out of 29 patients interviewed reported an increase in a sense of safety due to the home care services, they used less medication and their sense of isolation decreased from 30 per cent to 9 per cent. The economic assessment of the project indicated that consultations using remote technology were 50 per cent more economical for the health system.

Source: Teponnen and Hammar (2011b).

Similarly, the French example Geoloc (Naiditch, 2011a) uses a bracelet worn by Alzheimer patients that helps a call centre locate their exact position. It is efficient in finding a lost person and the service organises a network of supportive relatives, neighbours, health care and social care professionals (Box 12.6). This project has elements of the Finnish ISISEMD example (Box 12.7).

Box 12.6   IT applications at the interfaces between formal and informal care: Geoloc (France)

Geoloc is a national service managed by a French non-profit making organisation. It is targeted towards patients with Alzheimer's disease who wander. This problem of wandering leaves Alzheimer's patients in unfamiliar territory, feeling lost and unable to find their way back to their usual surroundings. This has negative consequences for their safety, as well as effects on quality of life for themselves and for those who are responsible for them. The service addresses this challenge by means of a call centre managed by well trained staff and equipped with electronic GPS/GPRD territorial software that allows users who get lost to be located. The system is coupled with a wireless phone and operates on a permanent basis (24hr/day all year). Users wear a special bracelet with software built in that will send repeated signals as soon as they are moving out of their 'usual' walking zone. As signals are sent to the call centre, staff will rapidly locate the wandering user and call one of the defined contact people. In addition to family members, neighbours and other volunteers are organised for each patient as contact people. Therefore it can be guaranteed that there will always be somebody who can be reached in order to bring the user back safely to his or her place of residence (own home or care home). The system can also be actively used by members of the family or volunteer groups to find out where the user is at any point in time. This system provides an important cost-effective and safe method of patient monitoring.

Source: Naiditch (2011a).

Box 12.7   IT solutions and new welfare technology facilitating integration for people with dementia: ISISEMD (Finland)

The aim of the ISISEMD project (Intelligent System for Independent living and SElf-care of seniors with cognitive problems or Mild Dementia) is to develop and pilot an innovative and intelligent set of services that support the independent living of older people in general and groups of older people with cognitive problems or mild dementia in particular, while at the same time supporting formal and informal caregivers. The ISISEMD pilot includes home safety equipment (e.g. intelligent front door, bed sensor, fire alarm and cooking monitor) and touch screens with reminders with date and time display. In addition a tracker device is available for outside activity. This tracker also includes a falling sensor which sends an alarm to informal/formal caregivers if a fall occurs. A voice connection can be made via the device to the older person to assess the situation, and it can also provide coordinates on a map to locate the older person. Informal caregivers have been trained to use an internet portal where they can control the ISISEMD system, edit sensor settings and see what kind of events home safety sensors have detected at older people's homes. The cost is relatively low given its preventive potential. The equipment cost between €2000 and €3000, depending on what equipment had been installed. Equipment costs cover touch screen computer, sensors and wires and installing costs varied between €200 and €400, depending on the size of the older person's home.

The service has been evaluated and first evaluation results have shown that it supports older people with dementia and their relatives in their daily activities, and older people were able to manage at home longer. In addition, older people feel safer at home and informal caregivers benefit from reduced stress.

Source: Tepponen (2011).

There is significant similarity of technology between these examples and this may be an advantage for transferability to other regions or countries. Sustainability of the above examples is unpredictable as, being in different systems, there is a variety of funding mechanisms, from fully funded through sponsorship in Finland, to being a paid-for service in France. However as they both deal with the need for security and provide a sense of autonomy, they have a strong case for becoming a mainstream solution.

IT applications at the interfaces between formal and informal care

This includes three examples originating from Finland, and two from France (see previous sections for Finnish examples and the French Geoloc service). The additional French example Equinoxe (Naiditch, 2011b, see Box 12.8) is similar to the Slovak Smart Call monitoring system, only it is linked to three neighbours who check on the older person if called to do so on a voluntary basis and resolve the situation themselves if possible.

In the Finnish ISISMED example (Tepponen, 2011), IT solutions focus on the care of an older person from the perspective of the formal and

Box 12.8 IT applications at the interfaces between formal and informal care: Equinoxe (France)

This French example was established in 1986. Created by Equinoxe, a non-profit association, it is a nationwide home alarm system that aims to maintain frail people living at home alone and enhance quality of life. Equinoxe is based on a complex high-tech call centre operating 24 hours a day with specially trained people. In order to participate, each older person must nominate a 'neighbourhood committee' which consists of about three people living nearby, holding a key to the older person's house. They have a central role – when notified by the call centre after contact from an older person, they are sent to assess the situation, sort out the issue if possible, and/or help and support the older person if medical assistance is required. The call centre staff will continue monitoring until the issue is resolved and users benefit from feeling safe and secure. In spite of a high turnover of users (death; moving into residential care) and having to compete against many other providers, Equinoxe is steadily growing. At the end of June 2011, 40 paid employees managed the system with the help of 35 'staff' volunteers for 8756 users. On average, the users were 84 years old, called 13 times and used the service for about 3 years. They rated the service highly.

Source: Naiditch (2011b).

particularly the informal carer. The example shows how technology can facilitate the independence of both the informal carer and the older person through enabling constant communication, and automated alarms and tracking. Thus this solution eases the constant worry the informal carer may have for the frail older person and provides a sense of security for them both, as well as improving the management of the situation from the professional perspective. To a large extent, aspects within all of the examples in this section have this potential psychological benefit built into their models.

Although this one-year pilot project has been supported by five countries (Finland, Denmark, UK, Greece and Italy), no data are available as yet to ascertain continuation or sustainability of this project. As the cost of this application is about €3000–4000, serious consideration needs to be given to outcomes and cost–benefit information, so this is a potential weakness. Nevertheless, transferability seems to be evident in countries with different LTC systems as it adapts to the users' need and is both controlled and connected with formal and informal carers' devices.

So having identified some transversal themes within a small number of elaborated INTERLINKS examples, it can be seen that the analysis has enabled understanding of IT benefits to be broadened to include those associated with the links and interfaces between services as well as formal and informal care. What appears to be challenging in this environment however are the necessary steps and knowledge to facilitate the scaling up of IT devices in LTC for older people. While costs and investment are intrinsically linked to wider application, a singular challenge highlighted within the literature review and perhaps implicit within INTERLINKS examples is the lack of user orientation and perspective on design and implementation methods, which is crucial to success. Taking the interpretation of 'user' in its broadest sense, the following section will critically discuss this issue in more detail.

The relevance of the user perspective in the design and use of ICT solutions

While IT applications in health care are largely being embraced across Europe as creating potential benefits for those in need of long-term care, there is a need to consider the end user's perspective as even the most advanced technology will fail if there is poor understanding and limited use in the field. This section aims to focus on the relevance of the user perspective in its widest capacity in the design and use of IT solutions, seeking to determine the relative merits and shortcomings of applying technology in their environments. Hence user groups will include older people, informal carers, as well as health and social care professionals and administrative personnel.

Older people: challenges and facilitators to the use of technology

A primary concern relates to the fact that technology is largely placed within the older person's home. Recent research undertaken across Europe focuses upon this locational issue of the home environment in relation to the social and cultural impact of new technologies, exploring how they reconfigure the 'spaces of care' where care takes place (Vincent and Fortunati, 2009; Schillmeier and Domènach, 2010). One concern deals with the effects of social control that comes with the introduction of technological care devices. López (2010) argues that the use of telecare and technological devices such as GPS makes people visible and potentially subjected to practices of surveillance and control in the home, which is usually considered a safe and private environment. Using his ethnographical research findings on home telecare services in Catalonia in Spain, López suggests that telecare produces a regime that continuously gathers information about users and subsequently requires them to be disciplined at home, interfering with familiar routines. This interference heightens suspicion and becomes a barrier to convincing recipients that monitoring is used to reduce the risks associated with living independently at home, not to eliminate them.

So while the assumption exists that new technological devices will be able to support the care needs of frail older people in their homes, it is important to consider how they reshape the physical and affective nature of home, and the older people's experiences therein. Milligan *et al.* (2010) have examined this aspect further in the UK. They conclude that the implementation of new technologies is contributing to a new topology of care, where home comes to represent a 'theatre of operation' in which a multiplicity of care functions and practices take place. The balance between the concept of home as a protected place, with a front door that allows older people to decide whom to exclude and how to live their lives, and the use of technologies that enable care professionals and call centre operators to enter the house without a physical presence, is the very concern of new care technologies.

While the above arguments put forward an interesting perspective on the potential invasiveness of technology, Mestheneos (2012) provides a good overview of the nature of other challenges faced by older people when encountering IT applications. Primarily, she states that the issue of motivation is key. Those, for example, who have never used IT technologies remain unconvinced that it is essential to their lives; they have successfully used other modes of communication and can see little reason to change. A gap in knowledge often exists and relates to what needs to be understood in order to effectively use systems. When knowledge deficits are apparent, prejudices and resistance become evident as difficulties in using systems are experienced, especially by older people, many of whom across Europe are computer illiterate (Eurostat, 2011).

Often new technologies are sold to people on the basis of fear such as of falling, of burglary and their safety in their homes, and this can be seen as a poor motivator. A further issue is cost; while some older people have adequate incomes, others will have to rely on their families to meet the costs. Inadequate and underfunded health and welfare systems are however facing difficulties in meeting the capital and restructuring costs of new expensive technologies, even when they might save on staff and provide more support services in the long run. Mestheneos (2012) adds that private insurance policies will have to be radically altered if they want to cover some of the future risks and expenses for ageing customers.

Whatever forms technology take, there are issues for older people about the installation, maintenance and monitoring of the equipment. However good and reliable the equipment is, if it does not provide a quality service through professional support, it will not be used. Many older people do not have someone in their social network able to help them with the installation of new equipment, teach them the necessary skills and deal with problems. It is of interest that most users of digital technology rely on informal hints, support and help from friends and peer users (Etchemendy *et al.*, 2011).

Attention and careful preparation is therefore needed to enable the older person to adapt and incorporate the new technology into their everyday lives, and embrace IT as a solution to their problems (Eisma *et al.*, 2004). For example, given the practice examples cited in the previous section, it is noteworthy that it is in general the older people themselves who are expected to be physically attached to equipment in order to benefit from the sense of security, by being allowed to leave the house without becoming lost, or staying at home and enjoying a better quality of life by early detection of exacerbations of illness. Although this creates a dependency on equipment and may appear to be a barrier and personal intrusion, it is in reality no different from dependency upon other aids to daily living such as spectacles or walking frames. Fears could therefore be mitigated by aligning this situation to others that have already been socially adjusted to.

Importantly, it is essential to have insight into what it means to be a recipient of this technology in order to best understand the impact on the older person, and here, needs assessment is vital to ensure goodness of fit between the technology on offer and the specific situation of the client. Long-term conditions cannot exclusively be dealt with through new technologies and all situations have to be analysed in terms of their risk and problems for the older frail person and how it can most effectively be prevented. IT can play a role, but often there will be other and simpler solutions to be tried first and in other cases IT alone will not be enough. Several of the examples cited in the previous section included multidisciplinary assessment, the benefits of which are well documented with regard to care improvement (e.g. Billings and Leichsenring, 2005). Ensuring adequate training as well as providing personal support is valuable in ensuring that older people and their carers do not abandon the assistive technologies, and there is some evidence that this is effective (Chiu and Man, 2004).

Other factors need to be in place to ensure engagement. Older people with chronic or acute health conditions may find it difficult to follow complex instructions or procedures, and this may be overcome by the use of voice-activated solutions or Skype. Effective intermediaries for improving self-confidence in using systems seem to be those connected with the target community, for example, groups of motivated older people or older people's clubs and training or educational centres accessed through trusted sources, as well as extended families, particularly children and grandchildren (Etchemendy *et al.*, 2011).

A further consideration is connected to cosmetic issues. Mestheneos (2012) notes that IT solutions in the form of alarm pendants and bracelets in dull grey colours are unlikely to be acceptable to many users and there is no reason why they cannot be made attractive and not just clinically functional. Additionally there is a pride in not wanting to accept that they now need aids, so disguising or mainstreaming them so they can be seen as normal is important.

Once technology is in place however, there does appear to be acceptance and some evidence of personal improvement. Within the INTERLINKS examples there appeared to be high levels of satisfaction with devices and increases in quality of life, obtained through independent research sources (Cordero, 2011; Hammar and Finne-Soveri, 2011; Mastroyiannakis, 2011b; Naiditch, 2011a). Evaluative data obtained from older people has however to be viewed with caution, given the tendency for those in receipt of care to overestimate personal benefits and create a false positive bias in results (Atwal and Caldwell, 2005).

In addition to this, reflections from the French Geoloc example (Naiditch, 2011a) introduce some important ethical issues that should be considered when implementing technology of this nature, particularly for patients who have reduced mental capacity. The author stresses that consent from the older person and his or her family should be obtained. Family members must be informed that the use of this device does not come without risk – this relates to carrying the risk of the older person's new-found freedom. Only families who are able to accept this risk will be able to benefit from such a system. This has important political implications: while more financial and practical support should be offered to develop such initiatives to support family carers and prevent institutionalisation, only providers who respect the above 'ethical code' should be supported. Selling and marketing the technology alone, without the necessary social support from neighbours and family, should be discouraged as without these prerequisites, there is a high risk for its inappropriate use in terms of control.

Informal carers: challenges and facilitators to the use of technology

Moving now to a greater consideration of the informal carer, it could be argued that their perspectives as users of technology could be similar to the older people's perspectives, in terms of knowledge gaps and potential prejudices. However, caution is needed as INTERLINKS work on informal carers has highlighted the dangers of assuming that their perspectives 'naturally' coincide with those of the older people they care for (Triantafillou *et al*, 2010). This was also noted in the CARICT project, funded by JRC-IPTS and DG CNECT of the European Commission, where impact assessment methodology was designed and employed to evaluate the effect of IT for the support of informal carers, necessitating not only an assessment of the effects of the technology on both end users (older person and informal carer), but also a clarification of whose needs were being addressed in the original design of the technological intervention. The CARICT project examined and evaluated 53 IT projects for the support of informal carers (including a subgroup of migrant care workers). This took place in 12 European countries, spanning 5 care regime models, from which 10 good practices were selected and supplemented by 2 North American examples and used for the development of an impact assessment methodology (IAM), based on identified dimensions, subdimensions and related indicators (Barbabella *et al.*, 2011).

Informal carers can undoubtedly benefit from IT solutions that have been designed to support them in improving their caring work, reducing the negative impact of caring and improving both informal carers' quality of life, as well as that of the older people they care for (Carretero *et al.*, 2012b).

Professional and administrative staff: challenges and facilitators to the use of technology

As mentioned at the beginning of this section, it is important to consider the perspective and needs of the professionals who are exposed to new technologies in addition to daily practice. In examples with more complicated equipment such as E-Health (Mastroyiannakis, 2011b) and AMICA (Cordero, 2011) concerning the monitoring of patients with COPD, considerable skills and competencies are necessary. While cooperation and acceptance of the technology are required on the part of the user and informal carer, the programme would not be successful without health care professionals' input regarding assessment, monitoring and measurement of vital signs, ensuring prompt responsiveness and timely information exchange, and understanding how the systems are generally defined and operated to ensure patient improvements.

Returning to the CARICT programme, some interesting points relate to the professional perspective. Inflexibility of service providers make for inefficient care provision, creating gaps in care and fragmentation of services, however the programme has established that IT can facilitate the adoption of a user-centred philosophy and more seamless provision of care. Timing is crucial in sharing information, making real-time online services a necessity for ensuring quality. Formal carers as well as informal carers should be included in IT training and should be dealt with as a group needing support. But user-friendliness of new IT applications is not on its own sufficiently incentivising to facilitate changes in professional practice and to overcome traditional ways of caregiving. Thus ongoing support and professional reflection on practice is needed alongside technological implementation.

Furthermore, new technologies demand new types of training outside the close professional boundaries of knowledge. New aspects of care including ergonomics or architecture of buildings, assistive technology, how to train older people and their carers in the use of ever changing high-tech gadgets and how to choose the most appropriate IT system to meet needs should be part of the new prerequisites. A health professional will need to change his or her practice to become more efficient, and with this comes the potential for resistance to change, particularly if resources are scarce and there is little managerial support and acknowledgement of the potential complexities involved. Such resistance was found within the Finnish example ADaL (Hammar and Finne-Soveri, 2011), where it came in the form of lack of money and insufficient time to get accustomed to the new technologies. In addition, evaluation revealed prejudices and objections to technological change. Conversely however, within other INTERLINKS examples, there mostly appeared to be a general acceptance of new technologies from health care staff, evidenced by increasing satisfaction, efficiency, ease of care, reduction of stress, and cost-effectiveness, and this is encouraging. An example of this is an E-training and certification programme, the Elderly Care Vocational

Skill Building and Certification (ECVC), which is a comprehensive national Greek project, based on practical and theoretical training materials for both formal and informal carers, developed, piloted and currently implemented in five countries and languages (Triantafillou, 2011).

Within this professional perspective one must not forget the administrative staff or the IT personnel who are responsible for the deployment, installation, creation, maintenance and change of the IT equipment. As implementation and technology acceptance will often depend on technical staff, it would be useful to also provide them with an introduction to care needs and to coping with older people in LTC in order to adjust their knowledge to a specific health or social care environment.

Involving users in the design and implementation of technology

These factors aside, the importance of user involvement in IT solution design must be considered, in order to maximise the opportunity of acceptance. It is now a prerequisite in proposals, and older people's participation should be included at all levels of a project, from planning, to implementation and importantly, as objective evaluators (HelpAge International 2009; see also Age Platform examples: www.age-platform.eu/en). Mestheneos (2012) states that too often older people are designated as users of a technology or beneficiaries of a policy over which they have had no control nor has their opinion been solicited. In asking for opinions from users it is important to involve their informal and formal carers to discuss their needs where appropriate. Making them more visible will increase awareness among policymakers about the need to design and deliver more tailored and carer-centred support services, as well as to increase their ability to respond efficiently and effectively to their specific needs (Schmidt *et al.*, 2011).

Although within INTERLINKS there were not many examples of user involvement in technology design and application (e.g. Naiditch, 2011b), there are several examples within the wider literature over the past decade that demonstrate how older people and their carers are being consulted and involved. Eisma *et al.* (2004) and Cassim (2007), for example, stress the importance of early involvement in the development of technology-related products for older people and relate the results of projects and case studies using workshops and group discussion for how designers and users can be mutually empowered within the design process. This is particularly important for mobile devices which can be intrusive and unattractive as previously discussed, and such technology was the focus of a project conducted successfully by Goodman *et al.* (2004).

While many projects focus on the involvement of older people and their carers only, a project conducted by Gil *et al.* (2010) concerning the development of a domestic well-being indicator system also involved physiotherapists, occupational therapists, researchers, technologists and a policymaker. The inclusive design process used workshops, brainstorming and focus groups. In addition, scenarios and paper-based prototypes were also incorporated and an interactive version was built to provide an early model for the design team and to provide a system which could then be refined. A final version of the system was tested with ten older people and ten carers, and results showed that the system was helpful to enhance the dialogue of care between older people and carers.

Newell *et al.* (2007), through their extensive research in this area, provide an important overview of discrete issues that must be taken into consideration when dealing with older people and offer some solutions to overcoming challenges. For example, the method of selecting a user group is important. Because of the diversity of the sensory, motor and cognitive characteristics of older people, as well as in their education and technical background, a representative sample of the user group will rarely be achieved. Thus a purposefully selected sample is permissible in the early stages of technological design, provided that wider testing is planned for (Lines and Hone, 2002).

A central consideration when including older disabled people is the quality of communication (Newell and Gregor, 2000). There can be significant problems between users and designers due to hearing deficits, and to lack of understanding of computer jargon (Eisma *et al.*, 2004). Visual conventions can also cause confusion – Newell *et al.* (2007) refer to scroll bars being an example of a whole repertoire of 'widgets' of which older people have limited or no experience.

Other studies also indicate problems when running focus groups with older people and give ideas as to how information can be elicited from this age group (Gheeraw and Lebbon, 2002; Lines and Hone, 2002). For example, older people tend to be very positive about the prototypes presented to them, feeling inclined to praise the developers rather than give an objective view, or blame themselves rather than poor design (Eisma *et al.*, 2004). Age-related factors can also make self-reporting inaccurate, in questionnaires for example, and difficulties may arise because older people tend to tire more quickly (Kayser-Jones and Koenig, 1994) and this can limit the length of sessions.

Added to this, Lines and Hone (2002) have found that it is not easy to keep a discussion group of older people centred on the subject at hand. They suggest that many informants see these groups as vehicles for socialising as well as providing information to the researchers, so it is important to provide a social gathering as part of the experience of working with IT researchers rather than treat them solely as participants. This aspect is endorsed by Eisma *et al.* (2004) who recognise significant benefits in treating collaboration in design and testing as a pleasurable occasion, which in turn can overcome the communication and psychological challenges previously mentioned. The authors stress that in order for meaningful involvement to take place from the start, both older people and researchers need to be willing and able to interact in language understandable to each other, and always to respect the other's contribution and expertise. This can be facilitated by making focus groups or other activities into social events, by providing refreshments and crucially, time for social engagement (Mival, 2002).

Newell *et al.* (2007) add that other methods to improve involvement include hands-on sessions where older people experience new technology, which have proved more successful than verbal explanations, and these can often lead to suggestions for improvements or for new products. Hands-on sessions also allow researchers to observe the difference between what people report and what actually happens (Mival, 2002). Newell *et al.*'s (2006) particular approach to technology design, experimentation and evaluation with older people has centred on the use of a studio theatre with actors and live performances, which can be used for very novel design briefs where an entirely new technology is being developed. This approach encourages open dialogue between the participants and the researchers, and the use of actors removes the ethical problems of 'protecting the users'. Actors can also present a more generic picture of a user and can change their personae in response to requests from the designers (e.g. What would happen if you were older or if your sight/hearing was impaired?). Finally theatre encourages a creative approach to design, involving users as well as designers, rather than the traditional view of focus groups and usability testing being the only method of eliciting users' views and opinions, and to determine their abilities to use specific interfaces and systems.

Conclusions: the future of IT in long-term care provision

In order to fully promote, gain investment for and integrate sustainable IT solutions into mainstream LTC activity, it is clear that there are a number of factors that must be in place. Fundamentally, the evidence base must be firmly established using rigorous research methods and reputable funding sources that are independent from industries seeking to promote their products. The literature review suggested that evidence is emerging; at present however, the quality of the evidence varies and its diversity in terms of technology used, patient group supported, and outcome measures assessed makes comparison highly problematic.

The transversal INTERLINKS analysis highlighted that although the number of elaborated examples appeared to be small, there were signs of additional IT activity within others to support the notion of an emerging yet growing industry, even though some activity is still in the pilot phase. Importantly, these examples provided additional perspectives of how IT can be of benefit, moving away from a specific disease orientation towards a focus on integration in LTC and the links between formal and informal care. For example, it would appear that the common element of having a call centre through which the older person obtains immediate assistance by the push of a button seems to be an effective way of mobilising IT solutions in a community set-ting. There are considerable benefits in terms of the use of professional time and reduced costs, as well as opportunities to enhance the mental well-being and independence of informal carers and the older people themselves.

The different types of funding, as well as the range of cost of the different examples, will also be influential in the sustainability and transferability of using IT solutions within LTC, and this remains unpredictable. It must be also be noted that transferability of an example may depend on a number of variables, such as the national or local characteristics, cultures or habits, and the surrounding health and welfare system or the existing suppor-tive services. In general, the evidence of effectiveness in many cases in the INTERLINKS examples was absent however, or not fully disclosed, there-fore the strength of the benefits mentioned needs to be viewed with some caution as comparability was not possible.

From a broader perspective, what seems to be clear from the innovations that are evolving and being presented at places such as international forums is that the research evidence seems to lag behind technological innovation, and care systems are slow to adopt new ways of working. What is needed is 'simultaneous innovation' where the system changes to keep pace with technology. The result of these difficulties means there is a lack of under-standing of the impact of new technologies and how to successfully apply and adopt them in practice (Goodwin, 2012).

Despite this lack of understanding and research evidence, it could be argued that it not only seems 'morally right' to offer such care, but that it is also the intelligent thing to do in terms of reducing the cost burden on the care system in the long term, as well as appearing to have an impact on user and carer independence, and their health and social well-being. This application must be coupled with a clear pathway to meaningful user involvement in the widest sense in the development and evaluation of initiatives, particularly with regard to issues of training, user-friendliness and technical support. There is the issue of a more democratic approach to participation that must be embraced (Beresford, 2003b; Newell *et al.*, 2007) since improving people's quality of life involves them making active, con-sidered reflection on their current and future needs and experiences. Their preferences need to be incorporated into current innovations and policy development and this in turn will change the social realities facing the different and emerging generations of older people (Vaarama *et al.*, 2008).

All this requires appropriate organisational change, and recourse to the literature provides us with some distinct pointers to what kinds of fac-tors must be taken into consideration for growth and sustainability when implementing IT solutions. Giordano *et al.* (2011) examined the progress and impact of telecare and telehealth interventions across 12 Whole System Demonstrator sites in the UK between 2008 and 2011. Four themes emerged as particularly important areas for consideration when adopting telehealth and telecare: leadership; working practices, skills and development; and data management. Key characteristics for growth and sustainability exhibited across the 12 sites included: collaboration within and across organisations; leadership; developing alliances and partnerships; identifying critical services; developing a shared vision; cultivating participation; building capacity; exploiting funding opportunities; and working across professional boundaries.

While these factors appear necessary to sustain and expand telecare and telehealth services, the authors state that they are insufficient on their own. Other areas that need to be addressed include: fostering fundamental service redesign; supporting professional development and staff training; analysing and designing the infrastructure prior to equipment being deployed; applying recognised standards; making decisions based on good interpretation of available data and evidence; and developing governance arrangements at national level to avoid regional variations in services.

Reflecting on these sentiments and from the 2012 International Congress on Telehealth and Telecare, Goodwin (2012) identified a number of key 'laws' that are more orientated towards the user perspective for successful telehealth adoption, most of which resonate with INTERLINKS findings:

- *keep it simple* for patients and carers to use, and for professionals to adopt
- *tailor the service* to the specific needs of the user; consider how they might best use and accept new technology
- *enhance human contact* by better connecting patients to family, friends and care professionals; users must feel safe, secure and empowered
- *embed an IT infrastructure* to act as the bedrock of better care through integrated information systems
- *build relationships and networks* to influence behaviours, build alliances, and overcome the significant mismatch of motives that exist between patients, carers, professionals, commissioners and industry.

As can be seen from this chapter, the development of assistive technology and digital training of older people is high on the agenda across the EU. Many countries are also committed to the goals of better health and greater participation of older people in society. Particular attention needs to be given to frail older people who already experience social and economic exclusion, by adding the dimension of technological exclusion to their existing vulnerability.

These analyses have demonstrated that IT continues to be in the implementation 'testing' phase in selected countries, with varying degrees of evidence. For now, it is necessary to concentrate on bridging the gaps between the use of new technologies and the needs and characteristics of older people, and firmly establishing the economic impact of its use to convince governments further of the benefits in widening social policy deployment tuned to these innovations.

13
Palliative Care within Long-Term Care Systems: Beyond End of Life?

Laura Holdsworth and Georg Ruppe

Introduction

Long-term care (LTC) is often discussed and presented in the context of prevention and rehabilitation, empowerment, dignity and self-autonomy and mechanisms that support holistic care. However, ultimately LTC for older people culminates with caring for people at the end of their lives. This entails that LTC in all its dimensions is regularly confronted with sensitive and sometimes even ethically precarious situations. A balanced mix and proper coordination of social, medical, formal and informal services as well as the integration of long-term, palliative and medical care concepts is of need in such situations. Decisions have to be taken regarding medical and psychosocial interventions, care approaches and appropriate support under different individual circumstances in the very last moments of a person's life.

The focus of this chapter is on the challenging relationship and interdependency of LTC and palliative care for older people in their very final stage of life. This phase of life is often characterised by chronic conditions and multimorbidities which may be exacerbated by acute episodes of ill health. This chapter will first discuss end-of-life and palliative care in the context of LTC including terminology, the historical context of palliative care and challenges faced in delivering palliative care to older people. This will set the scene for analysing the links between palliative and LTC within a European context. Various examples of initiatives across Europe that demonstrate good practice at the integration between palliative care and LTC will be presented including a critical commentary of their strengths and weaknesses. Finally, the chapter will argue that appropriate LTC for dependent older people necessarily incorporates the ideology and approach of good palliative care, but that LTC would benefit from further, explicit integration with palliative care to improve end-of-life experiences.

Terminology and the development of palliative care

There is a wide range of terminology used within and across countries to describe the care given at the end of life, including but not limited to: palliative care, end-of-life care, hospice care, continuing care, comfort care, and supportive care. Confusing the issue further is that there is no single accepted definition of the end-of-life period; it may span hours, days, weeks, months or even years (Radbruch and Payne, 2009). Many people with chronic or irreversible conditions may be in receipt of palliative care for years and, while they may be aware that they are life-limited, they may hardly consider themselves to be at the end of their lives. On the other hand, older people often recognise that they are in the final stage of their life (Janlöv *et al.*, 2005), although this period is often indeterminate and may refer to an entire decade or more of a person's life. Though older people are often treated palliatively for their chronic conditions, they are still not seen as a central target group for specialist palliative care as they approach the end of their lives (Hall *et al.*, 2011). It is not the aim of this chapter to define these terms, as this has been done elsewhere (see Radbruch and Payne, 2009; WHO, 2011), but rather to understand their complexity within the LTC system. Especially for increasingly dependent older people, who are still not widely recognised as a core target group for palliative care, long-term and palliative care are intrinsically intertwined and could be merged as approaches to care when considering appropriate care provision (Hallberg, 2006).

An important basis for linking palliative care and LTC is provided by the World Health Organization's (WHO) definition of palliative care (see Box 13.1), which describes a typology and philosophy of care which is relevant throughout the course of LTC and particularly towards the end of a person's life, younger or older.

Box 13.1   WHO's definition of palliative care

Palliative care is an approach that improves the quality of life of patients and their families facing the problems associated with life-threatening illness, through the prevention and relief of suffering by means of early identification and impeccable assessment and treatment of pain and other problems, physical, psychosocial and spiritual. Palliative care:

- provides relief from pain and other distressing symptoms;
- affirms life and regards dying as a normal process;
- intends neither to hasten or postpone death;
- integrates the psychological and spiritual aspects of patient care;
- offers a support system to help patients live as actively as possible until death;

- offers a support system to help the family cope during the patient's illness and in their own bereavement;
- uses a team approach to address the needs of patients and their families, including bereavement counselling, if indicated;
- will enhance quality of life, and may also positively influence the course of illness;
- is applicable early in the course of illness, in conjunction with other therapies that are intended to prolong life, such as chemotherapy or radiation therapy, and includes investigations needed to better understand and manage distressing clinical complications.

Source: WHO (2011).

The WHO's definition of palliative care illustrates a holistic system of care which is focused around the physical, emotional and spiritual needs of the patient and his or her family. The need for an integrated multidisciplinary approach to care is evident in order to fully achieve the aims of palliative care. Medical, psychosocial, nursing, social and spiritual professionals must work with families, informal carers and patients together to achieve the aims of palliative care. Indeed the definition of palliative care fits well with the WHO's definition of LTC; both emphasise the importance of quality of life through a multidisciplinary approach (see Box 13.2).

Box 13.2   WHO's definition of long-term care

LTC is the system of activities undertaken for people requiring care by informal caregivers (family, friends and/or neighbours) and/or professionals (health, social and others) to ensure that a person who is not fully capable of self-care can maintain the highest possible quality of life, according to his or her individual preferences, with the greatest possible degree of independence, autonomy, participation, personal fulfilment and human dignity. Important elements include, but are not limited to:

- maintenance of involvement in community, social and family life;
- environmental adaptations in housing and assistive devices to compensate for diminished function;
- assessment and evaluation of social and health care status, resulting in explicit care plans and follow-up by appropriate professionals and paraprofessionals;
- programmes to reduce disability or prevent further deterioration through risk reduction measures and quality assurance;
- care in an institutional or residential setting when necessary;

- provision for recognising and meeting spiritual, emotional and psychological needs;
- palliative care and bereavement support as necessary and appropriate;
- support for family, friends and other informal caregivers;
- supportive services and care provided by culturally sensitive professionals and paraprofessionals.

Source: WHO (2000).

The definition of palliative care describes an approach to care which is applicable across settings, including LTC and is included as an element within the WHO's definition of LTC. Palliative care has its origins in the 1960s when the nurse, physician and social worker Dame Cicely Saunders envisaged a new form of holistic care for the terminally ill and thus began the hospice movement (Council of Europe, 2003). Her endeavours for this alternative form of care were driven by her concerns about insufficient medical care for terminally ill cancer patients. Within this concept she integrated religious, spiritual, social and psychological aspects of care alongside symptom-oriented medical interventions (Saunders, 2006). The hospice movement quickly spread worldwide, underwent a series of amendments and regional adaptations and was, finally, also transformed into the concept of palliative care. The publication of the WHO's *Cancer Pain Relief and Palliative Care* (WHO, 1990) and the recognition that palliative care and curative medical interventions are not mutually exclusive represent important milestones in the development and establishment of palliative care (Pleschberger, 2010). Throughout its development, the focus of palliative care has always been on tailoring all forms of care to the individual needs of the patient and his or her family. This user-centred approach to care has messages which are applicable across care paradigms, including LTC.

Challenges in accessing palliative care and views on death

With its origins in cancer care, palliative care is still largely perceived as relating to the treatment of end-stage cancer (Kellehear, 1999). Indeed the WHO's definition of palliative care is found within the cancer section of its website. Despite this cancer context, it is defined broadly as applicable to all people with life-threatening illness. If we conceive of the very old to be life-limited, as old age inevitably threatens life though is not in itself an illness, then all people reaching old and very old age should be enabled to benefit from a palliative care approach. Certainly all those nearing the end of life, regardless of their health condition, would benefit from the psychosocial and spiritual support inherent in palliative care.

Palliative care services are most often initiated once a person enters a clear end phase of their life; a patient's condition becomes irreversible or terminal and thus measures seeking full remission or cure cease. Sometimes the 'surprise' question: 'Would you be surprised if this patient were to die in the next 6–12 months?' is used among health care professionals in order to determine the suitability of patients for palliative care services (GSF, 2012). This is more easily identifiable in cancer patients, but much less so for frail older people with chronic conditions who usually experience a gradual decline without an obvious stage for when such triggers for palliative care should be assessed. Additionally, people in this latter category may be perceived as 'users of LTC services', rather than 'patients', which may further hinder their access to specialised services as they may already be seen to be part of a system of care which is complex. Ageism among professionals may also be a barrier for referral to palliative care, particularly in systems which ration care, such as the UK (Gott *et al.*, 2011). These are largely medical views based on physical functioning as the boundaries of the end-of-life and palliative care services. Palliative care services are reactionary, responding to the needs of patients as they become 'terminally ill', with death constructed as a medical event.

However, an alternative view is to conceptualise dying as a social event which is a part of everyday life for which people should be prepared in order to be proactive in their own self-determination (Kellehear, 1999). Of course this is not to diminish the seriousness of death and that many will require intense medical intervention, but rather that it might be normalised as an event that everyone will experience. Kellehear argues for a concept of health promoting palliative care, in which the main objectives 'are to enhance a sense of control and support for those living with a serious life-threatening illness' (1999, p. 77). The core points of health promoting palliative care are presented in Box 13.3. Within this model, end-of-life issues and the principles of palliative care might be integrated within LTC settings and choices for end of life discussed as part of the processes that take place as the health of older people deteriorates.

Box 13.3 Health promoting palliative care

- provides information and education about health, dying and death;
- provides a social foundation for the idea of 'informed consent' when people are confronted with treatment decision-making;
- helps with interpersonal reorientation, to assist, facilitate and enable those living with chronic illness to adjust to life and lifestyle changes;
- encourages a reorientation of palliative care services;
- alters community attitudes to a wide range of ideas which have attracted, and continue to attract, negative imagery and discriminating behaviour;
- encourages a participatory process in which clients identify and direct their own information and education needs.

Source: Kellehear (1999).

However, a pervasive challenge in LTC, and indeed within wider society, is the taboo of death. While discussing end-of-life issues may be relatively standard practice among specialist palliative care nurses and doctors, in the wider society and even in geriatric care, end-of-life scenarios and dying may still be dealt with as taboo issues (Zimmermann, 2007; Bytheway, 1995; Aries, 1974). At the most basic level, medical professionals are trained and encouraged – by medical laws or professional ethics – to cure disease and prevent death. This ideology and professional principle may partly trace back to times when acute illnesses, such as infectious diseases, were more frequent and life threatening due to lack of treatment options and thus considered more perilous than the chronic and multiple diseases present in old and very old age. However, against medical paradigms the death of a patient may be regarded as a scientific and professional failure and dying in itself even as something inhumane. For these reasons professionals may find it difficult to raise the prospect of death with an older person as they may be fearful of being perceived as sending any kind of message to a patient that she or he may be dying (Dharmasena and Forbes, 2001). Older people themselves may find the issue taboo and may prefer not to discuss it, while at the same time fearing related issues such as becoming a burden to their family, losing their autonomy, or suffering from severe pain, which they may fear more than their own death (Ruppe, 2006). Much emphasis in LTC is put on preventative and rehabilitative care to prolong independence; relatively less emphasis seems to be on the acceptance of death as an outcome of a long and productive life even though palliative care should be a component of LTC (WHO, 2000). This possibly reflects anti-ageing views within society (Vincent, 2011) which contributes to the notion of death as a taboo and may be making it harder to identify opportunities to discuss the end of life. Thus while death is normalised within the academic and medical literature (e.g. Kellehear, 1999) and within the palliative care paradigm, the practice of discussing death is still a challenge.

Increasingly among palliative care professionals, advance care planning is being used for planning care at the end of life in a sensitive and practical manner. Advance care planning is a process of discussing with patients and/ or their representatives the goals and desired direction of the patient's care, particularly end-of-life care, in the event that the patient is or becomes incompetent to make decisions (MeSH Database). The content as well as the format of advance care planning can be variable and usually contains a statement of preference or refusal of future medical interventions, such as artificial nutrition or respiration, resuscitation, etc. Discussions and achievement of preferred place of care and death have also been given prominence within policy documents and initiatives (DH, 2008). Such statements may be oral or written declarations of will, may have different legal value depending on their form, and should be decided in consultation with a health professional who can provide additional advice on appropriate care options. Written declarations for such treatment are generally referred to as advance directives or living wills (Miles *et al.*, 1996). Advance care planning is particularly relevant to those in the early stages of dementia. Indeed, there is some evidence that engaging older people with dementia in advance care planning earlier on in their illness can increase the satisfaction with end-of-life care and improve the experience of family members (Birch and Draper, 2008). GPs and even social workers can initiate end-of-life discussions; it does not need to be led by palliative care specialists. Beginning end-of-life discussions earlier may also help to normalise end-of-life issues which may be of benefit to older people as they will have the option of expressing their wishes through advance care planning, thus giving them a degree of control at the end of their lives (Kellehear, 1999). Certainly in taking a health promoting palliative care approach, older people should be encouraged to participate in, or even initiate advance care planning discussions or advance directives even if they are not terminally ill. How the end-of-life and palliative care are handled is clearly complex and subject to social and cultural norms, interpersonal relationships, and the structure of health and social care systems.

Palliative care in European long-term care systems

Thus far palliative care has been presented as a holistic approach to care which addresses the full range of physical, social and spiritual needs that older people, and certainly those with LTC needs, may face towards the end of their lives. However, older people face a number of challenges in accessing palliative care, including professional bias, organisational barriers, and a societal taboo of discussing death. Possible solutions have also been discussed as methods of increasing older people's access to palliative care, including adopting a health promoting palliative care view which aims to normalise death, advance care planning to improve discussing end-of-life choices, and using triggers to identify people who may benefit from a palliative care approach. One of the most fundamental ways to increase access to palliative care is to improve its integration within LTC. Though palliative care is identified as a part of LTC, it is advised for use as 'appropriate and necessary' (WHO, 2000, p. 7). But as discussed, identifying for whom it is necessary and initiating end-of-life discussions is a challenge. It may be argued that in order to ensure that all those who might benefit from palliative care actually receive it, seamless integration of palliative care into LTC needs to be achieved. Therefore it may be useful to consider how countries within the INTERLINKS network have addressed the palliative care needs of LTC populations, and whether they have achieved seamless integration of palliative and long-term care, or whether palliative care continues to be treated as a separate sector called upon only when all other LTC options are exhausted.

The right to palliative care within long-term care

It may be most appropriate to start with an example that presents a view of palliative care as a right of people with LTC needs. The German Charter of Rights for People in Need of Long-term Care and Assistance was developed from a round table of some 200 experts from all areas of care in old age and is a detailed catalogue of rights covering eight distinct areas (see Box 13.4).

Box 13.4   German Charter of Rights for People in Need of Long-term Care and Assistance

The Charter is designed as a guideline for people with LTC needs, and institutions and organisations providing LTC, support and treatment. The Charter describes the rights of people with respect to eight areas:

1. Self-determination and support for self-help;
2. Physical and mental integrity, freedom and security;
3. Privacy;
4. Care, support and treatment;
5. Information, counselling, informed consent;
6. Communication, esteem and participation in society;
7. Religion, culture and beliefs;
8. Palliative support, dying and death.

In 2007, the Charter was disseminated to 27,000 care providers across Germany and is supported by a government funded agency which provides support and advice to health professionals and the public. The German Charter has been used as the basis for the development of the 'European Charter of the Rights and Responsibilities of Older People in Need of Long-Term Care'.

Source: Stiehr (2011a).

As with palliative care, to achieve good care in all eight areas requires multidisciplinary working. It also makes allowances that people should be able to express their preferences and choices and that these should be respected. This is particularly relevant to end-of-life care as people might have strong wishes for where and how their care should be carried out. The Charter places emphasis on older people having ownership of their care; this could be achieved through the use of advance care plans. However, ensuring this for all people is made difficult given that some older people may lack adequate mental capacity at the end of life due to dementia which becomes more prevalent as people age. The importance of advance directives are highlighted in the document, but are treated as optional. Treating such discussions about future care as optional acknowledges that not all people wish to discuss the possibility of death. While this is a legitimate preference which should be respected, it should be recognised that only through open discussion can people be enabled to have control and choice over their future care (Kellehear, 1999).

Users of LTC are the focus of the Charter rather than their informal carers. However, as in every stage of LTC, informal carers are an integral part of care at the end of life, but tensions can arise if family and/or informal carers have differing views on where and what care and interventions should be provided at the end of life. Additionally, carers may want more detailed information than patients about the dying process as death approaches (Parker *et al.*, 2007), and ensuring that carers have access to the information that they want, to reassure them in their caring role without compromising the wishes of the patient, may be difficult to manage. Within the definition of palliative care (WHO, 2011), informal carers and family are viewed as service users, but to what degree this affords them control over end-of-life events and their bereavement is unclear, as often care choices are patient centred rather than informal carer centred. Certainly the system of LTC, and indeed palliative care, is heavily reliant on the care provided by informal carers (Triantafillou *et al.*, 2010). The needs of informal carers and family are recognised within the WHO's (2011) definition of palliative care which 'offers a support system to help the family cope during the patient's illness and in their own bereavement'. The Charter thus falls short in this respect.

The Charter outlines values that should be followed in all care settings; however adhering to the Charter is not a legal requirement. In 2010, a Charter for the Critically Ill and Dying was created. This may be seen as an advancement of the rights and needs of people at the end of their lives, but consequently creating a separate Charter for the dying draws a line between what is seen as LTC and the end of life, though in reality these lines are often blurred within the frail elderly population. Patients can often receive different treatment types concurrently as is advocated by the WHO (Sepúlveda *et al.*, 2002), i.e. palliative and maintenance treatments, even up until death. Thus delineating where living ends and dying begins may not be clear to all those receiving care and their families. Waiting until the dying phase to take action may mean that opportunities to plan for future care will be missed. Additionally, such kind of separation necessarily supports an attitude and thinking which does not regard a phase of dying and death as an integrated part of human life.

Dying at home, rather than in institutional settings

The modern European conception of palliative care policy and services, as with much of the LTC sector, is largely to care for older people in their own homes in the community, rather than in institutional settings. Within a pilot project in the region of Umbria in central Italy, protected hospital discharge and palliative care teams (Di Santo, 2011a; see Box 13.5) are able to securely transfer older people from hospital to home, hospice or other facilities for end-of-life care. The teams aim to coordinate care with community-based teams for continuous care as well as prevent inappropriate admissions to hospital, reduce aggressive treatments, improve pain control, and support informal carers and families. Clearly for older people this means that they can be cared for in a preferred or more appropriate setting rather than what is often a chaotic hospital environment. However, currently the palliative teams in this region organise standard community care for end-of-life care at home and paid carers from these services may not have any kind of palliative training and, thus, insufficient skills to care for dying patients. Basic skills in communication or coping strategies might be lacking as well as sufficient team support by specialists. This is important as, for example, pain management provided by palliative specialists is well evidenced to be superior to that provided by GPs (Mitchell, 2002). The Umbrian pilot project demonstrates that care organised this way can be more cost-effective as the location of care is transferred from a costly hospital environment to the home setting, but it seems both necessary and meaningful to supplement the respective home care teams with sufficient palliative care specialists and to provide basic palliative training to all formal carers in such teams. Also, multidisciplinary working between specialised palliative care teams and existing community services can help spread the ethos of palliative care to other services.

Box 13.5 Protected hospital discharge and palliative care teams (Italy)

This pilot project in Umbria, Italy is run by a local health authority involving a variety of care providers and organisations including hospitals, hospices and home care services. These organisations have set up a small, multiprofessional team to facilitate an integrated care pathway to protect the discharge process for terminally ill patients so they can have a smooth transition from hospital to home. The team consists of a palliative care doctor, the hospice's head nurse, a hospital doctor, the hospital's head nurse, and if necessary the hospice's social worker and psychologist. The team meets once a week in hospital to assess patients and plan for care at home or in a hospice. The service has achieved:

• Lower rates of calls to emergency services and hospitalisation in the last 48 hours of life;
• Shorter length of stay in hospital;
• Increased use of opiate drugs;
• Support for families in managing the patient's illness and bereavement.

Source: Di Santo (2011a).

Likewise in Austria, mobile palliative teams (Ruppe, 2011; see Box 13.6) provide support to people with chronic and terminal illnesses living in the community. They can also support primary care and LTC professionals in issues relating to palliative care, thus potentially improving the palliative care delivered by GPs (Mitchell, 2002). As in the above example, older people are enabled to express preferences for care to be provided at home rather than in institutional settings. However, while the focus of 'palliative care teams' in the Italian example is mainly on the coordination and continuity of care between hospitals, traditional home care services, hospices and other specialists, the Austrian mobile palliative teams additionally represent an autonomous entity and team directly providing multiprofessional care in the form of pain management, nursing care, counselling, small medical interventions, social work, etc. Mobile palliative teams consist of medical doctors and nurses trained in palliative care, physio, psycho, speech and occupational therapists, as well as dieticians and chaplains. The multidisciplinary team thus collaborates and coordinates their interventions with the older person's existing formal and informal network. The Austrian mobile palliative teams are supported by voluntary work and administrative staff and provide an example of how specialists from different professions can deliver appropriate care directly to people in their own homes. However, these teams are often positioned in close proximity or even organisational affiliation to hospitals or large social care providers (e.g. non-profit organisations (NPOs) such as Caritas or Red Cross). Some mobile palliative teams also act as palliative consultant teams for the affiliated hospital, providing specialist support and counselling in difficult cases and filling respective gaps as needed. Often these gaps pertain to knowledge about pain management and medication and, in particular finding solutions to achieve a good quality of life in home settings where pharmacotherapy can be difficult to manage.

Box 13.6   Mobile palliative teams (Austria)

Services delivered by Mobile Palliative Teams (MPTs) can be taken up by people with chronic and terminal illnesses living at home as well as by professionals, who work in mobile or residential LTC services and need support with issues of palliative care or treatment. Composed of a multiprofessional group (always consisting of medical doctors, nurses, physio, psycho, speech and occupational therapists, as well as dieticians and chaplains), an MPT collaborates and coordinates interventions with the client's existing formal and informal network. In Austria, it was estimated in 2010 that one MPT was needed for every 140,000 inhabitants (approximately 57 MPTs for all of Austria).

Together with the registered clients and their caregivers, MPTs arrange individually adapted plans for home visits including all kinds of specific services provided by the multiprofessional teams. Telephone counselling and acute (medical) care are available over a 24-hour period. Additionally, support to coordinate admission to or discharge from hospitals as well as transfer to hospices is offered by MPTs.

MPTs represent a promising role model for mobile community care services for frail older people in general. The multiprofessional composition of MPTs as well as the individuality and flexibility of provided services appear as important success factors.

Source: Ruppe (2011).

The concept of providing care direct to patients at home is not new however, as in France, hospital at home (Com-Ruelle and Naiditch 2011; see Box 13.7) has been provided for over 50 years. Initially this was implemented to reduce the number of cancer 'bed-blockers', but this now extends to all patients who can be cared for at home to avoid or delay hospitalisation. Hospital at home is able to deliver complex hospital care to people with co-morbidities in their own home, something which many home care providers cannot do. Palliative care at home is one facet of hospital at home, though as it started with cancer care then perhaps it was initially conceived to achieve similar aims to the palliative care teams already discussed. Hospital at home is only organised for patients through a contractual agreement with a GP, which may serve to prevent duplication in care as there is a single line of care management. However, hospitals are not generally thought of as ideal places for end of life, in part because of the generalist skill mix of professionals which in this scenario would also be providing care at home. Thus perhaps the Austrian mobile palliative care teams offer a better at-home solution as palliative care encompasses more than just medical pain management; it is an holistic treatment of problems; physical, psychosocial and spiritual (Sepúlveda *et al.*, 2002).

Box 13.7   Hospital at home for older people (France)

The aim of hospital at home is to shorten the length of stay in hospital and to avoid or delay hospitalisation in acute or rehabilitation wards. Hospital at home for older people targets older people who have serious acute or chronic progressive conditions which require intensive or complex medical, nursing or social care. Hospital at home operates as independent not-for-profit organisations or as autonomous departments within hospital settings. Care delivered at home by hospital at home teams can only be implemented with agreement from the patient and their family and with a contractual arrangement with the patient's GP. Hospital care at home is supervised by a physician, implemented by a coordinating nurse, and the team social worker arranges access to cash benefits or in-kind services. The coordinating team can be reached 24 hours a day, and the patient is visited at least once a week by members of the team and the GP who assess the patient's condition. The care team provides: information to patients to comply with treatments, training for families in specific caring tasks and respite care for families. Hospital at home primarily addresses palliative care needs and cares for those undergoing cancer treatment. Research reports that satisfaction levels are high among users of the hospital at home service.

Sources: Com-Ruelle and Naiditch 2011; Com-Ruelle and Raffy (1994).

Supporting family carers at their loved one's end of life

Apart from different kinds of professional mobile and institutional services, caring for older and chronically ill people at the end of life still remains to a large extent the duty of family members and informal carers. Care at the end of life is demanding physically and emotionally and is usually a full-time responsibility. For working age people, it can be a difficult decision to choose between continuing to work and looking after a loved one. The German example of the Care Leave Act (Mingot, 2011; see Box 13.8) offers carers an opportunity to care for their loved ones with assurances of maintaining employment. The legislation allows a family member to take time off work, up to ten days in an emergency or up to six months with ten days' notice, to care for a needy relative. Employees must give their employers notice, but cannot be dismissed for taking this time off. Though it is not specific to end-of-life care, it may be of great benefit to carers of people dying from chronic or terminal illnesses as such high intensity of caring is usually limited to days, weeks or months. Caring for older people as they become more and more frail may take up more time, but it is not likely to be constrained to only ten days or even six months. Care for people with long-term conditions may not be adequately addressed by the Care Leave Act. The legislation does imply the value of family caring and that dying relatives may want their families to care for them. However, family carers who take time out of work have to make their own arrangements for health insurance during this period. It is helpful that informal carers can have ten days off in case of emergency, but resolving a crisis within this period so that the carer

Box 13.8 Care Leave Act (Germany)

The Care Leave Act was created to protect employees from dismissal in situations where they need to take a leave of absence from working to care for a relative. Family members can use one of two options, depending on the situation:

- In cases of emergency, carers can take leave for up to ten days to organise care or to care themselves for a relative. They are not paid during this period, unless they can claim a benefit under other legal regulations and no social insurance contributions occur during this period. There is no notice period for taking this time.
- In cases where care needs can be anticipated, employees can take up to six months' leave or partial leave from work, but they must give their employer a minimum of ten days' notice. They are not paid during this period and no social insurance contributions occur during this period. Companies with fewer than 15 employees are exempt. Care needs must be assessed and classified into one of three categories as set out in legislation.

There are no regulations about how often a person can use the right to care leave. The Care Leave Act is not widely known, therefore many family carers who could benefit may not do so.

Source: Mingot (2011).

can return to work will be dependent on the speed with which community services can respond to the person's needs. Because carers can be enabled to keep their jobs, it may make the transition from caring back to work easier.

Palliative care in institutional settings

While these examples address the palliative needs of people living at home in the community, older people living in countries with a strong institution-based LTC system (e.g. Austria, Finland) will have many of the same palliative needs which may not be met in nursing or residential care homes (Reynolds *et al.*, 2002). This issue has been addressed in certain regions of Austria through 'hospice and palliative care in nursing homes' (Beyer *et al.*, 2011; see Box 13.9). A similar programme exists in the UK with the Gold Standards Framework (GSF, 2012). Both programmes have achieved greater levels of nursing home staff confidence in caring for people at the end of their lives and achieve better standards of palliative care as viewed by residents and family members (Badger *et al.*, 2009). However, uptake of the GSF is voluntary and in many cases nursing homes that participate already have a basic knowledge of palliative care and therefore see the value in increasing their skills. Thus nursing homes which lack a palliative care ethos are unlikely to adopt the system, furthering the divide between those that 'do' palliative care and those that do not.

Box 13.9   Hospice and palliative care in nursing homes (Austria)

Hospice Austria, which represents 250 hospice and palliative care organisations in Austria, has developed guidelines for the delivery of high quality hospice and palliative care in nursing homes. There are three steps recommended for the implementation process of palliative care in nursing homes:

- A nursing home staff member is appointed as the palliative care representative for the home and is required to have basic training and education in palliative care to guide palliative care provision among the nursing home team;
- The organisation of the nursing home must meet national standards which may involve multidisciplinary coordination with outside professionals;
- Palliative care provision in the nursing home must be made sustainable through organisational structures.

Source: Beyer *et al.* (2011).

Palliative and long-term care integration – possible pathways for the future

It has been argued so far that seamless integration between palliative care and LTC is necessary. In many of the examples discussed, there are common themes of multidisciplinary working, holistic and person-centred care guiding working arrangements, and integration of family into the care of older people. These are also the tenets of LTC and in this way palliative care represents how LTC might be conceptualised – and the other way round. While there are clearly a number of links between these models of care, there are number of ideas from within palliative care that are applicable within LTC to achieve a better life experience until the very end.

One of the key messages from palliative care approaches for LTC is the normalisation of dying. While it may seem contradictory, normalisation and acceptance of death may enhance quality of life, particularly for those nearing the end of their lives. Froggatt *et al.* (2011) found that residents dying in a care home setting were often separated from the living which reduced their life world. The authors state that this has a negative impact on the experience of those dying as well as those still living in the care home: 'separation of living and dying residents can lead to isolation and an ending of relationships with fellow residents, hence creating particular social experiences for those people dying and also expectations in those residents yet to be in this state' (Froggatt *et al.*, 2011, pp. 265). The consequence of isolation in dying is in direct contrast to many LTC programmes which aim to improve older people's engagement in community, such as the Neighbourhood Solidarity project in Switzerland (Repetti, 2011). Dying is indeed a personal experience which some people may prefer to experience in private. However, this should be the decision of the person dying, rather than the decision of carers and health professionals, as may be the case in institutional settings and possibly reflect society's unease with death. Advance care planning and initiatives such as the Charter of Rights for People in Need of Long-term Care (Stiehr, 2011a) may be ways to ensure that people are given choice and control over how their dying experience is handled. The Charter states that self-determination and participation in society are rights of older people – why should this end for those who are in a dying phase? In this case, LTC would benefit from the values of palliative care, in which dying is regarded as a normal process where people should be respected as individuals who are still a part of a community.

Safety and security are also key components of palliative care services. This is particularly evident in the example on protected hospital discharge; the potential instability at the end of life is emphasised within the service name in that the pathway of care must be 'protected'. Dying is a frightening prospect for most people, but particularly for those with terminal illnesses who experience pain and other distressing symptoms. Specialist palliative care services are shown to be more effective in managing such symptoms compared to conventional care and can increase patient and informal carer satisfaction (Hearn and Higginson, 1998). However, evidence from a systematic review shows that older people are less likely to be referred to or access specialist palliative care services (Burt and Raine, 2006). Burt and Raine (2006) argue that this possibly reflects less of an inequity in service provision and more of unequal need for these services from older people as their needs are possibly being met by other LTC services. Many examples illustrating the INTERLINKS framework aim to address the isolation and insecurity that older people face living in the community (see examples by Overmars-Marx, 2011; Huijbers, 2011; Naiditch, 2011b), which may be sufficient for many people dying in the community. Feelings of security are influential in shaping people's preferences for where and how they want to spend the end of their lives (Thomas *et al.*, 2004).

The examples of palliative care services presented in this chapter largely appear medically orientated, as they are often organised from within clinical settings, such as hospitals or hospices. Access to palliative care services is also generally predicated on knowing that a person's life is limited, again reflecting a clinical rather than social need. Prognostication of death using the 'surprise' question remains a challenge, particularly for frail older people and people with dementia, as it may not be surprising if they died within the next year, nor would it be a surprise if they did not. Taking a health promoting palliative care approach (Kellehear, 1999) may negate the necessity of determining service accessibility based on prognosis which thus creates boundaries between living and dying. Specialist palliative care services, such as mobile palliative teams and hospital at home, are scarce resources, so it is not suggested that all people have access to such services if not needed, but rather that the values of holistic care which regards dying as a part of life form a natural progression within various stages of LTC, rather than being bolted onto 'the end'. Indeed what may be necessary is a whole systems rethinking, or possibly simply education among health and social care professionals and users, of the role of palliative care, which is often viewed as cancer related, towards the management of long-term conditions. A vision of seamless integration of palliative care into LTC programmes may help to increase the number of users who will benefit from a palliative care approach in the final phase of their life.

Conclusions

This chapter has set out to discuss the interrelationship of palliative and LTC concepts for old and very old people at their final stage of life. In depicting a rough snapshot of European palliative care and its particular relationship to developments in LTC for older people, a remarkable approximation of the two concepts and a number of similarities underlying these ideologies become obvious. Evident in good examples of both are multidimensional care, patient-centredness, and the integration of services and sectors. Indeed, Hallberg (2006) states simply that 'the concept of palliative care is above all based on the idea of reinforcing factors that improve quality of life, and decreasing the impact of factors that may reduce quality of life, which seems adequate as a concept for the long-term care of older people' (p. 226). However, a seamless integration of both concepts in everyday practice of comprehensive LTC provision for older people remains a major challenge.

Emerging as an alternative form of care for terminal cancer patients, palliative care still seems to be widely labelled as alternative care which comes into play only when other approaches and services are exhausted. Different forms of palliative care provision usually enter the care scene as soon as the individual has crossed a certain borderline where it is clinically or individually accepted that death is inevitable and medical interventions can at most delay death, but not prevent it. Such demarcation might be more easily defined among certain older people with specific types of illnesses, bringing us back again – probably not accidentally – to the group of terminal cancer patients. However, when talking about older people suffering from multimorbidities and frailty, this borderline is much more difficult to determine, if not a bit arbitrary in some cases, and yet palliative care interventions are still very appropriate (Boockvar and Meier, 2006).

In spite of endeavours by the WHO (Hall *et al.*, 2011; WHO, 2004; 2011) and the health and social sciences (see for example Arnold and Jaffe, 2007; Ruppe and Heller, 2007; Pleschberger, 2010) to continuously widen the scope of palliative care as a holistic and integrated concept, which 'affirms life and regards dying as a normal process' (WHO, 2011), it remains a difficult task to achieve or implement such kind of *normality* in practice.

In briefly analysing and discussing selected practice examples from different European countries, this chapter has demonstrated that a number of positive initiatives for collaboration between long-term and palliative care services for older people have emerged during the last years. However, it turns out that a seamless transition from one to the other or even a full integration of both approaches into a kind of natural continuity of services for older people is more difficult and less likely to be achieved. A reason for this might be that a necessary but difficult precondition for such kind of integrated service provision lies in the integration of human death into human life at the individual level. Accepting death as a part of life is probably one of the greatest challenges for humankind, let alone health and social care professionals. It would mean that an older person, and perhaps more importantly medical professionals, would need to alter their view of life as either *living* or *dying* towards a (more natural) transition from *life* to *end of life*.

In spite of this major challenge of integrating death and related precautious arrangements into human life, a number of valuable initiatives to bridge this gap and to integrate dying at least as a topic to be touched upon, have been made. Advance care planning, integration of palliative knowledge and specialists into multiprofessional LTC services, and a seamless inter-organisational coordination of services are some examples of approaches which allow for more palliative care expertise, preparedness and awareness for borderline situations in old age and for older people, who feel more or less close to the end-of-life phase or an acceptance of such a phase.

Developments of holistic palliative care as well as related LTC concepts have contributed to help make individual and professional discussions concerning end of life much less a taboo than a few years ago. However, we might also need to accept that it is not necessarily human or natural to openly plan our own death. Attitudes towards this topic might vary significantly at the individual level, and integrated palliative and LTC for older people will need to adapt respective approaches to best reflect this individual process and challenge.

Part IV
Conclusions

14
Improving the Evidence Base

Jenny Billings

Introduction

One characteristic of emerging practices in long-term care (LTC) for older people is the lack of evidence concerning specific interventions that have been applied. While evidence-based medicine has started to become a mainstream issue (Long *et al.*, 2006; Moriarty *et al.*, 2007), practitioners and policymakers in the realm of LTC seem to lack guidance and general openness for this kind of working. This is in part due to the still-emerging nature of LTC as a discrete system, but also due to the tensions that exist in how evidence is understood and used for both health and social care practice and policy (Klein, 2000).

This chapter will pursue these themes by briefly outlining the challenges and opportunities of more evidence-based approaches in health and social care evaluation and providing an account of and rationale for a more pluralist approach to gathering evidence in LTC, as adopted within the INTERLINKS project (see the Chapter 1 for elaboration of project approach and methods). The chapter will follow with a critical transversal analysis of those practice examples that reported project effects, providing an account of how they were evaluated, what sources of evidence were used, the project effects on users, carers and organisations, followed by an interpretation and critique of the extent to which INTERLINKS examples have contributed towards the evidence in LTC debates.

Evidence: The current situation

Evidence-based practice has now been recognised and embedded into health and social care policy and practice across Europe to varying degrees, with the understanding that with this label comes a set of assumptions about what constitutes evidence. The etymology of the word 'evidence' is rooted in the concept of 'experience', relating to what is manifest and obvious (Upshur, 2001) and there has been considerable debate on its epistemological derivation.

This discourse invariably embraces various permutations including not only evidence-based practice, but also evidence-based medicine, decision-making, and policymaking. The integrity of such concepts is invariably questioned within these discourses which does not seem to progress thinking in a manner that is practical (French, 2002; Olsson, 2007).

Despite this, evidence-based practice in the wider health and social care arena is generally understood as delivering care on information about what works best (Rycroft-Malone *et al.*, 2004). Further to this, through sources such as the Cochrane Collaboration, evidence and knowledge have come to be viewed as hierarchical and primarily objective, and as a consequence, this places randomised controlled trials and systematic reviews above evidence drawn from more qualitative sources reporting more in-depth levels of perceptions and experiences (www.cochrane.org; Scherer, 2009). This accepted view has been largely brought about by the interplay between the dominance of the medical academic field in determining what constitutes evidence and by what means, and clear policy imperatives (Glasby and Beresford, 2006). The promotion of this view of evidence has been powerful, and over recent years there has been a concentration upon and consequent financial investment in getting quantitative research evidence produced, synthesised, disseminated and used in practice (Stevens and Ledbetter, 2000).

For a while now, the need for a broader view of what constitutes evidence is being vocalised by health and social care academics, as traditional views of evidence do not account for the complexities of policy generation and practice needs (Glasby and Beresford, 2006; Rycroft-Malone *et al.*, 2004). Moriarty *et al.* (2007), for example, argue that policymakers in particular have always needed to draw upon diverse sources of information when making decisions about complex health and social care issues. This includes not only evidence from research, but also other more qualitative evidence stemming from value judgements such as public preferences for a particular intervention or approach to care (Mays *et al.*, 2005). Scientific evidence therefore is just one of the strands drawn upon for legitimate decisions about policy and implementation processes.

There are many aspects of health and social care where the research evidence for decision-making remains limited. In these cases, traditional definitions regarding what constitutes evidence is extended to include 'grey literature' in the form of small-scale research or policy reports and personal accounts. The prominence ascribed to the more 'hierarchical' view has however meant the relative neglect of other forms of evidence in terms of making them available for critical scrutiny and public review. More qualitative forms of evidence, for example, continue to be excluded from the majority of electronic databases and viewed as untrustworthy, even though their use may be high. Moriarty *et al.* (2007) even hint that these sources may be unwittingly or even deliberately excluded from many conventional systematic or narrative reviews of the literature.

Murray *et al.* (2009) extend these latter sentiments by arguing that there is an even more insidious explanation for the promotion of evidence as a form of hierarchy. Drawing on the philosophy of Foucault and Deleuze, the authors interrogate the constitution of evidence as it is defined in the evidence-based movement and question under what social and political conditions scientific knowledge appears to be 'true'. They draw on Foucault's thinking which describes these conditions as a 'state science', being a regime that privileges economic modes of governance and efficiency. Murray *et al.* state that the Cochrane taxonomy and research database are increasingly endorsed by government and public health policymakers who are support-ive of the evidence-based paradigm in promoting the 'noble' idea of true knowledge, free from bias of any sort. The authors argue that this apparent neutrality and objectivity is dangerous as it masks the methods by which power silently operates to inscribe rigid norms and ensure political domi-nance by, for instance, defining expert standards. Murray *et al.* urge scholars to use their positions to oppose such regimes and expose what hides behind the distortion and misrepresentation of 'evidence'.

While these arguments may lean towards the reactionary, the literature would indicate that academics have been responding. Prior to this 'rally call' there have been definite signs of a more eclectic understanding and acceptance of what constitutes evidence. For example, researchers have dem-onstrated that policy, practice and operational guidelines such as standards of care can be generated from a range of data and indeed, would be incom-plete without drawing upon diverse sources (Long *et al.*, 2006; Moriarty *et al.*, 2007). There is a consequent move towards recognising a 'pluralist' view that emphasises the importance of triangulation of data and recognises a degree of equality between data sources and the approaches used from which they were derived. Authors such as Rycroft-Malone *et al.* (2004), Glasby and Beresford (2006) and Murray *et al.* (2009) centre their arguments on concepts of objectivity and knowledge as being key when considering evidence.

While it can be argued that objectivity and knowledge are linked, in the following, the first consideration focuses more on the issue of objectivity. The second consideration takes up and expands on the issue of knowledge, with a particular focus on *practical knowledge*, which emanates from health and social care professionals, and *service-user knowledge*, which is derived from lay experiences.

Objectivity and evidence

Research evidence tends to be seen as providing watertight answers to questions posed, however as new processes and technological advances for practice come to light, it is becoming increasingly evident that evidence is seldom definitive and can be changed as new research emerges. Upshur (2001) states that research evidence needs to be viewed as provisional and evolving, particularly the evidence base for practice.

Rycroft-Malone *et al.* (2004) discuss an important objectivity 'paradox' concerning the relationship between the production of research and its use. They state that, while the producers of research attempt to achieve a level of objectivity, the use of evidence is embedded within a social as well as scientific process. The authors cite research undertaken by Dopson *et al.* (2002), which involved a cross-case comparison and synthesis of seven evidence into practice studies in health-related research concerned with professional nursing practice. A theme emerging from the secondary analysis related to the competing levels of evidence, which were capable of engendering a number of different interpretations, despite the fact that the individual studies looked at precise scientific topics amenable to scientific testing. Dopson *et al.* concluded that evidence is dynamic, eclectic and less value-free than is generally acknowledged.

Leading on from this, objectivity becomes even more removed when considering the potential for further differing interpretations from a multi-stakeholder perspective. This is even more so when it comes to LTC in terms of needs assessment and measuring outcomes for example, an aspect elaborated upon later. Whether for policy, practice or organisational requirements, evidence can potentially be 'flexed' to support a decision, depending upon the rationale needed to support that decision.

A further critical viewpoint is put forward by Glasby and Beresford (2006), who argue that the values of objectivity, neutrality and distance should now be seen as only one strand of thought among a variety of emergent approaches to research and the production of knowledge. Glasby and Beresford use the examples of the feminist, mental health and disability movements, who share a common view in rejecting the positivist assumptions behind objectivity by questioning the nature of the relationship between the researcher and the participants. It is argued that being less 'distanced' from participants achieves significant gains in research by eliciting fuller, more meaningful responses, and this knowledge has certainly raised our political understanding of the human and civil rights of service users.

From this position, the concept of neutrality in research can be challenged, so that what is traditionally seen as a merit may actually be a deficiency (Beresford, 2003a). Being distant from what is being investigated and interpreted, for example, can lead to a distortion and misunderstanding of the phenomenon under study. Glasby and Beresford emphasise that being 'closer' to the research issue is vital and indeed beneficial in generating knowledge that changes policy and practice in a relevant way. Importantly, it follows that an acceptance of this perspective negates the need for a 'hierarchy' of evidence, and the adoption of a pluralist approach is therefore more fitting.

Knowledge and evidence

In relation to practical knowledge, over the years, a number of academics have explored the nature of knowledge, and have substantiated the important contribution of different sources of knowledge beyond the technical and propositional (e.g. Benner, 1984; Higgs and Jones, 2000). Commentators argue that tacit and experiential forms of knowledge are persuasive and have a reciprocal mutually enhancing relationship with forms of research that are considered more scientific (Rycroft-Malone *et al.*, 2004).

Despite this, there is still an assumption in the field of evidence-based care that such sources of knowledge are subject to bias and lack credibility. However, those supporting the value of practical knowledge argue that it is intuitive, embedded in everyday actions, often draws upon the experience of others and in addition is frequently verified by others. As such therefore, it becomes an evidence base in itself (Glasby *et al.*, 2007; Rycroft-Malone *et al.*, 2004). Harnessing this tacit knowledge through exploratory research processes has contributed significantly to professional understanding (Grypdonck, 2006). This is particularly evident through the publication of case studies of good practice.

With reference now to service-user knowledge, Barker (2000) acknow-ledges that 'good practice' throughout an organisation cannot be separated from the unpredictable ways in which individuals and their families respond to concepts of health, illness and their social circumstances. The personal knowledge, experiences and preferences of clients and carers should there-fore be central components in the practice of evidence-based care. Over the past decade there has been a considerable international growth of published qualitative research that has contributed towards our understanding of the wide variety of user experience in many different areas of health and social care (see Leys, 2003).

In addition, there has been an increase in user involvement at policy level, and in collaboration with organisations that produce guidelines such as the National Service Framework for Older People in the UK (Moriarty *et al.*, 2007) and the Big Care Debate (HM Government, 2009). Alongside this, there has been a parallel growth in user-led research projects. Given that there is an increasing recognition that the actual progression of applying and developing research findings to practice is complex, the user knowledge gained from exploratory methods provides vital detail that can reveal why certain processes do and do not work in practice (Grypdonck, 2006).

Scepticism of the scientific merit of this knowledge as being worthy of use for evidence-based care has already been discussed, however others would argue that there is a definite shift towards its acceptance alongside a range of other data (Rycroft-Malone *et al.*, 2004). Pencheon (2005) suggests that the accepted meaning has changed from the use of best available evidence (preferably from peer-reviewed sources), to one that acknowledges the judicious mixture of evidence incorporating the experiential. He notes that this has resulted in a more healthy and critical approach to decisions and actions.

The case of long-term care

The discussion so far has focused on the relationship between evidence, policy and practice in its broadest sense, but it is important now to focus on how these discussions translate to LTC where there are a number of challenges. With the European move towards addressing LTC services for older people, there is a corresponding need for evidence of effectiveness upon which to base the newly developing services and enhance effectiveness of the more established ones. Over the last two decades there have been a number of studies (see Johri *et al.*, 2003; Ovretveit, 2011) that have sought to demonstrate the benefits of various approaches; however, the evidence agenda is distracted by a series of debates and tensions.

Taking the example of integrated care, it could be argued that such a service provision should benefit people at home; for a number of years, institutional care has been known to have a detrimental effect on health outcomes for frail older people (Challis *et al.*, 1995). Yet proving this perceived beneficial effect has been testing. Indeed it has become more commonplace to highlight that the evidence base for the beneficiaries is fragmentary (Cameron *et al.*, 2000; Dowling *et al.*, 2004; Ramsay and Fulop, 2008).

For example, Johri *et al.* (2003) undertook an international synthesis of the evidence for integrated care, choosing projects that used experiments. While the authors concluded that some frameworks were capable of providing more effective care for older people, they warned that overall the results should be viewed with caution. In general, the review demonstrated that certain frameworks of LTC could be cost-effective, but there was an issue with extension beyond the experimental phase and there were numerous problems with artefact and design effects. These anomalies are compounded by the fact that 'integrated care' is still often reduced to integration within the health care system, and therefore does not truly reflect a unified approach to care through health and social services (Singer *et al.*, 2011).

These conclusions typify a form of research that seeks to generate evidence for LTC outcomes. An experimental approach cannot always identify why aspects seem to work as they are unable to reveal important process variables. This means that, although the primary function of such research may not be to discover these processes, the transferability of evidence becomes difficult to establish when these are not made evident. Many 'successful' experiments fail to transfer to much larger populations of frail older people, or those with different cultural needs. Findings are often inconclusive and readers are often left not understanding what was at fault; possibly the design, the intervention, the context or the measures used.

With the latter, difficulties with measurement are inherent within some of the tools used for evaluation purposes. To gain evidence of effectiveness, researchers use an armoury of instruments to measure status, social and functional characteristics, all varying in complexity and reliability. Quality of life indicators, for example, measure a snapshot view that, when sample sizes are not adequate, can vary in people with multi-pathology and not necessarily relate to professional care received. Not only are there challenges in capturing reliable responses among the target group with these kinds of measures, but crucially, even with the most intensive input, one cannot expect significant changes in functional status among a population that is slowly declining anyway.

Added to this, evidence obtained through trials can act as a deterrent to innovative practice, where aspects may not fit into the established 'proof'. All the while that uniquely quantitative designs are relied upon, the evidence for LTC will remain elusive and incomplete. Therefore it could be argued that a wider spectrum of evidence including qualitative sources needs to be embraced in order to be able to develop methods that can fully describe and analyse the multifaceted and complex system that is LTC.

Cheetham *et al.* (1992) and Petch *et al.* (2005) add weight to this argument by stressing the importance of a multi-perspective approach to understanding what constitutes meaningful evidence and outcomes in LTC. This includes a move away from the narrow range of indicators towards organisational, professional and user inputs, making more visible important factors that contribute towards a more appropriate evidence base. This ranges from policy to case studies of good practice, and databases are currently being developed (Petch, 2009). Evidence from PROCARE (Billings and Leichsenring, 2005) and EUROFAMCARE (Mestheneos and Triantafillou, 2005) have also done much to contribute towards this.

INTERLINKS and evidence for long-term care

Thus the pluralist approach was seen to be the most fitting for the gathering of evidence in INTERLINKS, where the strengths of knowledge derived from different methodological approaches and perspectives could be triangulated. This permitted the project to adopt a broader, more diverse and more inclusive notion of what constituted valid evidence through a balanced interaction between the scientific and the experiential.

In order to enable partners to ensure that this approach was incorporated when describing practice examples, criteria were developed within the example template (see Appendix B) that included sections for elaborating on a broader range of materials. This included different types of research, insights from practice and the lived experience of service users, particularly important given efforts made within the INTERLINKS project as a whole to render the user and carer as visible as possible. Specific criteria were developed to guide example descriptions and analyses in terms of effects, strengths and limitations (Table 14.1); these guidelines encouraged a wide delineation of what constituted 'effects', and additionally permitted a critique

Table 14.1   Criteria to describe effects

What are/were the effects?

Use this section to describe the effects based on evidence: How was the project evaluated? How well did the example address the gaps and improve the interfaces? How were the benefits for older people and informal carers evaluated? What were the demonstrated or potential effects on older people, informal carers, formal carers, organisation/inter-organisational collaboration, costs? To what extent did the example change the way that LTC is provided across systems or services? Has the example proved to be sustainable, or mainstreamed? Has the example been implemented elsewhere?

What are the strengths and limitations?

Use this section to give your reflection on strengths, weaknesses, opportunities and threats, intended or non-intended outcomes/effects. Include any doubts whether the outcomes were caused by the measure/intervention/service. Consider the perspective of the person in need of care and the informal carer.

Source: http://interlinks.euro.centre.org (see also Appendix B).

of the information to be provided. This latter aspect was important, given the difficulties in generating LTC evidence as discussed.

Searching for evidence of effectiveness within the examples was undertaken therefore in the following four stages:

Stage 1: Examples that described implemented projects and project effects were selected only, as it was assumed that evidence of effectiveness would most likely be derived from these sources. This meant that the search was limited to all examples listed in Themes 3 (Pathways and Processes), 4 (Management and Leadership), 5 (Organisational Structures) and 6 (Means and Resources) of the INTERLINKS Framework for LTC (http://interlinks. euro.centre.org). Thus examples within 'Identity of LTC' (Theme 1) and 'Policy and Governance' (Theme 2) were excluded in this analysis, as these examples contained by and large descriptions of philosophical, value-driven and political approaches to integration in LTC, and few accounts of the effects of implementation (see Introduction for explanation of INTERLINKS framework). This resulted in a total of 59 examples (10 in Theme 3, 10 in Theme 4, 23 in Theme 5 and 16 in Theme 6).

Stage 2: A grid was then constructed using the key criteria/guidelines as outlined in Table 14.1. This featured an examination of evaluation methods which included a general description of research design and a specific analysis of those methods seeking user and informal carer views; a study of project effects regarding users and carers, organisations and costs; an assessment of sustainability and transferability; and commentary on the strengths and weaknesses of the evaluation. The selected examples represented different types of policy and practice driven examples: research projects, policy initiatives, mainstream service innovations, bottom-up and top-down initiatives, all of which had different methods and opportunities of producing evidence.

Stage 3: All 59 examples were subjected to analysis using the grid described above. Once analysis of the 59 selected examples was under way, it became evident that in a number of cases, the information in the examples was insufficient to conduct an analysis and critical review. Further work was undertaken compiling original published and unpublished research papers and reports of the examples. When these were not available, authors of the examples were contacted for further information. This was obtained either through interview with the authors or more commonly by the author liaising with their national contact linked to the example.

Stage 4: Once this information had been compiled in the grid, it was possible to conduct a transversal content analysis across all projects. Broad dimensions were applied to the projects that enabled them to be sorted into four clear categories. These categories are not intended to be hierarchical, but instead reveal the spectrum of evidence generation across INTERLINKS examples. Both dimensions and categories are described below:

- Category 1: Pluralist Evaluation (n = 8). These contained examples that included:
  - a pluralist mixed method evaluation of scientific merit, details, components and/or approaches;
  - reported effects and results contributed towards the evidence base;
  - results that could be generalised or used for local improvement;
  - projects that had been mainstreamed and/or transferred to other settings.
- Category 2: Other Evaluation Approaches (n = 17). Examples here were found that contained:
  - robust and predominantly single method approaches of scientific merit;
  - reported effects and results which contributed towards the evidence base;
  - results that could be generalised or used for local improvement;
  - projects that had been mainstreamed and/or transferred to other settings.
- Category 3: Developing and Employing Evaluation Tools (n = 4). This category consisted of:
  - examples which were based on developing and implementing evaluation/assessment tools and methods;
  - tools/methods that made a contribution towards the evidence base;
  - tools/methods that aimed to improve LTC practice by providing a direct and comparable evidence base for practice change;
  - tools/methods that had been mainstreamed.

- Category 4: Limited Evaluation (n = 30). These were examples with:
  - uncompleted, minimal or no evaluation, but the project had been rolled out or was ongoing;
  - evaluations that were in process.

An analysis of the type and nature of such projects in this last category was undertaken in order to isolate factors that compel investment and service implementation without accruing evidence of effectiveness.

A total of 25 examples therefore could be described as using evaluation methods sufficient to provide evidence for practice. In addition to this categorisation, further analysis was undertaken on those projects that had extracted user and/or informal carer perspectives in order to ascertain the methods used and the evaluation effects.

Category 1: Examples that included a pluralist evaluation of scientific merit

Examples within this category were selected if they had used mixed methodology (both qualitative and quantitative approaches) including traditional statistical analysis alongside capture of professional, managerial and/or user and carer experiences. Some may also have included an assessment of cost-effectiveness. Most produced evidence that could be generalised to a wider population.

A total of eight examples fell in this category. Two examples from Finland and one from the UK consisted of comparative process and outcome evaluations, two with cost analyses (Tepponen and Hammar, 2011b; Allen, 2011a) and one without (Hammar and Finne-Soveri, 2011); a further Finnish example used a cluster randomised trial, including user interviews and cost analysis (Hammar, 2011); a French example used a quasi-experimental approach with clinician interviews (Naiditch, 2011a); and three examples used post-project multi-method evaluations (Huijbers, 2011; Repetti, 2011; Stiehr, 2011c). The first two Finnish examples listed here were completed jointly with external experts. An innovative case is from the Netherlands (Huijbers, 2011) concerning 'Buurtzorg' which is a model providing care in the neighbourhood.

A range of instruments were used including quality of life, satisfaction surveys, semi-structured and group interviews. It was noticeable that some of the Finnish examples capitalised on the widespread use of the Resident Assessment Instrument (RAI), which appears to permit ease of comparison of outcomes. Boxes 14.1 to 14.3 provide illustrations.

Box 14.1   The PALKOmodel (Finland)

The purpose of this Finnish model was to improve on different aspects of integrated care, such as flow of information, cooperation across/within organizations, and coordination of services. The aim was to standardize practice and makes written agreements between hospital and home care.

Method: The effects and cost-effectiveness were tested using a cluster randomised trial with municipalities (n = 22). Data consisted of a documentary analysis of medical records and care registers, and questionnaires measuring changes in home care personnel's job, work satisfaction and the quality of services. In addition, client interviews took place at discharge, and at 3-week and 6-month follow-up.

Effects: The PALKOmodel improved the process: clarified and improved the transfer of information, defined roles and responsibilities, standardized practices, helped to integrate services and increased the proactive way of working. Clients' functional ability and quality of life did not improve however, except physical mobility at 3-week follow-up. This effect on clients was thought to be due to this changing practice. The use and cost of home care services as well as visits to a laboratory or doctor decreased. Clients were able to use the service less, which decreased costs due to reduction in overlapping/unnecessary work. So the model seemed to be a cost-effective alternative to customary care; it does not demand extra resources but is funded through existing ones and subsumed within normal working hours.

Sustainability: The practice has been implemented in 22 municipalities, and despite the uncertain outcomes for clients, some organisations have exploited the ideas of the PALKOmodel when developing their home care and discharge practices. The model is generic and goal-orientated, making it usable for all client groups in different settings and organisations.

Commentary: Despite the high interest, little progress has been made towards changing current practice. This may be largely due to the need for greater investment and time set aside for development of practice, as well as the view that the staff involved in developing and changing practice have different professional cultures and styles and attitudes which also need to be changed.

Source: Hammar (2011).

Box 14.2 POPPS: The Partnership for Older People Projects (United Kingdom)

This UK project focused on developing health promoting services for older people aimed at reducing social isolation and pressure on acute care. Partnership working was key; local authorities worked with health and third sector organisations, improving local interfaces and strengthening local networks. All local projects involved older people in their design and management.

Method: The 29 pilot sites used 15 different methods of comparative evaluation, drawing on national and local funding sources. The first phase consisted of baseline document analysis and questionnaires on all sites; second phase interviews and focus groups of staff and users on 5 case study sites; third phase interviews across all 29 sites. User involvement was a strong feature and included advisory panels, workshops, project design, staff interviews/selection, panel membership, as voluntary or salaried, as evaluators or interviewees in evaluation. Measures included EQ-5D health-related quality of life, comparing Partnership for Older People Projects (POPPS) participants with non-project control.

Effects: Evaluations demonstrated a high level of user participation in service design and governance, as well as increases in quality of life outcomes for exercise projects and falls prevention-focused projects. The impact on joint working was positive, especially where multi-agency staff worked together in one location, rather than in virtual teams. Better referrals across agencies was a linked benefit of improved partnership working. The impact of information and advice work was seen through increased receipt of state benefit. Cost reductions and efficiency of services appeared to be significantly improved in comparison with non-POPPS sites. The results show that for every £1 (€1.16) spent on POPPS, an average of £0.73 (€0.84) will be saved on the per-month cost of emergency hospital bed days.

Sustainability: Only 3% of projects closed down. Further incentives to fund came from the 'transformation agenda', the 2006 White Paper which had a similar focus to POPPS around person-centred approaches and integrated services. Evidence from local evaluations was also a key incentive for investment.

Commentary: As initially time-limited and localised pilot exercises, the impact of the POPPS programme is to some extent limited. Although measures were built into the programme in an attempt to ensure the sustainability of positive outcomes, evaluators identified several factors which may impede the mainstreaming process as it progresses, including financial constraints; inability to attribute positive gains to particular interventions; and changes in government policy.

Source: Allen (2011a).

Box 14.3   Neighbourhood Solidarity (Switzerland)

This Swiss example describes a method to help vulnerable older people to remain at home to improve the quality of their life, foster neighbourhood integration of older people and enable informal carers to cope with difficult situations.

Method: A multi-method process and outcome evaluation after a 4-year development period. It used indicators such as the extent to which certain social competences had increased and how the process developed. Research and evaluation was separately financed. Satisfaction of older people in their neighbourhood and degree of solidarity within the area were also measured from a user perspective, and the perspectives of informal carers regarding coping with difficult situations.

Effects: Solidarity and participation of older people in neighbourhood life increased; social contacts improved, older people were more involved in social activities and personal welfare, and the overall quality of life of the community improved. The project enhanced people's pleasure with meeting each other and raised self-confidence. Informal carers reported more shared care and both formal and informal carers felt there was an alleviation of tasks carried out.

Sustainability: Institutions are increasingly integrating the programme and introducing other activities such as conferences and debates to improve interaction between inhabitants and professionals. The project created sustainable social links in a neighbourhood and older people were more involved in community. Project leaders should however be cautious about replicating interventions and actions in similar communities and ensure that specific issues unique to individual neighbourhoods are addressed.

Commentary: While the aim of the project is to enable inhabitants of the neighbourhood to sustain the project, it is at risk of failing when key personnel such as a social worker leave if the local dynamic is not strong enough. The many data sources meant that results could be open to different interpretations and used in different ways by stakeholders.

Source: Repetti (2011).

Category 2: Examples with strong element(s) of evaluation employing predominantly single methods

There were 17 examples in this category, and they included mostly reliable methods of evaluation employing either quantitative measures (Dieterich, 2011a, 2011b; Leichsenring, 2011a; Mak, 2011a, 2011b; Mastroyiannakis, 2011b), extensive audits which in some cases involved user satisfaction (Di Santo, 2011a; 2011b; Møller Andersen, 2011) or qualitative methods (Tepponen, 2011; Weigl, 2011). In some cases there were a good range of mixed methods (Glasby, 2011, Naiditch, 2011b); often these were smaller studies and/or targeting limited project stakeholders (Beyer *et al.*, 2011

Kagialaris, 2011a; Kümpers, 2011), or were projects in response to government initiatives that had provoked several small-scale evaluations (Holdsworth, 2011a). Results were generalisable in some examples, in others applicable to the local service being evaluated, and provided relevant evidence for practice improvement. In the case of Holdsworth (2011a) however, the opportunity to develop a joined-up approach to evaluation to create a body of evidence across a national initiative had been missed. Three illustrations are provided in Boxes 14.4 to 14.6.

Box 14.4 Meeting Centres for people with dementia (The Netherlands)

There are 60 Meeting Centres across the Netherlands where people with dementia and caregivers get support. Users can participate in recreational and social activities and training such as reminiscence. Their informal carers can visit discussion groups and get help in practical, emotional and social problems.

Method: Several quasi-experimental studies were conducted to compare support to carers and users in two matched groups of people with mild to moderately severe dementia and their family carers. Measurements were at baseline, three and seven months. Measures of health and social well-being were used.

Effects: The clients visiting centres showed more activity, less antisocial behaviour, fewer depressive symptoms and a higher self-esteem than clients in regular day care. Informal carers had more support, felt competent, less burdened and less lonely. The time to nursing home admission was much longer in the Meeting Centres group. Longer time to nursing home admission meant reduced costs.

Sustainability: Some of the effects particularly occur after seven months of participation, which is an argument in favour of long-term support.

Source: Mak (2011a).

Box 14.5 Care at home (Italy)

This is a Social Services project to provide older people in need of LTC with vouchers to directly purchase home care services, such as registered migrant care workers, home care services provided by private registered organisations, and respite services provided by residential facilities. It aimed to integrate public and private resources to support continuity of care at home for older people by supporting migrant care workers.

Method: Evaluation research focused on user satisfaction of older people and informal carers with respect to levels of care vouchers and audit of uptake.

Effects: In 58 per cent of cases, care vouchers improved the quality of life for older people and their informal carers. This rate increased to 71 per cent who chose care provided by live-in migrant care workers. Positive aspects of the scheme were economic benefit and improved care arrangement. Users were satisfied as they received clear information. Access to and use of the care vouchers was clear and simple. Sixteen per cent of older people stated that the care voucher led to limited changes only. From about 90,000 older people in this province, 133 had applied for care vouchers. On the basis of resources, needs and means-testing, 72 requests (54 per cent) were accepted. However, older people who were not eligible for care vouchers could also use the counselling service and the register to employ a qualified migrant care worker with their own resources.

Sustainability: The project has demonstrated that this kind of intervention can be successfully reproduced in other areas, if there are adequate human and financial resources available. The project has become a mainstream service for the consortium catchment area, but it is also becoming a mainstream service throughout Torino province.

Commentary: Migrant care workers continue to be chosen through other channels such as word-of-mouth recommendations among neighbours, friends, acquaintances and parish members.

Source: Di Santo (2011a).

Box 14.6   Equinoxe (France)

This French example has been ongoing since 1986. The nationwide home alarm system aims to enable frail people to live at home alone and enhance quality of life. It is based on a complex, high-tech call centre operating 24 hours a day with specially trained people. Each older person must nominate a 'neighbourhood committee' which consists of about three people living nearby.

Method: Continuous evaluation divided into parts: internal analysis and audit headed by an external expert: monthly meetings of an internal pilot group to report on functioning or malfunctioning, and to suggest changes to be put in place. External observation study (internal communications, behaviour of the call centre worker); data analysis using documents, statistics, internal reports. External representative user survey.

Effects: The average length with the service was three years; withdrawal from the service was due to death or institutionalisation, and for 5 per cent due to dissatisfaction with the service. Ninety-eight per cent of users were satisfied or very satisfied. Urgent calls were mainly linked to falls and anxiety during the day, and to respiratory problems at night. Other frequent 'emergency issues' were pain, fever, feeling ill. Non-emergency demands were linked to information, feeling alone, but also to issues linked to fear of burglary. In, 2009, 112,020 overall calls received. Response to clients' calls resulted in 69,265 monitoring calls by centre staff. Eighty per cent of issues resolved by local committee when triggered by call centre. A committee member could be contacted on average after three phone calls and in less than two minutes.

Commentary: Despite its longevity, the organisation suffers from lack of recognition of the high quality of the support provided and has not as yet received acknowledgment. Service is entitled to public reimbursement through attendance allowance; its use as a preventative measure cannot be reimbursed. There are higher organisation costs which are linked to the skills needed for the service to operate at a high level.

Source: Naiditch (2011b).

Category 3: Examples to develop and employ evaluation or assessment tools and methods

In this category, there were four examples of ways in which instruments had been developed and tested that sought to assist practitioners in measuring levels of care with the intention of contributing towards the evidence base or aiming to improve practice, providing a direct and comparable evidence base for practice change (Diermanse, 2011; Hammar, Niemi and Finne-Soveri, 2011; Leichsenring, 2011b; Weritz-Hanf, 2011; see Boxes 14.7 to 14.9).

Box 14.7   Outcome indicators (Germany)

The aim of the project was to identify appropriate and scientifically valid outcome indicators for care homes.

Method: An examination of which quality criteria/outcome indicators worked well in research and practice and how they were recorded or rated was carried out. This was the basis on which a new set of tools was developed. Field-tested in 46 care homes, experiences of providers, users and carers were gathered and improvements made. The satisfaction levels of relatives regarding care provided by institutions are included in the questionnaire. It also includes indicators for identifying quality of life and measurement of informal carers' rating of the institution.

Effects: The focus of attention is placed on the person in need of care. Such a tool will result in better quality of health and social care. Outcomes of the project may promote development of quality in institutions and provide a basis for quality assessments and reporting by external institutions which allows outcomes to be compared across institutions and quality of care measured re the links and interfaces of services.

Commentary: Area-wide implementation has not yet happened and will need to be taken forward in the future in order to sustain momentum. It is unclear how this can be supported at present.

Such an indicator-based system requires completed surveys of residents and informal carers, using techniques that are in line with specific rules and conducted at regular intervals; and software developments are required for their evaluation. The necessary skills are also needed and training in this area has not yet been made clear.

Source: Weritz-Hanf (2011).

Box 14.8   E-Qalin (Austria, Germany, Luxembourg, Slovenia)

E-Qalin is a quality management system based on LTC care home staff training to become Process Managers and Group Facilitators. An organisational self-assessment is undertaken using 66 structures and processes criteria, and, 25 results-based ones. Quality management projects are implemented in the care homes with all stakeholders according to assessment outcomes.

Method: Evaluation of training and the applicability of the system in care homes were conducted. This included structure and process of the test-phase by means of questionnaires for participants in training (Process Managers, Facilitators, Consultants)

Effects: Inter-professional relationships have improved in terms of teamworking and information sharing. On the organisational level, stakeholders involved in the assessment process have increased their perception of residents' and their families' needs.

Sustainability: E-Qalin has expanded in Germany and Luxembourg, but particularly in Austria and Slovenia. Specific versions for community care services and services for people with disabilities have been developed as well as an external certification procedure. In Slovenia, E-Qalin has become the general quality management standard among larger care home providers, enabling comparisons and benchmarking.

Commentary: No assessment of residents but this poses its own difficulties with measuring perceptions of quality improvements. E-Qalin is currently being implemented in to other countries, but it will depend on national regulatory frameworks, leadership issues and traditions of further education in the care sector to create acceptance and a systemic impact.

Source: Leichsenring (2011b).

Box 14.9   Integrated care programmes: quality management of stroke-related services (The Netherlands)

Based on a previous study about improving stroke outcomes through integrated care, a Dutch Stroke Network has been set up and several projects started. A set of performance indicators for integrated stroke care was developed. This describes a project using indicators and criteria for stroke services between two hospitals and a care home.

Method: The criteria were implemented using an action research approach where they were implemented and assessed at various intervals. The performance indicators refer to the following criteria: hospitals have earmarked beds; professionals work with evidence-based guidelines; specialised nurses in hospitals and rehab centres; weekly multidisciplinary consultations; structured way of transferring patient information; after-care for patients and their carers. User and carer satisfaction is also included.

Effects: Using the action research approach, the measurement of performance against indicators resulted in rapid changes and improvements. In the pathway for stroke, satisfaction has improved among patients and carers with faster access to care and less uncertainty for patients and carers. Patients get quicker and more appropriate care. Following implementation of indicators, changes included: fast thrombolytic treatment, stroke units in hospitals, triage in nursing homes, stroke pathway monitoring with stroke coordinator. Outcomes include: reduction of hospital stays from an average of 2–6 months in 2000 to a maximum of 10 days in 2006; quicker rehabilitation; improved functional recovery; reduction in regional differences. Shorter stays in hospitals reduce hospital costs, but costs are higher in other places (residential care).

Commentary: Sometimes, patients stay too long in hospital, because there is no available place in the rehabilitation centre of a nursing home. Hospitals are not sufficiently able to provide good care for a longer stay, because there are insufficient opportunities for therapy (e.g. physiotherapy). This has a negative effect on the condition of the patient.

Source: Diermanse (2011).

Category 4: Examples with incomplete, limited or no evaluation

This was the largest category (n = 30). It included those projects where evaluations were incomplete or ongoing at the time of reporting and were not able to describe a full range of effects. A number of these projects contained robust, multi-method designs that will contribute towards providing valuable evidence (Di Santo, 2011a; Tepponen, 2011; Wagner, 2011). For others, the limited evaluation contributed towards insecurities about future sustainability and was a problem. An example of this was in Austria, where evidence for the effectiveness of the Mobile Palliative Care Teams was reliant upon a national audit of palliative care services not specific to these teams (Ruppe, 2011).

In several examples however, there was little evidence of effectiveness, with data (if any) being drawn from small-scale investigations, informal feedback, screening tools, audits, or a range of different uncoordinated sources (e.g. Brichtova, 2011b). But they had nevertheless been or were being rolled out, or parts of them were being replicated elsewhere, as they were seen as innovative, significant and relevant, intrinsically transferable and there was an urgency to address perceived need.

Examples of this included the Integrated Access Point for older people in Italy (Ceruzzi, 2011), Social Security and Informal Carers (Garcés, 2011) and a Needs Assessment Model in Slovakia described more fully in Box 14.10 (Brichtova, 2011a). A further example is from Greece that addresses the relevant issue of training. The Elderly Care Vocational Skill Building and Certification (ECVC) project (Triantafillou, 2011) was developed to address the needs of the Care of Older People sector in Europe for vocationally skilled workers with a common level of competencies, and has been extended beyond a pilot phase to provide training across the EU.

An interesting 'collection' of three examples from the UK also fall into this category, and are largely concerned with describing national government initiatives. Hospital Discharge Lounges, Direct Payments to Informal Carers (Holdsworth, 2011b, 2011c) and Section 75 Partnerships (Allen, 2011b) offer basic audit data or unfocused, often process-oriented evaluation as evidence, but despite this are ongoing, even though informal analysis is sceptical of its value. Holdsworth (2011c) is of particular interest as it demonstrates that attaching a performance indicator to an imperative can change outcomes quite dramatically (see Box 14.11).

Box 14.10   Needs assessment model (Slovak Republic)

Within this new model, a health and social care needs assessment was organised and processed in an integrated manner by various professionals cooperating under a common organisational umbrella.

Method: The implementation period of this project was too short to conduct a full evaluation. Therefore some process information existed that had been collected by towns and municipalities. This included the number of health and social assessment reports and some basic unofficial feedback obtained from users, social services providers and government officials.

Effects: This process information indicated that there was a heavy administrative burden (too much paperwork) which was slowing down the assessment process, resulting in a delay to entitlements. Some non-public representatives called for a change to the legislation that simplifies the whole assessment process. On the basis of this informal feedback and audit, the model was reformed and simplified.

Commentary: The opinion with this model was that there was insufficient time to wait for the results of a full evaluation as action needed to be taken as the needs were perceived to be so great.

Source: Brichtova (2011a).

Box 14.11   Direct payments (United Kingdom)

These are cash payments made by local authorities to carers of older or disabled people living in the community. The aim is to increase choice, flexibility and independence for carers so that they can continue caring. The payment can be used to purchase services or equipment and also to help them maintain their health and well-being; it is not a source of income.

Method: Audit.

Effects: Low uptake of direct payments among carers at first in 2003, then became a social services quality indicator, whereupon it dramatically increased: from 7350 in 2007/2008, to 21,014 in 2008/2009.

Commentary: Carers who can maintain older people at home save local councils money as they do not have to pay for residential care; therefore initiatives such as direct payments which support carers who might otherwise give up caring are viewed as cost-effective.

Carers may not consider themselves to be carers and therefore would not seek an assessment of their needs which is a prerequisite for a direct payment. There is very little evidence of the impact that direct payments have on carers of older people.

Source: Holdsworth (2011c).

The User and Informal Carer Perspective

Approximately 23 of the 59 examples contained some identifiable methods that included the user perspective. As far as could be ascertained in the information provided in the examples, 13 of these included informal carers. The user/informal carer input ranged from total involvement throughout the project (Allen, 2011a; Di Santo, 2011a) to opinion seeking (e.g. Brichtova, 2011a). Methods were largely either purely quantitative (Mastroyiannakis, 2011b) or qualitative (Stiehr, 2011c; Weigl, 2011), with some including both approaches (Allen, 2011a; Glasby, 2011; Hammar, 2011).

The UK POPPS programme described in category 1 (Allen, 2011a; Box 14.2), for example, used mixed methods which included advisory panels of users and carers, involvement in project design, staff selection, and evaluation. EQ-5D health-related quality of life measures were additionally used for those taking part in health promotion interventions. EQ-5D is a generic measure of health status that provides a simple descriptive profile and a single index value that can be used in the clinical and economic evaluation of health care and in population health surveys. Equally, the ongoing Italian project ROSA (Di Santo, 2011b) has a strong involvement with all local stakeholders including families to monitor and evaluate progress.

From a quantitative perspective, evaluation of an E-health unit in Greece (Mastroyiannakis, 2011b) employed a range of established and specifically designed instruments to ensure user effects were captured. These included the SF 36 (a well-validated health survey focusing on physical, emotional and social functioning as well as vitality, pain and mental health) the St George's Respiratory questionnaire (a 50-item questionnaire designed to measure health impairment in patients with asthma and COPD), and questionnaires to measure autonomy and acceptance of IT. Many projects adopted health-related measures as the sole metric and the most common quality of life measure used was the SF 36. Finnish examples employed quality of life-related items within RAI (Hammar, 2011; Hammar and Finne-Soveri, 2011; Tepponen and Hammar, 2011b).

Methods for gaining perceptions of informal carers included qualitative interviews (Glasby, 2011; Repetti, 2011; Stiehr, 2011c), a Zarit burden structured interview (a popular 22-item caregiver self-report measure assessing personal strain and role strain) (Tepponen, 2011), informal carer ratings of care homes (Weritz-Hanf, 2011) and quality of life (Naiditch, 2011a). For both users and informal carers, a high proportion employed satisfaction levels either as stand-alone instruments or sections within other measures such as audits (Diermanse, 2011; Di Santo, 2011a,b; Mak, 2011a; Møller Andersen, 2011; Naiditch, 2011a; 2011b) or capitalised on national satisfaction surveys to extract information on their projects (Huijbers, 2011).

Regardless of methodology, it is important to ascertain perceived benefits and outcomes from a user/informal carer perspective regarding integrated health and social care services for those with LTC needs. Although effects were specific to each project, some trends can be seen and in general, the methods used assisted in establishing a variety of health improvements for those using the interventions described in the examples. A number of projects reported high levels of satisfaction and/or improved quality of life or health and wellness (Allen, 2011a; Beyer *et al.*, 2011; Diermanse, 2011; Dieterich, 2011b; Di Santo 2011a; 2011b; Mak 2011a; 2011b; Mastroyiannakis, 2011b; Naiditch, 2011b; Repetti, 2011); improved social contacts, less isolation, and from an informal carer perspective, feeling more supported and less burdened (Garcés, 2011; Hammar and Finne-Soveri, 2011; Mak, 2011a; Repetti, 2011; Stiehr, 2011c; Tepponen, 2011; Tepponen and Hammer, 2011); and improved care, access to care and/or care support (Campbell, 2011; Diermanse 2011; Di Santo 2011a; Sedlakova, 2011; Stiehr, 2011c).

Less beneficial outcomes included a lack of detailed information sharing between services about needs (Møller Andersen, 2011). Of interest is a finding in the Finnish PALKOmodel (Hammar, 2011) where there was physical improvement but no improvement in quality of life as measured by RAI. Similarly, in the French example COPA (Naiditch, 2011a), while outcomes for older people showed health and wellness improvement, no significant effect on satisfaction was found for informal carers.

Discussion

This discussion of the findings will focus on critically reviewing the extent to which the selected INTERLINKS examples contribute towards our understanding of evidence in LTC, with a particular examination of the methodologies used, the nature of the evidence, and its potential to provide practical knowledge for sustainable service improvement.

As previously argued, a key feature of INTERLINKS was to adopt a pluralist approach in order to gather knowledge derived from a triangulation of different methodological approaches and perspectives. The purpose of this was to allow the project to adopt a diverse and inclusive notion of what constituted valid evidence through a blend of scientific and experiential data. INTERLINKS examples showed that approximately a quarter had been evaluated in sufficient detail to provide some meaningful evidence, so while there is forward movement, evaluation of initiatives remains the exception rather than the norm. Within the database only a minority had used a mixed method approach, and traditional quantitative designs were evident such as potentially generalisable experimental designs (e.g. Mak, 2011a; Naiditch, 2011a) and large qualitative exploratory approaches, such as Weigl (2011) and Glasby (2011) that were amenable to analytical generalisability (Guba and Lincoln, 1992). It must be noted however that, despite attempts to seek further information on project evaluation, analysis was sometimes restricted to the information supplied through the examples, so that weaknesses due to artefact and design effects were not always evident.

The pluralist mixed method approach is however not without its challenges, in that the numerous data sources may generate competing evidence unless an analytical framework is used that blends data sources cohesively, and there was little evidence of this. Dopson *et al*. (2002) warn that compet-ing levels of evidence were capable of engendering a number of different interpretations. This was recognised in the Swiss example 'neighbourhood solidarity' (Repetti, 2011), where results were expressed in a cautious manner as they could be open to different meanings and bias in reporting.

But on the whole, adopting a broader approach to evaluation within INTERLINKS appeared justified. As previously stated, commentators have argued that experiential data in particular can best be obtained through qualitative methods, in order to provide the detail that can reveal why processes do and do not work in practice (Barker, 2000; Rycroft-Malone *et al.*, 2004; Glasby and Beresford, 2006). This is especially so when sample groups are frail, due to sensory, mechanical and cognitive difficulties in completing forms, and it was therefore encouraging to see some examples that embraced these approaches. However, the fact that a large proportion of examples relied quite heavily on satisfaction levels to obtain user and informal carer views is surprising, given the often negative commentary regarding the effectiveness of this approach in gaining a valid picture, as older people tend to be overly positive about services compared to other groups (Atwal and Caldwell, 2005). A factor at play may be that they are often still in receipt of services themselves and feel that a negative appraisal may jeopardise their care (Castle, 2007).

When considering the general lack of evaluation and subsequent evidence from integrated care initiatives in LTC, INTERLINKS examples may reflect the general situation as previously discussed; commentators have argued for example that it has become commonplace to highlight that LTC evalua-tion is testing and fragmentary (Johri *et al.*, 2003; Dowling *et al.*, 2004; Ramsay and Fulop, 2008). It must be reported however that INTERLINKS examples may not reflect the general trend and not all countries across the EU were included, and indeed some countries within the partnership were underrepresented. But it is worth speculating on reasons for this evaluation shortfall that lie outside of methodological challenges, and INTERLINKS data provides us with some clues.

In general, the priority and resources allocated for evaluation purposes appeared to vary across projects and countries. Without doubt, rigorous independent evaluation is costly and can sometimes compete with the resources needed to implement the initiatives when costs are contained, and this appeared evident in category 4. Evaluation can be prompted only when further investment is required, rather than being built into project design. The exception did appear to be those examples associated with information technology, telehealth and telemedicine for instance, where by and large evaluations had been conducted and comparisons could to some extent be made (e.g. Naiditch, 2011b). In addition, service providers cannot often wait for the outcomes of lengthy evaluation studies when there is a felt need for action.

Further to this, an interesting paradox concerning the prevailing argument for evidence-based practice through robust evaluation was situated within category 4, and begs the question whether there is always a need for evidence prior to implementation. This related to those projects that were being rolled out despite lack of evidence and a clear evaluation framework. This was particularly so regarding the 'cluster'of policy-focused UK examples (Allen, 2011b; Holdsworth, 2011b; 2011c), which provided instances of where political imperatives demanded prompt service responses in rolling programmes out regardless of their potential effectiveness. Any evaluations were based on simple audit measures or were conducted piecemeal across the country with no coordinated approach. Thus there is a spiralling problem in that it becomes difficult to compile a reliable body of comparable information to inform future implementation or to develop strategies for policy improvement.

However, there are explanations for this from a public policy perspective. Naughton (2005) states that despite the obvious common sense of evidence-based policy (EBP) – that sound research should inform all areas of public policymaking – since its emergence it has been subjected to a sustained attack on almost every level. Klein (2000) in particular argued that EBP rests on a gross misunderstanding of the policy process, which is already governed by 'evidence', but of a very different kind from the narrow scientific view of evidence. Policy decisions incorporate evidence as to whether a policy will be politically acceptable and implementable. If a policy cannot meet these requirements it is deemed to be not worth pursuing, whatever the research evidence says (Dopson *et al.*, 2003). To support this, within the examples, some initiatives came about through being driven by completely different intentions, such as policy imperatives and innovative local ideas, than by producing or creating evidence.

Aside from this, in other INTERLINKS examples, the rolling out of projects has been due to identified needs, but how these needs were established is not clear within the confines of the example descriptions (Brichtova, 2011a; Ceruzzi, 2011). With regard to innovative examples, it could be argued that examples such as this have a wide common sense appeal that often transcend the need for rigorous evaluation, so much so that other organisations appear to 'buy in' to the ideas at face value. This is a movement that is frequently present when implementing health promotion initiatives where the evidence base is equally difficult to build and demonstrate, and is referred to as the 'precautionary principle' (Foxcroft, 2006). This principle is based on the premise that it is better to do something than nothing provided you do no harm, and has been applied in fields such as interventions to reduce teenage alcohol misuse across Europe (Billings, 2009). Thus in answer to the question of whether evidence is always needed, there is an argument in favour of this, provided the risks are low.

But when it comes to creating credible, generalisable bodies of European evidence, evaluation within INTERLINKS examples did not appear to be of sufficient levels to achieve this. Category 3 however demonstrated that there are potential approaches and instruments that could permit such a comparison (Leichsenring, 2011b; Weritz-Hanf, 2011) and should be exploited. While there were similarities between examples across the database that could be grouped within the INTERLINKS framework, it was clear that projects were sufficiently different in approach and nature to disallow comparable methods of evaluation that would permit evidence to amass for discrete areas. This is of course unsurprising, as project objectives would be honed to the specific organisational, population and political requirements of the localities and it must be noted that creating bodies of evidence are not local imperatives. Even when initiatives were derived from clear national project specifications amenable to comparative evaluation across sites however, evaluations still tended to be operationalised in a less than joined-up fashion and according to local drivers, providing disparate evidence and missing an opportunity to add to a wider evidence base (e.g. Holdsworth, 2011b).

Further to this, it is equally important to comment on how practical knowledge emanating from the INTERLINKS examples has been used to inform, improve or sustain practice, and the INTERLINKS examples described here give some information on this. Individual project effects have been described in the previous section, and when it comes to service improvement at a local level, many project evaluations reported being able to provide sufficient information upon which to generate local service recommendations (e.g. Beyer *et al.*, 2011; Kagialaris, 2011a; Kümpers, 2011), even through audit alone (e.g. Ceruzzi, 2011), appearing to fulfil local needs. But it was interesting to note that despite having a seemingly robust pluralist evaluation that can provide a clear steer for how practice can develop, there still remains the issue of ensuring further investment, as well as changing practice and combatting cultural issues. With regard to the former, the home alarm system Equinoxe in France (Naiditch, 2011b) has some clear positive outcomes, but despite its longevity, the organisation suffers from lack of recognition of the nature of the support provided and has failed to receive mainstream funding.

With regard to changing practice, the PALKOmodel in Finland (Hammar, 2011) demonstrated a cost-effective alternative to customary care, where staff were involved in developing and changing practice, but the evaluation revealed that the different professional cultures, styles and attitudes acted as a barrier to progress and change. Similarly, a comprehensive evaluation can demonstrate problems that disable sustainability and investment for the long term and therefore be a negative outcome (Allen, 2011a) that may inhibit practice change in the longer term. These findings do resonate with Johri *et al.* (2003), whose review demonstrated that certain frameworks of

LTC could be cost-effective and have potential for service improvement, but there were many examples where extension beyond the research or pilot phase did not transpire. However the INTERLINKS analysis does to some extent verify the objectivity of the research process and the value of a multi-method approach to evaluation, permitting an understanding of all factors that constitute meaningful evidence and outcomes in LTC (Cheetham *et al.*, 1992; Petch *et al.*, 2005).

Conclusion

Thus the evidence debate remains a complex and long-standing issue. What is clear is that in LTC, evidence still struggles to move away from the 'informal', and there would even appear to be a movement away from seeking outcomes and effects towards 'precautionary' implementation due to service and political imperatives.

What is apparent is that there are no easy solutions or clear ways forward. What constitutes evidence within INTERLINKS and in the wider literature oscillates between the clear 'hierarchical' position to the almost anecdotal, and for LTC this results in a pathway that is far from transparent. This is particularly so when considering the differences and tensions that exist between how both health and policy accumulate and use evidence for implementation. INTERLINKS has however contributed towards unravelling the complexity and providing a critical understanding of this subject through this analysis of evidence within the examples, which allowed a delineation of the extent to which different methods can support an evidence base. In addition, by dynamically evolving the accumulated knowledge during the lifetime of the project, INTERLINKS has sought to promote an acceptance of different levels of evidence, and describe the extent to which it can influence policy and practice.

15
Improving Policy and Practice in Long-Term Care

Kai Leichsenring, Jenny Billings and Henk Nies

Meeting the challenges

When Peter Townsend published his landmark study about long-stay institutional care in England and Wales 50 years ago, he could not help calling these institutions 'The Last Refuge' (Townsend, 1962). He was 'both daunted and shocked' by what he had seen and heard, from overcrowded dormitories 'with ten or twenty iron-framed beds close together' (p. 4). He noticed 'isolated persons sitting alone in a wash-room, standing in a corridor and one looking out of a staircase window weeping silently' and he found heavily restricted privacy together with authoritarian matrons in uniforms and inappropriate staffing (p. 5f.). A 'revisiting study' by Johnson *et al.* (2010) found that 37 out of the 173 care homes visited by Townsend in the late 1950s were still providing long-term care (LTC) for older people. Results for the 20 homes that were sampled for revisiting revealed that:

> (...) residential care for older people, insofar as it is now catering for an older and more infirm population has been transformed into a radically different instrument of social policy when compared to the 1950s (...) While the physical environment may have improved, we recorded many institutional features which appear to characterise the twenty-first century care home. These features reflect not only the changed function of residential care but also an increasing concern with risk and safety.
>
> (Johnson *et al.*, 2010, p. 209f.)

Certainly, the framework conditions, legislation and safety regulations introduced during this time period have contributed to a greater degree of privacy and a more enhanced personal and friendly atmosphere, but resident's autonomy remains restricted, notably by the very same regulations.

Apart from residential care, changes and social innovation in LTC have happened quite quickly on a number of occasions. For instance, the phenomenon of live-in 'migrant carers' providing personal assistance to frail older people in their own homes has taken place during the past 10–15 years due to the proximity of low-wage countries, and partly fuelled by cash benefits for people in need of care. Another example is the transformation of residential care into facilities for nursing care where the majority of residents have cognitive impairments. Drivers such as quasi-markets, competition and public reporting of quality have further hastened the need for radical change within LTC over the past decades.

Considering European survey data, residential care still remains 'The Last Refuge' for the majority of citizens. Most citizens reject residential care as they prefer to live at home for as long as possible, even when frail and in need of care. Along with users, policymakers, governments and other relevant stakeholders have been looking for alternatives. These changing expectations and preferences have contributed to the emergence of different approaches and a broader range of services to meet needs, for instance intermediate care, day care facilities and various forms of care in the community. In terms of public policies, this has included initiatives such as guaranteeing more integrated health and social care services for frail older people, supporting informal carers, fostering prevention and rehabilitation as well as assuring and improving quality. Given the continuing importance of user contributions to care that are relied upon in most countries, families have been searching for less costly alternatives such as the employment of migrant care workers.

As the history of LTC is relatively recent, this book has set out to provide knowledge and insight into LTC as a sector that cuts across institutional and community care, across social and health care, and across formal and informal ways of supporting frail older people.

As has been acknowledged previously and within this book, there are a number of key demographic, cultural and political changes across Europe and indeed the whole industrialised world that have turned the issue of LTC for older people into a challenge that calls for proactive political strategies. Key challenges are addressed below, and include indications of how and where INTERLINKS has contributed to the knowledge base:

- Ageing of the population causes – in absolute terms – a significant increase of needs for health and social care. Contrary to general prejudices, however, ageing as such does not necessarily trigger rising expenditures on health and social care (see Chapter 9; for various scenarios see European Union, 2012). While factors such as increasing costs, medical technology, new options for treatment, and policy changes explain more than 50 per cent of rising expenditures, only 15 per cent can be attributed to ageing (RIVM, 2011; Berenschot and Van der Geest, 2012; Breyer *et al.*, 2010). Improvements and sustainability thus depend chiefly on political and organisational choices, rather than on components that are difficult to influence.

- Family structures and gender relations are changing in the direction of smaller households, fewer births and a rising proportion of women in paid employment (European Union, 2011). Arguably, one of the most valuable discoveries of capitalism is unpaid and mainly female household and care work. As this is insidiously reducing, new ways and patterns of support for informal care, intergenerational and societal solidarity will have to be 'invented' and promoted to counteract the increasing vulnerability of older people (see Chapters 3 and 11).

- While some countries have 'professionalised' female care work via an exceptional growth of health and social care sectors, there are rising problems in recruiting and retaining a sufficient number of care workers. This is partly due to poor payment, limited education and the generally low status of care work (OECD, 2008; Hasselhorn *et al.*, 2005), which is even more evident within the LTC sector. However, with new types of job profiles, multiprofessional working and a wide range of conceptual tasks to be fulfilled, there are also opportunities for the development of more appropriate professional identities (see Chapter 4).

- Complementing and replacing the workforce or informal carers with the aid of live-in migrant care workers has become common practice in many countries as a means of responding to unmet care needs (see Chapter 10). As a consequence, significant movement of care workers can be observed across national borders, and in many cases under conditions of unregulated legal status and a lack of social security.

- The changing character of formal and informal LTC has also been highlighted by showing that many countries have acknowledged LTC as a social risk. In the context of a market-oriented governance, this has included the introduction of funding mechanisms promoting user choice by increasing the purchasing power of people in need of LTC, such as by means of attendance and care allowances and personal budgets. For older people in need of care and their carers, such mechanisms might not always be the most appropriate solution, particularly if cash benefits do not compensate for market prices of care, if they are not supported by the development of a proper service infrastructure, and if older people are suffering from dementia (see Chapter 3). The blurred boundaries between paid and unpaid, and between formal and informal work will continue to produce hybrid forms of organising and rewarding activities in LTC. This will mean combining not only the advantages of state, market, household and third-sector principles (Evers and Ewert, 2010), respectively, but also constructing an appropriate mix of formal health *and* long-term care provision.

- The rising expectations in terms of user-centredness and associated values such as dignity, privacy and autonomy means that there needs to be a paradigm shift in how organisations fundamentally deliver services and engage with their clients. This includes addressing the poor coordination and integration between health and social care organisations, inter-professional working and networking around users' needs. The evidence of this happening within LTC can be identified, but it is only sporadic and limited to individual projects that are of short duration and not mainstreamed. There are many challenges with engaging users; emerging LTC systems have yet to fully embrace this philosophy (Chapters 2, 6, 7 and 14).

• Sustainability and quality of LTC will remain an issue in all countries. Political debates are heavily biased by existing funding models, welfare traditions and family ethics in individual countries with, in general, Northern countries expecting the state to take up responsibility for paying for LTC, while Southern and Eastern European countries still consider the family to be in charge (see Chapters 2 and 8; see also Costa-i-Font and Courbage, 2011). Regulatory frameworks and political strategies are therefore needed that look wider than the immediate area of care for older people and consider employment, housing, gender issues and social planning.

• In most countries, funding and service provision are profoundly characterised by rationales for social assistance schemes, rationing and targeting, which is in stark contrast to health systems that largely provide universal access and funding. As LTC care has developed between acute care and social care and between formal and informal care, this means that LTC operates between various (policy) systems, layers of governance, public and private provision, funding systems, organisational cultures, quality systems and, in particular, a vast array of stakeholders. From a managerial and governance point of view, LTC is an extremely complex field, which requires complex strategic, tactical and operational skills. It seems clear that coordination between policy sectors, regulatory mechanisms and the various stakeholders involved is a necessity, but sound evidence on the effectiveness of existing mechanisms and structures is still weak or indeed lacking (Chapters 2, 8, 9 and 14).

• By the sector's very nature, a recurrent issue in long-term care concerns coordination and integration of various stakeholders, but also of different rationales, values and professional ethics (Chapter 2). Respective coordination mechanisms have been developed, tested and evaluated in many places, using various approaches and with differing results (Chapters 4 and 5). But the general tendency towards networking and integration has not yet resulted in coherent mainstreamed strategies to plan, organise, provide and evaluate LTC for older people. There is still a long way to go to mainstream improvements and to tackle ongoing changes, particularly as major shifts in resource allocation, education and training will be at issue.

Learning from emerging practices

Against the backdrop of these challenges, this project has developed the INTERLINKS Framework for LTC to contribute to a systematic description and analysis of key issues at stake. This has been achieved through the search for 'emerging practices' that point the way towards feasible and sustainable solutions for one or more of these key issues. The individual chapters in this book have discussed and analysed these practices in LTC across selected European countries and from different perspectives. These contributions range from analysing tangible and innovative *policy and practice* on individual, organisational and systems levels to outlining *new theoretical approaches* and considerations of how basic concepts such as the definition of identity, boundaries, evidence, quality, prevention and informal care may be reconstructed from a 'long-term care perspective'. Transcending these two areas, important messages emanating from this book concern the need to focus on *systematic improvements* to policies and practice in LTC that involve all relevant stakeholders. These three themes will now be elaborated on.

Policy and practice solutions

Informal carers provide the vast majority of long-term care. Acknowledging and supporting this important societal resource entails a differentiated and well-designed set of interventions and policies that consider the varying but not always compatible interests of those who provide informal care and those who receive it (Chapter 3).

Although the introduction of quasi-markets and new private stakeholders has failed to provide clear evidence for lower costs, better quality and more effectiveness in terms of 'value for money', it seems that the area of LTC will continue to develop in a market-oriented environment. It is therefore necessary to continue the search for ways of regulating these quasi-markets and promoting mechanisms to improve quality within and across 'hybrid organisations' (Evers and Ewert, 2010), that combine the advantages of public, private and third-sector principles, and reduce respective inconveniences. Quality assurance across the whole spectrum of services and organisations needs to be developed further (Chapter 8).

The search for resources and affordable alternatives to formal care arrangements has revealed that LTC, with its ambivalent practices and outcomes, is a proficient area for societal experimentation and innovation. On the one hand, LTC offers important opportunities for volunteering at almost any age, spanning from help in the neighbourhood to 'paid volunteering' in the context of community care or in residential care settings (Chapter 11). On the other hand, the phenomenon of migrant live-in carers from Eastern Europe or other countries, where particularly women suffer from low wages or unemployment, has mushroomed. This has occurred notably in neighbouring countries (e.g. Germany, Austria) and in countries where the network of formal services does not seem to be in line with the growth of LTC needs (e.g. Italy, Greece, Spain; see Chapter 10). In both cases, there is a vast amount of initiatives to boost the advantages of specific arrangements and to overcome their inconveniences. Apart from awareness-raising campaigns, these initiatives include legal regulations to formalise care by private assistants, training, information and counselling. However, this type of 'social innovation' needs further investigation in terms of ethical considerations, quality assurance and sustainability.

A more promising perspective has become apparent through a consideration of opportunities for prevention and rehabilitation, even for older people who are already in need of LTC. Apart from preventative and health-promoting approaches across the life course, it seems appropriate to apply concepts, policies and practices that provide adequate support at the right time and at the right place. By avoiding further deterioration, unnecessary admissions to hospitals and other rather expensive health care facilities, a range of practice examples indicate the way forward. This includes in particular local initiatives that support users (and informal carers) in their normal life, in their self-care capacities and in the context of socially inclusive communities (Chapter 7). These notions should be encouraged as they seem to be beneficial when dealing with long-term conditions. While these initiatives may not be available for the majority of older people, such practice is a serious option for the future, where prevention in this arena will be crucial. Increasingly, theory and concepts relating to health promotion and illness prevention are being included into the education and training of practitioners across Europe, and it can be expected that this expansion of knowledge will improve opportunities for including prevention and rehabilitation interventions in care planning.

Decentralisation of public responsibilities has for a long time been a guiding principle for public sector reforms and thus has also had an important impact on emerging services and organisations dealing with LTC at the interface between social and health care. With new orientations in terms of 'rescaling social welfare policies' (Kazepov, 2010), some local initiatives are facing problems under new administrative environments (see Wagner, 2011a). At the same time, a new type of successful 'bottom-up' initiatives driven by professionals as pioneers can be identified in some countries (Huijbers, 2011; Leichsenring, 2011a; Stiehr, 2011b; Kümpers, 2011). These examples highlight the necessity of improving LTC in the local context and in the living environment of older people with care needs, taking into account the well-known concept of 'ageing in place' (Tinker, 1999) to enable older people to stay in the homes they currently live in for as long as possible.

Given the rather patchy and fragmented development of individual services that provide LTC, 'coordination' and 'networking' have probably been the most used catchwords in policy documents and recommendations over the past two decades. This has triggered the search for intermediaries resulting in new job profiles to carry out functions such as case management, 'bridge-building' and the facilitation of inter-organisational working.

Steps forward towards improving these connections should however be complemented by 'difference management, through which the system and its elements regulate their relationship to their environment' and by 'boundary redefinition processes' as exposed in Chapter 6. Despite huge quantitative and qualitative differences in terms of structures, cultures and funding mechanisms, European countries show very similar problems when it comes to coordination and networking between health and social care, and between formal and informal support. It can be observed that similar mechanisms are being used to overcome these bottlenecks, but different ways of working and implementing these methods are creating new types of divergences within and between countries.

A great hope for further improvements in health and LTC has been the application of information technology (IT), particularly in relation to data processing and connectivity between organisations, but also with respect to the reduction of workload and the provision of care and support at a distance. In this area there is still ample space for development, for example by means of stakeholder-driven design, as well as through better linkages between technological possibilities and human needs, and between professionals' and citizens' skills and expectations (Chapter 12).

Interestingly, it would seem that various specialisations and differentiated areas of LTC have contributed to further fragmentation, but have started to 'learn from each other'. This may be observed, for instance, in the context of policies for people with disabilities, but also in relation to palliative care. In both cases, policies and providers of LTC have been reluctant to acknowledge principles such as self-determination and user-driven care as promoted by the Independent Living Movement, or even how multiprofessional working has been achieved in specialised palliative care facilities. Yet integrating palliative care in LTC services and facilities should be seen as a logical step forward when considering the age and health conditions of users in LTC (Chapter 13). However, future developments in this area should not only focus on how links with the health system will be managed, but importantly on how the social characteristics of LTC can remain intact and not be subsumed within a medical model of care delivery.

New theoretical and conceptual thinking

Long-term care is not yet a system on its own. The challenge for the twenty-first century will remain how to design integrated networks to organise user-centred LTC in synergy with health and social care resources. This may not happen in the short term, but creating a unique LTC identity may support the blending of its multifaceted parts. This could be achieved through previously mentioned hybrid organisations with professionals that not only share core values, but fully embrace the conceptual understanding of what user-centredness means and how it can underpin working relationships and standards. Practice examples have shown that there are successful outcomes in individual projects expressed by anecdotal evidence or through systematic investigation. This is particularly so if professionals have been involved in developing a common understanding of organisational values, mission statements or quality standards. Specific importance must be given to strong leadership to reinforce these values around users' expectations. A proper LTC identity has the potential to help individual stakeholders and organisations to think and act across boundaries to carry forward the ongoing paradigm shift in LTC with respect to encouraging self-management.

More systematic improvement

To improve policies and practice in LTC it is necessary to adopt a systematic framework for setting priorities and working towards mutually agreed objectives. The INTERLINKS Framework for LTC contributes to such a systematic approach by determining themes, sub-themes and key issues, and by providing illustrative practice examples that show the feasibility and in many cases the sustainability of solutions. However, solutions that have been successful in some cases are not always transferable to another context. Frequently, individual solutions have been implemented and tested through time-limited projects, and while providing some inspirational ideas for practice improvement, they have neglected a range of other relevant issues, not least their own sustainability and how they can be scaled up. As a corollary, it has become clear that ready-made solutions and blueprints are neither available nor practical for all situations.

However, one way forward could be to exploit the wide and diverse illustrations of current practice both internationally and within the INTERLINKS Framework, encompassing those examples that are both pioneering and mainstreamed. Further work should be invested in extracting key success criteria from the examples, prototyping innovation, and developing 'social business cases'. According to Leatherman *et al.* (2003, p. 18) 'a social business case for a health care intervention exists if the entity that invests in the intervention realises a financial return on its investment in a reasonable time frame, using a reasonable rate of discounting'. This definition is equally valid for interventions in LTC improvement, where added value may be realised by direct profit, a reduction of losses, avoidable costs or through other indirect effects that contribute to sustainability. As the latter will depend on the intervention's impact on different stakeholders, social business cases require a stakeholder analysis and an impact map, as well as insight into the qualitative, and if possible quantitative, social benefits (Competentiecentrum transities, 2009). The presentation of social business cases can create a firmer foundation upon which to embed practice advancement into an organisation, through a more expedient consideration of viability, context and sustainability.

Despite these ideas, it must be stressed that implementation of innovations to achieve sustainable, high quality interventions requires long-standing commitments and investments in LTC organisations. In general there are hardly, if any, innovation budgets and structures for investments in LTC, other than those that can be obtained through competitive means. While this sector counts for up to 3 per cent of GDP budgets, research and development investments are negligible in most countries. Improvement on the scale required needs far-reaching investments of governments and other stakeholders, otherwise innovation will remain as a 'sideshow', limited to a temporary impact on practice development only.

Taking all this into consideration, systematic improvement of LTC practice therefore has to include the following features:

- *An emphasis on prevention and rehabilitation*: Prevention can be effective for older people as well, even if they are already in need of care and live in residential settings. Prevention as a concept needs to be redefined for people in need of LTC, and it needs to be conceived more in terms of social innovation and community development. Rehabilitation provides the option of regaining functional capabilities or reducing deterioration.
- *Greater attention to the capabilities and weaknesses of informal care*: It has been demonstrated that a broad range of support policies and services is strengthening the capacities of informal carers, rather than weakening their involvement. Enabling strategies thus contribute to conceiving informal carers not only as a resource, but as a positive element for societal well-being, contributing to social cohesion and solidarity in general.
- *A focus on quality of life*: LTC care is characterised by human relations that aim to improve the quality of life of frail older people – the specific challenge is to 'organise normal life' through regulations that allow for flexibility and continuity. While the traditional division of labour cannot contribute sufficiently to this kind of organisation, working with quality improvement mechanisms may help to guide reflective and dynamic improvement processes.
- *Enabling all stakeholders*: Both top-down and bottom-up strategies for improvement rely heavily upon the expectations and interests of various stakeholders. The rhetoric of 'putting the users and carers at the centre' calls for an enhanced dialogue, new types of job descriptions and related training.
- *Innovation is more than implementing information technology*: The introduction of new technologies in terms of 'gerontechnology' or 'ambient assisted living' has triggered enormous interest both at the level of manufacturing and software development. However, given the negligible input by users themselves, the uptake and use of these technologies will only

be successful if IT pioneers fully embrace user and carer involvement in device design, piloting and evaluation.

- Finally, innovative approaches to home care such as 'Buurtzorg' have shown that, within only a few years, quasi-markets may be taken over by new ideas if they are supported by professionals, and appreciated by users and purchasers. It must be noted however, that only a few of these changes have explicitly been initiated by policymakers. In many cases, political debate has reacted to ongoing changes or unintended effects of earlier interventions. This latter issue emphasises the weaknesses of the still rather shaky evidence base with regard to addressing new societal challenges.

Future research

Further research is certainly needed in an area that has only started to gain momentum in becoming publicly and politically acknowledged. The rising number of families that are confronted with the need for long-term care of older relatives has rendered this important topic more visible, because traditional family ethics alone are no longer enough to coerce mostly women into the moral responsibility of family care. Political debates and legal regulations have turned long-term care into an issue of acknowledged social risk that calls for solidarity, and its integration into respective welfare policies. The 'long-term care industry' from an economic perspective has developed into an area of added-value in terms of employment, as well as private and public budgets. With between one and three per cent of GDPs spent on long-term care, the sector cannot be neglected any more when it comes to debates about sustainability, health expenditures or the funding of the welfare state in general. Interdisciplinary research is therefore needed to search for synergies between health and social care, to address the 'moral political economy' of care and organisational issues of under- and overregulation.

- With respect to health, there is a myriad of links and interfaces that call for coordination and integration. But the respective strategies that are applied in Europe have yet to be evaluated and further developed by social innovation in the context of ageing societies (e.g. equal access, services and health inequalities, support of healthy ageing).
- With respect to social welfare services, this is the area from which LTC services have emanated with respective 'poor law' traditions, social exclusion and disempowering regulations. These are still calling for modernisation and innovation towards social inclusion and demand-driven policies.
- With respect to education, both education services and further training within the wider area of LTC demand innovative research and strategies

to evaluate suggested innovations. This is compounded by the fact that LTC has been and will remain one of the sectors with the highest growth.

With the European move towards addressing LTC services, which has been emphasised by INTERLINKS research, there is a corresponding need for evidence of effectiveness upon which to base newly developing and innovative services. Many 'successful' experiments and examples of good practice fail to transfer to other populations and findings are often inconclusive; readers are often left not understanding what was at fault, possibly the design, the intervention, or the context. Often there are too many idiosyncrasies that make transfer impossible.

Therefore, there is a need to determine different designs that can better capture the effects of innovative LTC developments. This may include, for example, a move towards a broader multidisciplinary orientation, with organisational, professional and user inputs, and pluralistic and technological approaches that incorporate a more lateral spectrum of data collection techniques and data sources. As also stipulated within the UN Convention on the Rights of People with Disabilities and other equal opportunity issues, it is high time to *conceptualise relevant and interconnected long-term care topics that apply to all age groups.* It will be important in the future to expand LTC perspectives on social services, ensuring links between health, education, housing and welfare services. For example, the characteristics of LTC services call for *specific training and education* which are still to be created as current education pathways and job profiles remain connected to either health or social care. Searching for more appropriate specialised and/or integrated education and training designs in LTC will thus become an important issue. Equally, there are other relevant transversal issues, mechanisms and governance aspects, including inter-generational tensions of how the 'identity' of LTC is perceived, that need to be considered within the sphere of LTC services for people with different needs and expectations across the age spectrum.

A wide range of projects and initiatives that have been identified by INTERLINKS and other European projects have certainly enlarged the knowledge base for improving the efficiency of health service delivery. However, these initiatives that were analysed and described through INTERLINKS have also shown that the *rolling out and scaling up of their experiences and research outputs* have not been fully exploited. Indeed, the implementation of research appears to be very demanding and it can take many years to get evidence into practice. For system innovations, where health care organisations are radically changed, the process may even be more complex.

Last, but not least there is a need for developing community processes in which older people and their carers have a much stronger role in managing their health. 'Health literacy', but also 'long-term care literacy' will be another crucial factor for improving the well-being and health of citizens across Europe in the future.

The authors' hopes and expectations are that INTERLINKS and this book will contribute to visible, enduring and sustainable improvements that will be clearly identified when researchers will revisit and analyse long-term care systems in another 50 years' time.

Appendix A: The INTERLINKS Framework for Long-Term Care – Themes, Sub-Themes and Key Issues

The *INTERLINKS Framework for Long-Term Care* consists of validated themes, sub-themes and key-issues to guide the future development of long-term care systems in Europe. This framework was translated into an interactive website (http://interlinks. euro.centre.org) that consists of background information, national and European overview reports as well as almost 100 practice examples from across Europe to illustrate the individual key-issues.

1 Identity of Long-Term Care

1.1 Values
   a) how key principles that characterise LTC are expressed; what values dominate for which stakeholder perspectives (e.g. economic and quality perspectives, citizenship)
   b) whether there are sets of values that are shaping political, organisational and individual choices in LTC (e.g. through surveys of older people's specific needs for care and support)
   c) how informal care and/or family ethics are addressed in legislative frameworks
   d) how issues of dignity, quality of life and empowerment are described in policy papers
   e) how values relating to prevention and rehabilitation are considered
   f) how values embrace the diversity of users and carers (according to gender, culture and social inequalities) and support the specific needs of hard-to-reach groups

1.2 Mission statements
   a) organisations that explicitly address problems at the interfaces between, for example, formal/informal and health and social care, prevention and rehabilitation or the use of migrant care workers
   b) a description of how the problems are addressed and what particular factors distinguish an organisation as an 'LTC organisation', rather than a health or social care organisation
1.3 Organisational definitions
   a) how LTC is defined within or between organisations
   b) how health care providers define their purpose in relation to LTC (cure vs care; needs vs supply; expectations vs preferences)
   c) how clients are defined and positioned

2 Policy and Governance

2.1 Policy
   a) policy barriers and opportunities in terms of linking social and health care (e.g. rhetoric as opposed to more achievable opportunities and results, workforce planning)

b) policies addressing continuity and mechanisms to overcome barriers at the interfaces between social and health care
c) policies addressing continuity and mechanisms to overcome barriers at the interfaces between formal and informal care, in particular to set the balance between caring informally and working
d) policies addressing continuity and mechanisms to overcome barriers at the interfaces between prevention/rehabilitation and cure/care
e) policies addressing continuity and mechanisms to overcome barriers at the interfaces between care at home and residential care
f) implementation strategies
g) policies addressing issues of diversity and equal access, i.e. considering differences according to gender, culture, disability and/or social inequalities

2.2 Legal framework
a) legislation which explicitly addresses LTC with respect to informal carers
b) legislation which explicitly addresses LTC with respect to interfaces between health and social care
c) legislation which explicitly addresses LTC with respect to workforce planning
d) legislation which explicitly addresses LTC with respect to authorisation, accreditation, quality systems
e) legislation which explicitly addresses LTC with respect to coordination or integration of facilities or services

2.3 Governance mechanisms
a) institution building to support and emphasize LTC as a specific area of concern (e.g. quality assurance agencies, insurance system, awareness campaigns)
b) incentives linked to contractual and financial mechanisms
c) incentives to provide LTC by addressing multi-level governance
d) a description of steering mechanisms concerning access and eligibility criteria

2.4 Visibility of key topics
a) Programmes and initiatives to promote prevention and rehabilitation
b) Programmes and initiatives to promote informal care
c) Programmes and initiatives to promote quality development
d) Programmes and initiatives to promote user empowerment

3 Pathways and Processes

3.1 Accessing services
a) case finding through routine screening services (e.g. preventative home visits)
b) transfer of information to users and carers and about users and carers between services or agencies
c) the older person's and carers' interests and involvement which should consider rights, information, choice and entitlements
d) how services deal with diversity and equality of access, considering culture, disability, gender, sexual orientation and class to counter discrimination
e) performance management/indicators that relate to service access
f) ethical guidelines

3.2 Assessing needs
   a) multidisciplinary assessment (protocols, tools and instruments)
   b) assessment tools and instruments (older people's and/or informal carers' needs), protocols (and electronic records)
   c) follow-up of needs assessment (transfer of information)
   d) older people's and/or informal carers' rights: information, shared decision-making, consent, privacy regulations, complaints, second opinion
   e) dealing with diversity (cultural and socio-economic inequalities, disability specific issues)

3.3 Discharge, terminating professional contacts
   a) how professional and/or informal follow-up is properly communicated, available and well prepared
   b) how older people's and carers' rights are ensured (user-friendly information, shared decision-making, consent to care, privacy regulations, complaints, second opinion)
   c) how information (files, care plans) and responsibilities are transferred (logistics issues)
   d) how information to and dialogue with older people and their informal network are facilitated, and how capacities are enabled and strengthened
   e) how funding of next stage care and service delivery is ensured
   f) how outcomes are assessed

3.4 Interdisciplinary work
   a) fostering a culture of collaboration (requirements, training, team building)
   b) inter-professional exchange/development/agreement about views on care and pathways
   c) transfer of information (joint care plans, registers/files)
   d) accountability, responsibilities, dealing with hierarchies and professional–cultural clashes
   e) new ways of involving older people and/or informal carers

4 Management and Leadership

4.1 Management and leadership competence and skills
   a) foster leadership and management using appropriate training
   b) establish management and leadership competences in organisations through mentorship, secondment and shadowing
   c) establish leadership competencies regarding the management of networks

4.2 Quality assurance at workforce level
   a) training of professionals in interdisciplinary/inter-professional working
   b) enabling inter-professional knowledge transfer
   c) fostering diversity-sensitive knowledge and attitudes of staff, promote and make use of multiethnic teams
   d) shaping job profiles, fostering and mutual understanding of comprehensive pathways
   e) establishing competence regarding preventive and rehabilitative LTC
   f) establishing competence and providing capacity for supporting and negotiating with older people and/or informal carers at their level

4.3   Contractual bases of pathway links
   a) using contracts or agreements to enable and sustain processes between services and/or organisations
   b) contracts or agreements that link between services professionally and managerially
   c) contracts or agreements that specify funding across services

4.4   Administrative support at interfaces
   a) how organisations foster enabling administrative patterns as well as processes between services or organisations, making administrative systems compatible, reducing administrative burden

4.5   Ensuring relationships with stakeholders
   a) enable participation of older people and carers' representatives in shaping pathways and appropriate linkages
   b) ensure conditions for older people's and carers' shared decision-making
   c) mobilise volunteers' organisations, and ensure their participation
   d) consider socio-economic, sociocultural, disability and gender differences of older people and carers

4.6   Quality management
   a) approaches for promoting and facilitating the quality of mechanisms in relation to linkage, networking, cooperation, coordination or integration of agencies and organisations
   b) approaches to ensure diversity-suitable structures and processes
   c) approaches to ensure high quality structures and processes involving users/ informal carers
   d) approaches to shape preventive and/or rehabilitative structures and processes
   e) approaches focusing on quality of structures, processes *and* results of LTC providers
   f) approaches to measure and consider user satisfaction

5   Organisational Structures

5.1   Nursing and residential care homes
   a) multidisciplinary teams
   b) structures that facilitate individual and multiprofessional care planning
   c) integrated access points (e.g. concerning referral, financial issues, payment regulation, one-stop shops)
   d) programmes integrating prevention/rehabilitation/reintegration
   e) facilities that help preserving and maintaining informal family relationships
   f) structures that facilitate free choice and access to additional external services including medical (own GP), social (own hairdresser, friendships) or voluntary services
   g) diversity-friendliness: recognition of the specific care needs of hard-to-reach groups, especially their specific needs for information, coordination and support to access available services and benefits

5.2   Care within a hospital setting
   a) multiprofessional teams for assessment, care and treatment
   b) flexible outpatient/outreach services/geriatric ambulatory teams

c) integrated prevention, rehabilitation/remobilisation/reintegration programmes
d) access points (referral, one-stop shops)
e) day clinic services
f) structures that facilitate communication and planning with existing formal care resources and informal carers
g) structures that facilitate integrated discharge and follow-up planning
h) diversity-friendliness: recognition of the specific care needs of hard-to-reach groups, especially their specific needs for information, coordination and support to access available services and benefits

5.3   Transitory care facilities
a) structures that facilitate communication and planning with formal care resources and informal carers
b) access points (referral, one-stop shops)
c) structures that facilitate reassessment and follow-up planning
d) structures that facilitate individual and multiprofessional care planning
e) integrated prevention/rehabilitation/remobilisation/reintegration programmes
f) diversity-friendliness: recognition of the specific care needs of hard-to-reach groups, especially their specific needs for information, coordination and support to access available services and benefits

5.4   Assisted living arrangements
a) structures that facilitate individual and multiprofessional planning of care and living arrangements, e.g. care communities, small units, service housing, sheltered housing (i.e. without care facilities)
b) access points (referral, one-stop shops)
c) structures that facilitate preserving and maintaining informal family relationships
d) structures that facilitate free choice and access to additional external services including medical (e.g. own GP), social (e.g. own hairdresser, friendships) or voluntary services
e) structures that facilitate coordination with formal care resources (e.g. prevention and/or rehabilitation)

5.5   Formal care in the home and community
a) access points (referral, counselling, one-stop shops)
b) flexible and adaptable services to suit individual needs and individual lifestyle
c) multiprofessional teams (e.g. preventive/rehabilitative measures)
d) structures that facilitate coordination and cooperation with other formal and/or informal care (including mobility and transport)
e) structures that facilitate communication, planning and care delivery with informal carers
f) practitioners in independent practice as gatekeepers and/or personal case and care managers
g) diversity-friendliness: recognition of the specific care needs of hard-to-reach groups, especially their specific needs for information, coordination and support to access available services and benefits

5.6   Specialised case or care management centres
a) access points (referral, counselling, one-stop shops)
b) structures that facilitate multiprofessional and inter-agency care planning and coordination

c) structures that are responsible for care planning and coordination between different kinds of services (as independent provider)
d) diversity-friendliness: recognition of the specific care needs of hard-to-reach groups, especially their specific needs for information, coordination and support to access available services and benefits

6   Means and Resources

6.1   (Shared) funding
a) types of funding (insurance-based vs tax-based; co-payments)
b) financial incentives between different levels of funding (national vs regional or local; health vs social; formal vs informal levels; the role of insurance companies/agencies)
c) commissioning and contracting with individual or pooled budgets

6.2   Enabling, allocating and funding human resources
a) recruiting staff (including ethical international recruitment)
b) levels of payment of staff in LTC
c) (new) job profiles in LTC
d) innovative education or career patterns

6.3   Supporting informal carers as a resource for LTC
a) financial support schemes for informal carers and their funding
b) funding of services, training and other in-kind support directed at informal carers
c) funding of initiatives to reconcile work and caring

6.4   Financial indicators
a) expenditures for LTC as a percentage of GDP
b) percentage of population over 65 with LTC needs in residential care as opposed to home care
c) percentage of population over 65 with LTC needs using community care services
d) number of staff working in LTC (as a percentage of total workforce)
e) percentage of funding by stakeholders (public, private other)

6.5   Outcome indicators
a) initiatives that strive to develop and implement outcome indicators
b) costs of different services in relation to number of users
c) methods and indicators used to measure quality of service

6.6   Role of information technology
a) IT solutions in ambient assisted living and smart housing
b) IT solutions in LTC management including electronic records
c) IT applications at the interfaces between health and social care professionals
d) IT applications at the interfaces between health and social care administrations
e) IT applications at the interfaces between formal and informal care

Appendix B: The INTERLINKS Template to Describe and Analyse Practice Examples

The following template was used to describe and analyse practice examples that illustrate the INTERLINKS Framework for long-term care, focusing on the themes, sub-themes and key issues stipulated in Appendix A. With its prompts (in italics) for each individual category, it is also the basis for any further contributions by users of the INTERLINKS website who want to add their own example to illustrate individual key issues.

Theme

Sub-theme

Key issue(s)
Which key issues are addressed?
Choose from key issues from the INTERLINKS Framework

Title

Summary of this example
Use this section for addressing the following aspects: Where is the
example taking place (local, region, country)? In what settings
(e.g. individual organisation, across organisations, whole system)?
Why was the example developed? What are/were the main results?
Max. 150 words, no headings

What is the main benefit for people in need of care and/or carers?

What is the main message for practice and/or policy in relation to this sub-theme?

Links to other INTERLINKS practice examples

Country

Status
Choose from the following: pilot project (terminated), pilot project
(ongoing), project (terminated), project (ongoing), implemented practice
(restricted areas), widespread practice/rolled out.

Keywords

Why was this example implemented?
Use this section for addressing the following aspects: What were the
reasons for starting this example? Which interfaces or gaps does the
example address? For which target groups?
No headings, max. 200 words

Description of the example
Use this section for addressing the 'what' of the example (the type of intervention, service, policy measure, policy paper, legislation). Who are/were the people involved and who is/was driving the example forward (initiator, supporter, promoter, users and carers)? How does/did it affect the person in need of care and/or the carer? Was there user involvement? How is/was it carried out (include also costs, resources), when, where?
Max. 300 words with hyperlinks, with more specific information

What are/were the effects?
Use this section to describe the effects based on evidence: How well did the example address the gaps and improve the interfaces? How were the benefits for older people and informal carers evaluated? What were the demonstrated or potential effects on older people, informal carers, formal carers, organisation/inter-organisational collaboration, costs? To what extent did the example change the way that LTC is provided across systems or services? Has the example proved to be sustainable, or mainstreamed? Has the example been implemented elsewhere?
Max. 300 words

What are the strengths and limitations?
Use this section to give your reflection on strengths, weaknesses, opportunities and threats, intended or non-intended outcomes/effects. Include any doubts whether the outcomes were caused by the measure/ intervention/service. Consider the perspective of the person in need of care and the informal carer.
Max. 200 words

Credits
Author and organisation, reviewer, example verified by ..., and organisation (if available, refer to key person of the example)

Author and organisation:
Reviewer 1:
Reviewer 2:
Verified by:
Organisation:

Bibliography

Abbott, A. (1992) 'From causes to events: Notes on narrative positivism', *Sociological Methods and Research*, 20, 428–55.

Abbott, A. (1995a) 'Things of boundaries', *Social Research*, 62(4), 857–82.

Abbott, A. (1995b) 'Boundaries of social work or social work of boundaries?', *Social Service Review*, 69, 545–62.

Abbott, A. (2007) 'Mechanism and relations', *Sociologica: Italian Journal of Sociology*, 2, 1–22.

Ageing Society – Osservatorio Terza Età (2009) *Rapporto nazionale sulle condizioni ed il pensiero degli anziani*: una Società diversa. Roma, Osservatorio Terza Età.

Åhgren, B. and R. Axelsson (2005) 'Evaluating integrated health care: a model for measurement', *International Journal of Integrated Care*, 5 (August), 1–9.

Åhgren, B., Axelsson, S. and R. Axelsson (2009) 'Evaluating intersectoral collaboration: a model for assessment by service users', *International Journal of Intergrated Care*, 9, e03.

Ala-Nikkola, M. and H. Valokivi (1997) *Yksilökohtainen palveluohjaus käytäntönä*. [Case/care management as a practice], Jyväskylä, Stakes (raportteja 215).

Alaszewski, A. and K. Leichsenring (2004) *Providing integrated health and social care for older persons. A European overview of issues and stake.* Aldershot, Ashgate.

Alaszewski, A., Billings, J. and K. Coxon (2004) 'Integrated health and social care for older persons: theoretical and conceptual issues' in K. Leichsenring and A. Alaszewski (eds) *Providing integrated health and social care for older persons: a European overview of issues at stake*, Aldershot, Ashgate, pp. 53–94.

Alber, J. (1995) 'A framework for the comparative study of social services', *Journal of European Social Policy*, 5(2), 131–49.

Aligon, A., Com-Ruelle, L. and N. Raffy-Pihan (2000) L'hospitalisation à domicile: un patient à satisfaire?, *Informations Hospitalières*, 52, 16–21.

Allen, K. (2011a) 'The partnerships for older people projects: POPPS'. http://interlinks. euro.centre.org/framework, date accessed 17 May 2012.

Allen, K. (2011b) 'Section 75 partnership', http://interlinks.euro.centre.org/framework, date accessed 17 May 2012.

Allen, K. (2011c) 'Intermediate care', http://interlinks.euro.centre.org/framework, date accessed 22 May 2012.

Allen, K., Bednárik, R., Campbell, L., Dieterich, A., Hirsch Durrett, E., Emilsson, T., Glasby, J., Gobet, P., Kagialaris, G., Klavus, J., Kümpers, S., Leichsenring, K., Ljunggren, G., Mastroyiannakis, T., Meriläinen, S., Naiditch, M., Nies, H., Repetti, M., Repková, K., Rodrigues, R., Stiehr, K., van der Veen, R., Wagner, L. and B. Weigl (2011) *Governance and finance of long-term care across Europe. Overview report*. Birmingham/Vienna: University of Birmingham/European Centre for Social Welfare Policy and Research (INTERLINKS Report #4), http://interlinks.euro.centre. org/project/reports, date accessed 27 July 2012.

Allen, K., Glasby, J. and C. Ham (2009) *Integrating health and social care: a rapid review of lessons from evidence and experience*. Birmingham, HSMC.

Alzheimer Europe (2008) *Dementia in Europe yearbook 2008*. Luxembourg, Alzheimer Europe.

Anderson, M.A., Clarke, M., Helms, L. and M. Foreman (2005) 'Hospital readmission from home health care before and after prospective payment', *Journal of Nursing Scholarship*, 37(1), 73–9.

Anttonen, A., Baldock, J. and J. Sipilä (eds) (2003) *The Young, the old and the state*. Cheltenham, Edward Elgar.

Anttonen, A., Sipilä, J. and J. Baldock (2003) 'Patterns of social care in five industrial societies: explaining diversity' in A. Anttonen, J. Baldock and J. Sipilä (eds) *The young, the old and the state*. Cheltenham and London, Edward Elgar, pp. 167–98.

Aries, P. (1974) *Western attitudes toward death: from the Middle Ages to the present*. Baltimore: Johns Hopkins University Press.

Arksey H. and C. Glendinning (2007) 'Choice in the context of care giving', *Health and Social Care in the Community*, 15(2), 165–75.

Arnold, R. and E. Jaffe (2007) 'Why palliative care needs geriatrics', *Journal of Palliative Medicine*, 10, 182–3.

Atwal, A. (2002) 'Nurses' perceptions of discharge planning in acute health care: a case study in one British teaching hospital', *Journal of Advanced Nursing*, 39, 450–8.

Atwal, A. and K. Caldwell (2005) 'Older people: the enigma of satisfaction surveys', *Australian Occupational Therapy Journal*, 52, 10–16.

Audit Commission (1997) *The coming of age. Improving services for older people*. London, Audit Commission Publications.

Audit Commission (2000) *The way to go home. Rehabilitation and remedial services for older people*. London, Audit Commission Publications.

Audit Commission (2009) *A means to an end: Joint financing across health and social care*. London, Audit Commission Publications.

Axelsson, R. and S. Axelsson (2006) 'Integration and collaboration in public health – a conceptual framework', *International Journal of Health Planning and Management*, 21, 75–88.

Badger, F., Clifford, C., Hewison, A. and K. Thomas (2009) 'An evaluation of the implementation of a programme to improve end-of-life care in nursing homes', *Palliative Medicine*, 23, 502–11.

Baldwin-Edwards, M., Kraler, A. *et al.* (2009) *REGINE (regularisations in Europe), study on practices in the area of regularisation of illegally staying third-country Nationals in the Members States of the EU*. Vienna, International Centre for Migration Policy Development (ICMPD).

Barbabella, F., Hoffmann, F., Rodrigues, R., Chiatti, C., Fry, G., Hanson, E., Magnusson, L., Socci, M., Stückler, A., Széman, Z., Widéhn, N. and G. Lamura (2011) *ICT-based solutions for caregivers: Assessing their impact on the sustainability of long-term care in an ageing Europe (CARICT). Final report on the Methodological Framework*, Vienna, European Centre for Social Welfare Policy and Research, http://is.jrc.ec.europa.eu, date accessed 24 July 2012.

Barker, P. (2000) 'Reflections on caring as a virtue ethic within an evidence-based culture', *International Journal of Nursing Studies*, 37, 329–36.

Barlow, J., Singh, D., Bayer, S. and R. Curry (2007) 'A systematic review of the benefits of home telecare for frail elderly people and those with long-term conditions', *Journal of Telemedicine and Telecare*, 13, 172–9.

Barnett, A.M. (2004) 'An inside look at long-term care nursing', *Maryland Nurse*, 6(2), 27.

Barroso, J.M. (2011) *Europe leading social innovation*, http://ec.europa.eu/enterprise/policies/innovation, date accessed 11 February 2012.

Barthel, D. (1965) 'Functional evaluation: the Barthel Index', *Maryland Medical Journal*, 14, 56–61.

Bauch, J. (Hg.) (2006) *Gesundheit als System. Systemtheoretische Betrachtungen des Gesundheitswesens.* Konstanz, Hartung-Gorre-Verlag (Konstanzer Schriften zur Sozialwissenschaft).

Bauer, J.E., Duffy, G.L. and R. Westcott (2006) *The quality improvement handbook.* Milwaukee, Quality Press.

Baxter, K. and C. Glendinning (2011) 'Making choices about support services: disabled adults' and older people's use of information', *Health and Social Care in the Community*, 19(3), 272–9.

Baxter, K., Glendinning, C. and I. Greener (2011) 'The implications of personal budgets for the home care market', *Public Money & Management*, 31(2), 91–98.

Beck, U. (2003) 'The theory of reflexive modernization: problematic, hypotheses and research programme', *Theory, Culture & Society*, 20(1), 1–33.

Bednárik, R. (2010) 'Podmienky a pracovná c̆innost' opatrovatel'ov na Slovensku' [Conditions and care work of carers in Slovakia] in R. Bednárik, H. Jerabek and L. Smekalova (eds) *Sociologie zdravotnictví a medicíny.* Brno, MCSS and LF UP, pp. 86–100.

Bednárik, R., Brichtová, K. and K. Repková (2009) *Informal care in the LTC system.* National Report Slovakia, Bratislava, ILFR (INTERLINKS National Report, http://interlinks.euro.centre.org), date accessed 18 May 2012.

Bednárik, R., Brichtová, K. and K. Repková (2010) *Governance and finance of LTC in the Slovak Republic.* Bratislava, ILFR (INTERLINKS National Report, http://interlinks.euro.centre.org), date accessed 18 May 2012.

Benatar, D., Bondmass, M., Chitelman, J. and B. Avitall (2003) 'Outcomes of chronic heart failure', *Archive of Internal Medicine*, 163, 347–52.

Beneken genaamd Kolmer, D.M., Tellings, A., Gelissen, J.P.T.M., Garretsen, H.F.L. and I.M.B. Bongers (2008) Ranked motives of long-term care providing family caregivers, *Scandinavian Journal of Caring Sciences*, 22(1), 29–39.

Benner, P. (1984) 'From novice to expert: Excellence and power in clinical nursing practice.* Menlo Park, CA, Addison-Wesley.

Bensick, M., Hailey, D. and R. Wooten (2006) 'A systematic review of successes and failures in home telehealth: preliminary results', *Journal of Telemedicine and Telecare*, 12, 8–16.

Benušova, K., Kovačova, M., Nagy, M. and M. Wismar (2011) 'Regaining self-sufficiency: Slovakia and the challenges of health professionals leaving the country' in M. Wismar, C.B. Maier, I.A. Glinos *et al.* (eds) *Health professional mobility and health systems. Evidence from 17 European countries.* Copenhagen, WHO/European Observatory on Health Systems and Policies, pp. 479–510.

Beresford, P. (2003a) *It's our lives: A short theory of knowledge, distance and experience.* London, Citizen Press (in association with 'Shaping Our Lives').

Beresford, P. (2003b) *User involvement in research: Connecting lives, experience and theory*, http://www.feantsa.org, date accessed 9 July 2012.

Berenschot, L. and L. Van der Geest (2012) *Integrale zorg in de buurt.* Utrecht, Nyfer.

Bettio, F., A. Simonazzi and P. Villa (2006) 'Change in care regimes and female migration: the "care drain" in the Mediterranean', *Journal of European Social Policy*, 16(3), 271–85.

Bewley, C., Branfield, F., Glynn, M. *et al.* (2011) *Person-centred support: A guide for service users.* London and York, Shaping Our Lives and Joseph Rowntree Foundation.

Beyer, S., Bitschnau, K., Pelttari, L. and A. Pissarek (2011) 'Hospice and palliative care in nursing homes', http://interlinks.euro.centre.org/framework, date accessed 9 March 2012.

Billings, J. (2009) *Tackling alcohol misuse in teenagers: What works and how can practice develop?* Canterbury, CHSS/University of Kent.

Billings, J. and K. Leichsenring (eds) (2005) *Integrating health and social care services for older persons. Evidence from nine European countries.* Aldershot, Ashgate.

Billings, J., Campbell, L., Glasby, J., Kümpers, S., Leichsenring, K., Ljunggren, G., Naiditch, M., Nies, H., Ruppe, G., Stiehr, K., Triantafillou, J., Wagner, L. *et al.* (2011) *The INTERLINKS Framework – themes, sub-themes, key issues and the template to describe and analyse long-term care,* Canterbury, Vienna, University of Kent, European Centre for Social Welfare Policy and Research *et al.* (INTERLINKS Working Paper #10), http://interlinks.euro.centre.org/project/reports, date accessed 27 July 2012.

Birch, D. and J. Draper (2008) 'A critical literature review exploring the challenges of delivery effective palliative care to older people with dementia', *Journal of Clinical Nursing,* 17, 1144–63.

Birks, Y.F., Hildreth, R., Campbell, P. *et al.* (2003), Randomised controlled trial of hip protectors for the prevention of second hip fractures, Short report', *Age and Ageing,* 32, 442–4.

Bland, J.M. (2004) 'Cluster randomised trials in the medical literature: Two Bibliometric surveys', *BMC Medical Research, Methodology,* 4(21), www.biomedcentral.com/1471–2288/4/21, date accessed 28 July 2012.

Blonski, H. (Hg.) (1999) *Qualitätsmanagement in der Altenpflege – Methoden, Erfahrungen, Entscheidungshilfen.* Hagen, Brigitte Kunz Verlag.

Bode, I. (2005) Alter(n) auf dem Markt der Möglichkeiten. Die Disorganisierung der Seniorenversorgung und ihre Folgen für die Strukturen sozialer Ungleichheiten, *DZA Diskussionspapiere,* 44, 1–27.

Bode, I. (2006) 'Disorganised welfare mixes: voluntary agencies and new governance regimes in Western Europe', *Journal of European Social Policy,* 16(4), 346–59.

Bode, I. and A. Dobrowolski (2009) *'Governance-Performance' bei gemeinnützigen Trägern der stationären Altenhilfe. Zwei Fallstudien und ihre Botschaften.* Wuppertal, Bergische Universität (www.3q.uni-wuppertal.de).

Bode, I., Gardin, L. and M. Nyssens (2011) 'Quasi-marketisation in domiciliary care: varied patterns, similar problems?', *International Journal of Sociology and Social Policy,* 31(3/4), 222–35.

Böhm, E. (1999) *Psychobiographisches Pflegemodell nach Böhm,* 2 Bde., Wien et al., Maudrich.

Bolin, K., Lindgren, B. and P. Lungborg (2008a) 'Informal and formal care among single-living elderly in Europe', *Health Economics,* 17, 393–409.

Bolin, K., Lindgren, B. and P. Lundborg (2008b) 'Your next of kin or your own career? Caring and working among the 50+ of Europe', *Journal of Health Economics,* 27, 718–38.

Bonjean, C., Markham, W. and P. Macken (1994) 'Measuring self-expression in volunteer organizations: A theory based questionnaire', *The Journal of Applied Behavioral Science,* 30, 487–515.

Bonsang, E. (2009) 'Does informal care from children to their elderly parents substitute for formal care in Europe?', *Journal of Health Economics,* 28, 143–54.

Boockvar, K.S. and D.E. Meier (2006) 'Palliative care for frail older adults: "there are things I can't do anymore that I wish I could ..."', *Journal of the American Medical Association,* 296(18), 2245–53.

Boot, J.M. and M.H.J.M. Knapen (2005) *De Nederlandse gezondheidszorg.* Houten, Bohn Stafleu van Loghum.

Boult, C., Reider, L., Leff, B., Frick, K. *et al.* (2011) 'The effect of guided care teams on the use of health services: results from a cluster-randomized controlled trial', *Archives of Internal Medicine*, 171(5), 460–6.

Boyd, C.M., Reider, L., Frey, K. *et al.* (2010) 'The effects of guided care on the perceived quality of health care for multi-morbid older persons: 18-month outcomes from a cluster-randomized controlled trial', *Journal of General Internal Medicine*, 25(3), 235–42.

Bracci F. and G. Cardamone (2005) *Presenze. Migranti e accesso ai servizi socio-sanitar.* Milano, Franco Angeli.

Breyer, F., Costa-i-Font, J. and S. Felder (2010) 'Ageing, health, and health care', *Oxford Review of Economic Policy*, 26(4), 674–690.

Brichtova, L. (2011a) 'A new national model of needs assessment implemented by local governments', http://interlinks.euro.centre.org/framework, date accessed 11 May 2012.

Brichtova, L. (2011b) 'Smart call', http://interlinks.euro.centre.org/framework, date accessed 27 April 2012.

Broese van Groenou, M. and T. van Tilburg (2010) 'Six-year follow-up on volunteering in later life: A cohort comparison in the Netherlands', *European Sociological Review*, 28(1), 1–11.

Brozmanová Gregorová, A. et al. (2012) *Dobrovoľníctvo na Slovensku – výskumné reflexie.* Bratislava, Iuventa – Slovenský inštitút mládeže.

Bryan, K., Gage, H. and K. Gilbert (2006) 'Delayed transfers of older people from hospital: causes and policy implications', *Health Policy*, 76, 194–201.

Buckner, L. and S. Yeandle (2006) *Who cares wins: the social and business benefits of supporting working carers.* London, Carers UK.

Bundesministerium für Gesundheit & Bundesministerium für Familie, Senioren, Frauen und Jugend (Hg.) (2011) *Entwicklung und Erprobung von Instrumenten zur Beurteilung der Ergebnisqualität in der stationären Altenhilfe (Abschlussbericht).* Berlin, BMG/BMFSFJ.

Bundesministerium für Gesundheit, Familie und Jugend (2005) *Schnittstellenmanagement zwischen ambulanter und stationärer Versorgung. Endbericht des Projekts 'MedTogether'.* Wien, BMGFJ.

Bundesministerium für Soziales und Konsumentenschutz/BMSK (2010) *Österreichischer Pflegevorsorge-Bericht 2009.* Wien, BMSK.

Bundesnetzwerk Bürgerschaftliches Engagement (2011) *Wandel des Sozialstaats. Bürgerschaftliches Engagement in der Pflege*, http://b-b-e.de/index.php?id=14317, date accessed 21 August 2011.

Bundesregierung (2010) *Stellungnahme der Bundesregierung zum Bericht der Sachverständigenkommission für den Sechsten Altenbericht 'Altersbilder in der Gesellschaft'.* Berlin, Deutscher Bundestag.

Burau, V. (2007) 'Comparative health research' in M. Saks and J. Allsop (eds) *Researching health. Qualitative, quantitative and mixed methods.* London, Sage, pp. 368–83.

Bureau, V., Theobald, H. and R.H. Blank (2007) *Governing home care: A cross national comparison.* Cheltenham, Edward Elgar.

Burgio, A., Battisti, A., Solipaca, A., Colosimo, S.C., Sicuro, L., Damiani, G., Baldassarre, G., Milan, G., Tamburrano, T., Crialesi, R. and W. Ricciardi (2010) *La relazione tra offerta di servizi di long-term care e i bisogni assistenziali dell'anziano*, Roma, Istat (Contributi Istat, n. 4).

Burt, J. and R. Raine (2006) 'The effect of age on referral to and use of special palliative care services in adult cancer patients: a systematic review', *Age and Ageing*, 35, 469–76.

Busse, R., Geissler, A., Quentin, W. and M. Wiley (2011) *Diagnosis-related groups in Europe: Moving towards transparency, efficiency and quality in hospitals*. Maidenhead, Open University Press and WHO Regional Office for Europe.

Buyck, J.F., Bonnaud, S., Boumendil, A., Andrieu, S., Bonenfant, S., Goldberg, M., Zins, M. and J. Ankri (2011) 'Informal care giving and self-reported mental and physical health: Results from the Gazel Cohort study', *American Journal of Public Health*, 101(10), 1971–9.

Bytheway, B. (1995) *Ageism*. Buckingham, Open University Press.

C.A.R.D.O. (2010) 'Promotion of senior volunteering through international exchange. Practical and policy recommendations', http://www.cardo-eu.net/, date accessed 3 July 2011.

Callahan, C.M., Boustani, M.A., Unverzagt, F.W. *et al.* (2006) 'Effectiveness of collaborative care for older adults with Alzheimer disease in primary care: a randomized controlled trial', *JAMA*, 295(18), 2148–57.

Cameron, A., Lart, T., Harrison, L., Macdonald, G. and R. Smith (2000) *Factors promoting and obstacles hindering joint working: A systematic review*. Bristol, School for Policy Studies/University of Bristol.

Campbell, L. (2011) 'Home rehabilitation', http://interlinks.euro.centre.org/framework, date accessed 17 May 2012.

Campbell, L. and L. Wagner (2011) 'As long as possible in one's own life – sub-project: home-rehabilitation', http://interlinks.euro.centre.org/framework, date accessed 22 May 2012.

Cangiano, A., Shutes, I., Spencer, S. and G. Leeson (2009) *Migrant care workers in ageing societies, report on research findings in the United Kingdom*. Oxford, University of Oxford/Centre on Migration, Policy and Society (COMPAS).

Cano Arana, A., Martín Arribas, M.C., Martínez Piédrola, M., García Tallés, C., Hernández Pascual, M. and A. Roldán Fernández (2008) Eficacia de la planificación del alta de enfermería para disminuir los reingresos en mayores de 65 año, *Atención Primaria*, 40(6), 291–5.

Capitma, J., Leutz, W., Bishop, C. and R. Casler (2005) *Long-term care quality: Historical overview and current initiatives*. Washington, National Commission for Quality Long-term Care.

Caritas Italiana and Fondazione Migrantes (2008) *Dossier statistico immigrazione*. Roma, Edizione Idos.

Carnwell, R. and A. Carson (2008) 'The concepts of partnership and collaboration' in R. Carnwell and J. Buchanan (eds) *Effective practice in health, social care and criminal justice: A partnership approach*. Maidenhead, Open University Press, pp. 3–21.

Carretero, S., Garcés, J. and F. Ródenas (2007) 'Evaluation of the home help service and its impact on the informal caregiver's burden of dependent elders', *International Journal of Geriatric Psychiatry*, 22(8), 738–49.

Carretero, S., Garcés, J., Ródenas, F. and V. Sanjosé (2009) 'The informal caregiver's burden of dependent people: Theory and empirical review', *Archives of Gerontology and Geriatrics*, 49, 74–9.

Carretero, S., Garcés, J. and I. Monsonis (2012) Technologies pour les personnes âgées et politiques européennes, *Les Politiques Sociales*, 1 & 2, 80–91.

Carretero, S., Stewart, J., Centeno, C., Barbabella, F., Schmidt, A., Lamontagne-Godwin, F. and G. Lamura (2012) *Can technology-based services support long-term care challenges in home care? Analysis of evidence from social innovation good practices across the EU. CARICT Project Summary Report*. Seville, JRC-IPTS.

Casale, O., Di Santo, P. and P. Toniolo Piva (2006) Il ritorno del sociale in sanità?, *Animazione sociale*, 4, 18–28.

Cassim, J.D. (2007) 'Empowering designers and users: case studies from the DBA inclusive design challenge' in R. Coleman, J. Clarkson and H. Dong (eds) *Design for Inclusivity: A practical guide to accessible, innovative and user-centred design*. Hampshire, Gower.

Castle, N.G. (2007) 'A review of satisfaction instruments used in long-term care settings', *Journal of Aging & Social Policy*, 19(2), 29–41.

Cattan, M., White, M., Bond, J. and A. Learmouth (2005) 'Preventing social isolation and loneliness among older people: a systematic review of health promotion interventions', *Ageing & Society*, 25, 41–67.

Censis (2008) *Il sociale non decifrato*. Roma, Censis.

Ceruzzi, F. (2011) 'Integrated access point for older people', http://interlinks.euro. centre.org/framework, date accessed 10 May 2012.

Ceruzzi, F. and F. Tunzi (2003) *Professioni sociali. Governo del mercato del lavoro regionale*. Roma, Ediesse.

Challis, D., Darton, R., Hughes, J., Stewart, K. and K. Weiner (2001) 'Intensive care-management at home: an alternative to institutional care?', *Age and Ageing*, 30, 409–13.

Challis, D., Darton, R., Johnson, L. *et al.* (1995) 'Clients and the outcomes of care' in D. Challis, R. Darton, L. Johnson *et al.* (eds) *Care management and health care of older people: The Darlington community care project*. Aldershot, Ashgate, pp. 181–4.

Cheetham, J., Fuller, R., McIvor, G. and A. Petch (1992) *Evaluating social work effectiveness*. Buckingham, Open University Press.

Chiu, C.W.Y and D.W.K Man (2004) 'The effect of training older adults with stroke to use home-based assistive devices', *Occupation, Participation and Health*, 24(3), 113–20.

Cittadinanza attiva (2006) *Rapporto sull'Assistenza domiciliare integrata*, http://www. cittadinanzattiva.it/files/sanita/rapporti/Rapporto_adi.pdf

Cleland, F.G.F., Lousi, A.A., Rigby, A.S., Janssens, U. and H.M. Balk (2005) 'Non-invasive home telemonitoring for patients with heart failure at high risk of recurrent admission and death', *Journal of the American College of Cardiology*, 45(10), 1654–64.

Cochrane Collaboration (2012) 'The Cochrane Collaboration – working together to provide the best evidence for health care', www.cochrane.org, date accessed 27 July 2012.

COCIR (2011) *COCIR E-Health Toolkit for an accelerated deployment and better use of eHealth*, Brussels, COCIR.

Colombo, F. and J. Mercier (2011) 'Help wanted! Balancing fair protection and financial sustainability in long-term care', *Eurohealth*, 17, 3–6.

Colombo, F., Llena-Nozal, A., Mercier, J. and F. Tjadens (2011) *Help wanted? Providing and paying for long-term care*. Paris, OECD Health Policy Studies.

Com-Ruelle, L. and M. Naiditch (2011) 'France: HAH-OP: Hospital at home for older people', http://interlinks.euro.centre.org/framework, date accessed 22 May 2012.

Com-Ruelle, L. and N. Raffy (1994) *Les patients hospitalisés à domicile en 1992*. Paris, CREDES (Rapport CREDES n° 1007).

Comondore, V.R., Devereaux, P.J. *et al.* (2009) 'Quality of care in for-profit and not-for-profit nursing homes: systematic review and meta-analysis', *British Medical Journal*, 339(b2732), 1–15.

Competentiecentrum transities (2009) 'Social business case (care)', www.transitiepraktijk. nl/en/experiment/example/social-business-case-care, date accessed 23 July 2012.

Conclave, M., Facco, M. and G. Casanova (2012) *Il Ministero del Welfare a sostegno del lavoro di cura*. Roma, Italia Lavoro.

Cordero, L. (2011) 'AMICA – Telemonitoring system for COPD patients', http://interlinks.euro.centre.org/framework, date accessed 10 May 2012.

Costa-i-Font, J. and C. Courbage (eds) (2011) *Financing long-term care in Europe: Institutions, markets and models*. Basingstoke, Palgrave Macmillan.

Council of Europe (2003) Recommendation Rec (2003) 24 of the Committee of Ministers to member states on the organisation of palliative care, http://www.coe.int/t/dg3/health/Source/Rec(2003)24_en.pdf, date accessed 3 April 2012.

Council of the European Union (2008) *Council conclusions on public health strategies to combat neurodegenerative diseases associated with ageing and in particular Alzheimer's disease. 2916th Employment, Social Policy, Health and Consumer Affairs Council meeting*. Brussels, European Union.

Cour des comptes (2009) 'La prise en charge des personnes âgées dépendantes' in Cour des comptes (ed.) *Rapport public annuel 2009*. Paris, Cour des comptes, pp. 255–303.

Cox, R. (2006) *The servant problem: Paid domestic work in a global economy*. London, I.B. Tauris.

Cromwell, J., Trisolini, M.G., Pope, G.C., Mitchell, J.B. and L.M. Greenwald (eds) (2011) *Pay for performance in health care: Methods and approaches*. Research Triangle Park, NC, RTI Press.

Cuijpers P. and H. Nies (1997) 'Supporting informal caregivers of demented elderly people: psychosocial interventions and their outcomes' in B.M.L. Miesen and G.M.M. Jones (eds) *Care-giving in Dementia. Research and applications, Volume 2*. London and New York, Routledge, pp. 168–77.

Culyer, J. (1995) 'Need: The idea won't do – But we still need it', *Social Science & Medicine*, 40(6), 727–30.

Cylus, J. and R. Irwin (2010) 'The challenges of hospital payment systems', *Euro Observer*, 12(3), 1–4.

Da Roit, B. and B. Le Bihan (2010) 'Similar and yet so different: cash-for-care in six European countries' long-term care policies', *The Milbank Quarterly*, 88(3), 286–308.

Da Roit, B. and C. Castegnaro (2004) *Chi cura gli anziani non autosufficienti? Famiglia, assistenza privata e rete dei servizi per gli anziani in Emilia-Romagna*. Milano, Franco Angeli.

Dar, O., Riley, J., Chapman, C., Dubrey, S.W., Morris, S., Rosen, S.D., Roughton, M. and M.R. Cowie (2009) 'A randomized trial of home telemonitoring in a typical elderly heart failure population in North West London: results of the Home-HF study', *European Journal of Heart Failure*, 11, 319–25.

Darkins, A. (2012) 'International perspective on telehealth and telecare'. Keynote address at the *International Congress on Telehealth and Telecare*, 6–8 March 2012, London.

De Prins, P. and E. Henderickx (2007) 'HRM effectiveness in older people's and nursing homes: The search for best (quality) practices', *Nonprofit and Voluntary Sector Quarterly*, 36, 549–71.

Decruynaere, E. (2010) *The personal budget (PGB) in the Netherlands*. Gentbrugge, Expertise Centre Independent Living (www-en.independentliving.be).

Dekker, P. and L. Halman (2003) 'Volunteering and Values: An Introduction' in P. Dekker and L. Halman (eds) *The Values of Volunteering. Cross-Cultural Perspectives*, New York, Kluwer/Plenum Publisher, pp. 1–17.

Del Favero, A.L. (2011) *Secondo rapporto sulla non autosufficienza in Italia*. Roma, Ministero del Lavoro e delle Politiche Sociali.

DelliFraine, J.L. and K.H. Dansky (2008) 'Home-based telehealth: a review and meta-analysis', *Journal of Telemedicine and Telecare*, 14(2), 62–6.

Delnoij, D.M.J. *et al.* (2006) 'Made in the USA: the import of American consumer assessment of health plan surveys (CAHPS) into the Dutch social insurance system', *European Journal of Public Health*, 16(6), 652–9.

Denis, J.-L., Lamothe, L., Langley, A. and A. Valette (1999) 'The struggle to redefine boundaries in health care systems' in D. Brock, C.R. Hinings and M. Powell (eds) *Restructuring the professional organisation*. London, Routledge, pp. 105–30.

Department of Health (2004) *Making partnership work for patients, carers and service users: A strategic agreement between the Department of Health, the NHS and the voluntary and community sector*. London, HMSO.

Department of Health (2006) *Our health, our care, our say: a new direction for community services*. London, HMSO.

Department of Health (2008) *End of life care strategy – promoting high quality care for all adults at the end of life*. London, HSMO.

Department of Health (2009) *Prevention package for older people*. London, HSMO.

Dharmasena, H.P. and K. Forbes (2001) 'Palliative care for patients with non-malignant disease: will hospital physicians refer?', *Palliative Medicine*, 15, 413–18.

Di Santo, P. (2004) 'Costruire la filiera di servizi per la domiciliarietà', *Animazione Sociale*, 1, 162–7.

Di Santo, P. (2010) 'Immigrazione e promozione della salute. Il progetto Healthy Inclusion: sfide, opportunità e strategie per il cambiamento', *Welfare on line*, 5–10.

Di Santo, P. (2011a) 'Protected hospital discharge and palliative care teams', http://interlinks.euro.centre.org/framework, date accessed 12 April 2012.

Di Santo, P. (2011b) 'R.O.S.A. – Network for employment and care services: Promoting the regulation of undeclared work and increasing' quality of care work', http://interlinks.euro.centre.org/framework, date accessed 14 May 2012.

Di Santo, P. (2011c) 'Care at home by integrating formal and informal care', http://interlinks.euro.centre.org/framework, date accessed 14 May 2012.

Di Santo, P. and F. Ceruzzi (2009) *Quality in long-term care: Italy*. Rome, Studio Come (INTERLINKS National Report, http://interlinks.euro.centre.org/countries).

Di Santo, P. and F. Ceruzzi (2010) *Migrant care workers in Italy. A case study*, Rome/Vienna, European Centre/Studio Come (INTERLINKS Report #3a, http://interlinks.euro.centre.org/project/reports).

Di Santo, P. and O. Casale (2008) *Piano sociale di zona. Strumenti per il Welfare locale*. Roma, Ediesse.

Dickens, A.P., Richards, S.H., Greaves, C.J. and J.L. Campbell (2011) 'Interventions targeting social isolation in older people: a systematic review', *BMC Public Health*, 11, 1–22.

Dickinson, H. (2006) 'The evaluation of health and social care partnerships: an analysis of approaches and synthesis for the future', *Health & Social Care in the Community*, 14(5), 375–83.

Dickinson, H. and J. Glasby (2009) 'Introduction' in J. Glasby and H. Dickinson (eds) *International perspectives in health and social care*, Chichester, Wiley-Blackwell, pp. 1–9.

Dickinson, H. and J. Glasby (2010) 'Why partnership working doesn't work', *Public Management Review*, 12(6), 811–28.

Diermanse, I. (2011) 'Integrated care programmes: quality management of stroke related services', http://interlinks.euro.centre.org/framework, date accessed 14 March 2012.

Dieterich, A. (2011a) 'Family doctors contracted as staff members in care homes', http://interlinks.euro.centre.org/framework, date accessed 10 May 2012.

Dieterich, A. (2011b) 'Implementation of everyday assistance in institutional dementia care', http://interlinks.euro.centre.org/framework, date accessed 10 May 2012.

Directorate-General for Economic and Financial Affairs (2002a) 'Incorporating the sustainability of public finances into the Stability and Growth Pact', *European Economy*, 3, 62–74.

Directorate-General for Economic and Financial Affairs (2002b) 'The long-term sustainability of public finances', *European Economy*, 3, 32–6.

Dixon, A. and E. Mossialos (eds) (2002) *Health care systems in eight countries: Trends and challenges*. London, The London School of Economics and Political Science.

Dockray S. and A. Steptoe (2010) 'Positive affect and psychobiological processes', *Neuroscience & Biobehavioral Reviews*, 35, 1, 69–75.

Donabedian, A. (1966) 'Evaluating the Quality of Medical Care', *Milbank Foundation Quarterly*, 2, 166–206.

Donabedian, A. (1980) *The definition of quality and approaches to its assessment. Vol. 1. Explorations in quality assessment and monitoring*. Ann Arbor, Health Administration Press.

Dopson, S., Fitzgerald, L., Ferlie, E., Gabbay, J. and L. Locock (2002) 'No magic targets! Changing clinical practice to become more evidence-based', *Health Care Management*, 27(3), 35–47.

Dopson, S., Locock, L., Gabbay, J., Ferlie, E. and L. Fitzgerald (2003) 'Evidence-based medicine and the implementation gap, *Health: An Interdisciplinary Journal for the Social Study of Health, Illness and Medicine* 7(3), 311–30.

Dowling, B., Powell, M. and C. Glendinning (2004) 'Conceptualising successful partnerships', *Health and Social Care in the Community*, 12(4), 309–17.

Downey, J. (2011) 'Old age and new speak', *Journal of Care Services Management*, 5(1), 23–7.

Drummond, M., O'Brien, B. Stoddart, G. and G. Torrance (2001) *Methods for the economic evaluation of health care programmes*. Oxford, Oxford University Press.

Du Tertre, C. (2005) 'Services immatériels et relationnels: intensité du travail et santé', *Activités*, 2(1), 37–49.

Du Tertre, C. (2009) 'Modèle industriel et modèle serviciel de performance', *Economies et Sociétés, Série 'Économie et Gestion des Services'*, 10(4), 643–62.

Duffy, S. (2005) 'Will "in Control" at last put people in charge of their lives?', *Community Living*, 13(1), 8–16.

Dulac, A.-V., K. Falk, Giraud, O., Kümpers, S., Lechevalier, A., Lucas, B., Purschke, K. and L. Sgier (2012) *Policy learning and innovation in local regimes of home-based care for the elderly: Germany, Scotland and Switzerland*. Berlin, Geneva, Centre Marc Bloch, CNRS, Université de Genève (IRSP) et Haute Ecole Travail Social (HETS), Social Science Research Center (WZB).

Dupuis, A. and L. Farinas (2010) 'La gouvernance des systèmes multi-organisationnels. L'exemple des services sanitaires et sociaux au Québec', *Revue française d'administration publique*, 135, 549–65.

Economist Intelligence Unit (2011) *The future of healthcare in Europe*. A report from the Economist Intelligence Unit. Geneva *et al.*, EIU (http://www.managementthinking. eiu.com).

Eichler, F. and B. Pfau-Effinger (2009) 'The "consumer principle" in the care of elderly people: Free choice and actual choice in the German welfare', *Social Policy & Administration*, 43(6), 617–33.

Eigmüller, M. (2006) 'Der duale Charakter der Grenze. Bedingungen einer aktuellen Grenztheorie' in M. Eigmüller and G. Vobruba (eds) *Grenzsoziologie. Die politische Strukturierung des Raumes*. Wiesbaden, VS Verlag für Sozialwissenschaften, pp. 55–73.

Eisma, R., Dickinson A., Goodman, J., Syme, A., Tiwari, L. and A. Newell (2004) 'Early user involvement in the development of Information Technology-related products for older people', *Universal Access in the Information Society*, 3(2), 131–40.

Elfering, J. and R. Scherpenzeel (2009a) 'Intensive respite care. Complex care supported voluntarily', www.movisie.nl, date accessed 22 August 2011.

Elfering, J. and R. Scherpenzeel (2009b) 'Beside the family carer. A companion for family carers', www.movisie.nl, date accessed 22 August 2011.

Elfering, J. and R. Scherpenzeel (2009c) 'Crossing thresholds together. Buddies support non-western family carers', www.movisie.nl, date accessed 22 August 2011.

Elfering, J. and R. Scherpenzeel (2009d) 'Voluntary coaches. Intensive coaching of care volunteers', www.movisie.nl, date accessed 22 August 2011.

Emilsson, T. (2011) 'Municipal LTC obligations to support informal carers', http://interlinks.euro.centre.org/framework, date accessed 10 May 2012.

Emilsson, T. (2011) 'Dementia guidelines and informal carers', http://interlinks.euro.centre.org/framework, date accessed 10 May 2012.

Empirica, WRC and European Commission (2010) 'ICT & Ageing. European study on users, markets and technologies. Final report', http://www.ict-ageing.eu, date accessed 27 July 2012.

Employers for Carers (2010) '"Business case" webpage', www.employersforcarers.org/BusinessCase, date accessed 1 August 2012.

Engel, H. and D. Engels (2000) *Case management in various national elderly assistance systems*. Kohlhammer, Köln.

Engels, D. und F. Pfeuffer (2007) *Die Einbeziehung von Angehörigen und Freiwilligen in die Pflege und Betreuung in Einrichtungen*. Köln, Institut für Sozialforschung und Gesellschaftspolitik e.V.

Equality and Human Rights Commission (2011) *Close to home: An inquiry into older people and human rights in home care*. Manchester, Glasgow, Cardiff, Equality and Human Rights Commission (www.equalityhumanrights.com).

Esping-Andersen, G. (1990) *The three worlds of welfare capitalism*. Cambridge, Polity Press.

Esping-Andersen, G. (1997) *Welfare states at the end of the century: the impact of labour market, family and demographic change*. Paris, OECD (Social Policy Studies nr. 21).

Esping-Andersen, G. (1999) *Social foundations of postindustrial economies*. Oxford, Oxford University Press.

Etchemendy, E., Baños, M., Botella, C., Castilla, D., Alcañiz, M., Rasal, P. and L. Farfallini (2011) 'An e-health platform for the elderly population: The butler system', *Computers & Education*, 56, 275–9.

Eurobarometer (2007) 'Health and long-term care in the European Union', http://ec.europa.eu/public_opinion/archives/ebs/ebs_283_en.pdf, date accessed 27 July 2012.

EUROCARERS (2009) 'Family care in Europe', www.eurocarers.org/library_factsheets.php, date accessed 27 July 2012.

EUROFAMCARE (2006) 'Services for supporting family carers of dependent older people in Europe: the trans-European survey report (TEUSURE)' – http://www.uke.de/extern/eurofamcare/deli.php, date accessed 17 May 2012.

Eurofound EWCO (2008) 'Work-life balance policies benefit working parents', www.eurofound.europa.eu/ewco/2008/06/DK0806059I.htm, date accessed 18 April 2012.

European Centre for Social Welfare Policy and Research (ed.) (2010) *Measuring progress: Indicators for care homes*, Vienna *et al.*, European Centre for Social Welfare Policy and Research, City University London, E-Qalin Ltd., MDK, Stichting Vilans, TU Dortmund.

European Commission (2006) *Modernising labour law to meet the challenges of the 21st century, Green Paper, COM (2006)708 final*. Brussels, European Commission.

European Commission (2007) *A lead market initiative for Europe*. Brussels, EC.

European Commission (2008a) *Long-term Care in the European Union*. Brussels, DG Employment, Social Affairs and Equal Opportunities.

European Commission (2008b) *A better work-life balance: Stronger support for reconciling professional, private and family life, COM(2008) 635 final*. Brussels, European Commission.

European Commission (2009) *Telemedicine for the benefit of patients, healthcare systems and society*. Commission Staff Working Paper, http://ec.europa.eu/information_society/ activities/health/docs/policy/telemedicine/telemedecine-swp_sec-2009-943.pdf, date accessed 11 April 2012.

European Commission (2010a) *Overview of the European strategy in ICT for ageing well*. Brussels, European Commission, Information Society and Media.

European Commission (2010b) *A digital agenda for Europe. Communication from the commission to the European parliament, the council, the European economic and social committee and the committee of the regions*, Brussels, European Commission.

European Commission (2011a) *Europe for citizens. Education, audiovisual and culture executive agency (EACEA)*, http://eacea.ec.europa.eu/citizenship/index_en.php, date accessed 21 May 2012.

European Commission (2011b) *Roadmap: reconciliation between work, family and private life*, http://ec.europa.eu/governance/impact/planned_ia/docs/2011_empl_016_work_ family_private_life_en.pdf, date accessed 4 May 2012.

European Commission (2012) *Active ageing, special eurobarometer 378*, http://ec.europa. eu/public_opinion/archives/ebs/ebs_378_en.pdf, date accessed 21 May 2012.

European Union (2009) *2009 Ageing report: Economic and budgetary projections for the EU27 Member States (2008–2060)*. Luxembourg, Publications Office of the European Union (European Economy 2).

European Union (2010) *This is European social innovation* (co-ordinated by the Social Innovation eXchange (SIX) at the Young Foundation, Euclid Network, and the Social Innovation Park), www.youngfoundation.org, date accessed 11 February 2012.

European Union (2011) *Demography Report 2010. Older, more numerous and diverse Europeans*. Luxembourg, Publications Office of the European Union.

European Union (2012) *The 2012 ageing report. Economic and budgetary projections for the 27 EU Member States (2010–2060)*. Brussels, DG ECOFIN (European Economy 2).

European Union Agency for Fundamental Rights – FRA (2011) *Migrants in an irregular situation employed in domestic work: Fundamental rights challenges for the European Union and its Member States*. Luxembourg, Publications Office of the European Union.

Eurostat (2009) *Reconciliation between work, private and family life in the European Union*. Luxembourg, Publications Office of the European Communities.

Eurostat (2011) Information society statistics, http://epp.eurostat.ec.europa.eu/portal/ page/portal/information_society/data/main_tables, date accessed 15 February 2012.

EUSTaCEA (2010) 'European charter of the rights and responsibilities of older people in need of long-term care and assistance', www.age-platform.eu/images/stories/ 22204_AGE_charte_europeenne_EN_v4.pdf, date accessed 27 July 2012.

Evers, A. (1990) 'Shifts in the welfare mix – Introducing a new approach for the study of transformations in welfare and social policy' in A. Evers and H. Wintersberger (eds) *Shifts in the welfare mix. Their impact on work, social services and welfare policies.* Frankfurt a.M./Boulder, Campus/Westview, pp. 7–30.

Evers, A. and H. Wintersberger (eds) (1990) *Shifts in the welfare mix: Their impact on work, social services and welfare policies.* Frankfurt a.M./Boulder, Campus/ Westview.

Evers, A. und B. Ewert (2010) 'Hybride Organisationen im Bereich sozialer Dienste. Ein Konzept, sein Hintergrund und seine Implikationen' in T. Klatetzki (ed.) *Soziale personenbezogene Dienstleistungsorganisationen. Soziologische Perspektiven.* Wiesbaden, VS Verlag für Sozialwissenschaften, pp. 103–28.

Evers, A., Haverinen, R., Leichsenring, K. and G. Wistow (eds) *Developing Quality in Personal Social Services. Concepts, Cases and Comments.* Aldershot: Ashgate (Series 'Public Policy and Social Welfare', Vol. 22), pp. 9–24.

Evers, A., Pijl, M. and C. Ungerson (1994) *Payments for care.* Aldershot, Avebury.

Falk, K. (2012) 'Selbstbestimmung bei Pflegebedarf im Alter – wie geht das? Kommunale Handlungsspielräume zur Versorgungsgestaltung' in S. Kümpers and J. Heusinger (eds) *Autonomie trotz Armut und Pflegebedarf? Altern unter Bedingungen von Marginalisierung.* Bern, Huber, pp. 39–75.

Family Caregiver Alliance (2003) *Fact sheet: Women and caregiving: Facts and figures.* San Francisco, Family Caregiver Alliance.

Faß, R. (2009) *Helfen mit System. Systemsteuerung im case management.* Marburg, Tectum Verlag.

Feil, N. (2002) *The validation breakthrough: Simple techniques for communicating with people with Alzheimer's-type Dementia.* Baltimore, Health Professions Press.

Flick, U. (1995) *Qualitative Forschung. Theorie, Methoden, Anwendung in Psychologie und Sozialwissenschaften.* Reinbek bei Hamburg, Rowohlt.

Fontaine, R., Gramin, A. and J. Wittwer (2009) 'Providing care for an elderly parent: interactions among siblings', *Health Economics*, 18, 1011–29.

Forder, J., Knapp, M. and G. Wistow (1996) 'Competition in the mixed economy of care', *Journal of Social Policy*, 25(2), 201–21.

Foster, M., Harris, J., Jackson, K., Morgan, H. and C. Glendinning (2006) 'Personalised social care for adults with disabilities: a problematic concept for frontline practice', *Health and Social Care in the Community*, 14, 125–35.

Foxcroft, D. (2006) *Alcohol misuse prevention for young people: A rapid review of recent evidence.* WHO Technical Report. Geneva, WHO.

Franzoni, F. and M. Anconelli (2003) *La rete dei servizi alla persona. Dalla normativa all'organizzazione.* Roma, Carocci Faber.

Freiwilligen-Agentur Halle-Saalkreis e.V. (2006a) Abschlussbericht zum Modellprojekt 'Ehrenamt in der Pflege': beziehungsweise – Ehrenamtliche Besuchsdienste in der ambulanten und stationären Altenhilfe, http://www.gkv-spitzenverband.de, date accessed 21 May 2012.

Freiwilligen-Agentur Halle-Saalkreis e.V. (2006b) Ehrenamtliche Besuchsdienste für ältere Menschen. Praxishilfe für lokale Koordinierungsstellen, www.gkv-spitzenverband.de, date accessed 31 August 2011.

Freiwilligen-Agentur Halle-Saalkreis e.V. (2006c) Ehrenamtliche Besuchsdienste für ältere Menschen. Praxishilfe Freiwilligen-Management für Pflegeeinrichtungen, www.gkv-spitzenverband.de, date accessed 31 August 2011.

French, P. (2002) 'What is the evidence on evidence-based nursing? An epistemological concern', *Journal of Advanced Nursing*, 37(3), 250–7.

Frerichs, F., Leichsenring, K., Naegele, G., Reichert, M. und M. Stadler (2003) *Qualität Sozialer Dienste in Deutschland und Österreich*. Münster, LIT Verlag.

Frič, P., Pospíšilová, T. *et al.* (2010) *Vzorce a hodnoty dobrovolnictví v české společnosti na začátku 21. Století.* Praha, AGNES.

Fried, L.P., Tangen, C., Walston J. *et al.* (2001) 'Frailty in older adults: evidence for a phenotype', *Journal of Gerontology*, 56A, M1–M11.

Froggatt, K., Hockley, J., Parker, D. and K. Brazil (2011) 'A system lifeworld perspective on dying in long term care settings for older people: contested states in contested places', *Health and Place*, 17, 263–8.

Fuchs, P. (2006) 'Das Gesundheitssystem ist niemals verschnupft' in J. Bauch (Hg.) *Gesundheit als System. Systemtheoretische Betrachtungen des Gesundheitswesens*. Konstanz, Hartung-Gorre-Verlag, pp. 21–38.

Fujisawa, R. and F. Colombo (2009) *The long-term care workforce: Overview and strategies to adapt supply to a growing demand.* Paris, OECD (Health Work Paper, No. 44).

FUTURAGE (2011) *A road map for European ageing research*, http://futurage.group.shef.ac.uk, date accessed 11 February 2012.

Garcés, J. (2011) 'Social security and informal carers', http://interlinks.euro.centre.org/framework, date accessed 6 July 2012.

Garcés, J. and F. Ródenas (2012) 'Sustainable social and health care transitions in advanced welfare states' in J. Broerse and J. Grin (eds) *Towards system innovations in health systems: Understanding historical evolution, innovative practices and opportunities for a transition in healthcare, Part II. Innovating practices: Experiences and lessons.* New York, Routledge.

Garcés, J., Carretero, S. and F. Ródenas (2011) *Readings of the social sustainability theory*. Valencia, Tirant lo Blanch.

Garcés, J., Ródenas, F. and V. Sanjosé (2004) 'Care needs among the dependent population in Spain: an empirical approach', *Health and Social Care in the Community*, 12(6), 466–74.

Garcés, J., Ródenas, F. and V. Sanjosé (2006) 'Suitability of the health and social care resources for persons requiring long-term care in Spain: an empirical approach', *Health Policy*, 76, 121–30.

Garcés, J., Carretero, S, Ródenas, F. and C. Alemán (2010) 'A review of programs to alleviate the burden of informal caregivers of dependent persons', *Archives of Gerontology and Geriatrics*, 50, 254–9.

García-Armesto, S., Abadía-Taira, M.B., Durán, A., Hernández-Quevedo, C. and E. Bernal-Delgado (2010) 'Spain: Health system review', *Health Systems in Transition*, 12(4), 1–295.

Genton, A. (2008) *Quartiers solidaires: exploration d'un pari communautaire.* Rapport de recherche. Lausanne, Pro Senectute Vaud, Fondation Leenaards.

German Federal Ministry of Family Affairs etc. (2007) *Charta der Rechte hilfe- und pflegebedürftiger Menschen*, Berlin, German Federal Ministry of Family Affairs, Senior Citizens, Women and Youth, and German Federal Ministry of Health (www.bmfsfj.de).

Gérvas, J. (2008) 'Case and disease management and improved integration of healthcare services in Spain', *Gaceta Sanitaria*, 22(1), 163–8.

Gheeraw, R.R. and C.S. Lebbon (2002) 'Inclusive design – developing theory through practice' in S. Keates, P.M. Langdon, P.J. Clarkson and P. Robinson (eds) *Universal access and assistive technology.* London, Springer, pp. 43–52.

GHK (2010) 'Volunteering in the European Union – final report', http://www. eyv2011.eu, date accessed 31 August 2011.

Gieryn, T.F. (1983) 'Boundary-work and the demarcation of science and non-science: Strains and interests in professional ideologies of scientists', *American Sociological Review*, 48, 781–95.

Gil, N.M., Hine, N.A. and J.L. Arnott (2010) 'Interactive and inclusive design for older people and carers'. *Latin American conference on networked and electronic, edia*, 8–10 September 2010, Cali – Colombia, http://www.match-project.org.uk/resources/ project_publications.html, date accessed 17 May 2012.

Gilligan, C. (2001) *In a different voice. Psychological theory and women's development.* Cambridge MA, Harvard University Press.

Giordano, R., Clark, M. and N. Goodwin (2011) *Perspectives on telehealth and telecare. Learning from the 12 whole system demonstrator action network (WSDAN) sites. WSDAN briefing paper.* London, The King's Fund.

Glasby, J. (2003) *Hospital discharge: integrating health and social care.* Abingdon, Radcliffe Medical Press.

Glasby, J. (2007) *Understanding health and social care.* Bristol, The Policy Press.

Glasby, J. (2011a) 'Care Trusts: structural integration of health and social care', http:// interlinks.euro.centre.org/framework, date accessed 3 March 2012.

Glasby, J. (2011b) 'Modernisation of older people's services in Birmingham', http:// interlinks.euro.centre.org/framework, date accessed 3 March 2012.

Glasby, J. (2011c) 'The "big care debate" – public engagement in LTC funding', http:// interlinks.euro.centre.org/framework, date accessed 3 March 2012.

Glasby, J. (2011d) 'Reimbursement for delayed hospital discharges', http://interlinks. euro.centre.org/framework, date accessed 10 May 2012.

Glasby, J. and P. Beresford (2006) 'Who knows best? Evidence-based practice and the service user contribution', *Critical Social Policy*, 26(1), 268–84.

Glasby, J. and R. Littlechild (2009) *Direct payments and personal budgets: putting person-alisation into practice (second edition).* Bristol, Policy Press.

Glasby, J., Walshe, K. and G. Harvey (2007) 'What counts as "evidence" in "evidence-based practice"?', *Evidence and Policy*, 3(3), 325–7.

Glasl, F., Kalcher, T. and H. Piber (eds) (2005) *Professionelle Prozessberatung. Das Trigon-Modell der sieben OE-Basisprozesse.* Bern, Haupt.

Glendinning, C. (2003) 'Breaking down barriers: integrating health and care services for older people in England', *Health Policy*, 65(2), 139–51.

Glendinning, C. (2010) 'Continuous and long-term care: European perspectives' in D. Dannefer and C. Phillipson (eds) *The SAGE Handbook of Social Gerontology.* London, Sage Publications, pp. 551–62.

Glendinning, C. and P.A. Kemp (eds) (2008) *Cash and care: policy challenges in the welfare state.* Bristol, Policy Press.

Glendinning, C., Challis, D., Fernández, J. *et al.* (2008) *Evaluation of the individual budgets pilot programme.* York, Social Policy Research Unit.

Glendinning, C., Tjadens, F., Arksey, H., Moree, M., Moran, N. and H. Nies (2009) *Care provision within families and its socio-economic impact on care providers.* Report for the European Commission DG Employment. York and Utrecht, SPRU, University of York, Vilans Centre of Expertise for Long-term Care.

Gobet, P. (2011a) 'Case management for patients of lower socio-economic status experiencing complex somatic and psychosocial problems (KOMPASS)', http:// interlinks.euro.centre.org/framework, date accessed 3 May 2012.

Gobet, P. (2011b) 'Improvement of discharge planning through formal collaboration between hospital and home care organisation: Express service', http://interlinks. euro.centre.org/framework, date accessed 10 May 2012.

Gobet, P., Hirsch Durrett, E. and M. Repetti (2009) *Quality in LTC: Report on quality in Switzerland*. Lausanne, HESS/EESP (INTERLINKS National Report, http://interlinks. euro.centre.org/countries).

Gold Standards Framework (GSF) (2012) 'The gold standards framework: Enabling a gold standard of care for all people nearing the end of life', http://www. goldstandardsframework.org.uk, date accessed 3 May 2012.

Goodman, J., Dickinson, A. *et al.* (2004) *Gathering requirements for mobile devices using focus groups with older people. Second Cambridge workshop on universal access and assistive technology.* London, Springer.

Goodwin, N. (2012) Nick Goodwin on International Congress on Telehealth and Telecare, http://www.kingsfund.org.uk/multimedia/goodwin_telehealth12.html

Gori, C. (a cura di) (2002) *Il welfare nascosto. Il mercato privato dell'assistenza in Italia e in Europa.* Roma, Carocci.

Gott, M., Ibrahim, A.M. and R.H. Binstock (2011) 'The disadvantaged dying: ageing, ageism, and palliative care provision for older people in the UK' in M. Gott and C. Ingleton (eds) *Living with ageing and dying: Palliative and end of life care for older people.* New York, Oxford University Press, pp. 52–62.

Graafmans, J.A.M. and A. Brouwers (1989) 'Gerontechnology, the modelling of normal aging' in Human Factors and Ergonomics Society (ed.) *Proceedings of the 33rd Annual Meeting*, Denver CO, USA, pp. 187–90.

Gramain, A. and L. Malavolti (2004) 'Evaluating the effect of care programs for elderly persons with dementia on caregiver's well-being', *The European Journal of Health Economics*, 5(1), 6–14.

Gravelle, H., Dusheiko, M., Sheaff, R., Sargent, P., Boaden, R., Pickard, S., Parker, S. and M. Roland (2007) 'Impact of case management (Evercare) on frail elderly patients: controlled before and after analysis of quantitative outcome data', *British Medical Journal*, 224, 31–4.

Greaves, M. and C. Rogers-Clark (2009) 'The experience of socially isolated older people in accessing and navigating the health care system', *Australian Journal of Advanced Nursing*, 27, 5–11.

Groenewoud, H., Egers, I., Pool, A. and J. de Lange (2008) *Evaluatieonderzoek van de pilot casemanagement dementie in regio Delft Westland Oostland 2005–2007, eindrapport.* Rotterdam, Hogeschool Rotterdam.

Groenewoud, S. (2008) It's your choice. A study of search and selection processes and the use of performance indicators in different patient groups. Rotterdam, Erasmus University (PhD Thesis).

Grunfeld, E., Coyle, D., Whelan, T. *et al.* (2004) 'Family caregiver burden: results of a longitudinal study of breast cancer patients and their principal caregivers', *Canadian Medical Association Journal*, 170 (12), 1795–801.

Grypdonck, M. (2006) 'Qualitative health research and the era of evidence-based practice', *Qualitative Health Research*, 16(10), 1371–85.

Guba, E.G. and Y.S. Lincoln (1992) *Effective evaluation: Improving the usefulness of evaluation results through responsive and naturalistic approaches.* San Francisco, Jossey-Bass.

Hall, S., Petkova, H., Tsouros, A.D., Costantini, M. and I.J. Higginson (eds) (2011) *Palliative care for older people: better practices.* Copenhagen, WHO Regional Office for Europe.

Hallberg, I.R. (2006) 'Palliative care as a framework for older people's long-term care', *International Journal of Palliative Nursing*, 12(5), 224–9.

Ham, C. (2012) 'Competition and integration in health care reform', *International Journal of Integrated Care*, 12.

Hamerman, D. (1999) 'Toward an understanding of frailty', *Annals of Internal Medicine*, 130, 945–50.

Hammar, T. (2008) *Palvelujen yhteensovittaminen kotihoidossa ja kotiutumisessa: kotihoidon asiakkaiden avun tarve ja palvelujen käyttö sekä PALKO-mallin vaikuttavuus ja kustannus-vaikuttavuus* [Integrated Services in the Practice of Discharge and Home Care (PALKO) – Home-care clients' use of services and need for help, and the effectiveness and cost-effectiveness of the PALKO model – abstracts in English and Swedish]. Helsinki, Stakes (Research Reports 179).

Hammar, T. (2011) 'Integrated home care and discharge practice for home care clients (PALKOmodel)', http://interlinks.euro.centre.org/framework, date accessed 10 May 2012.

Hammar, T. and H. Finne-Soveri (2011) 'Activating daily life (ADaL)-programme together with technology solutions', http://interlinks.euro.centre.org/framework, date accessed 13 March 2012.

Hammar, T., Finne-Soveri, H., Mikkola, H.M., Noro, A. and T. Hujanen (2009) *Developing and ensuring quality in LTC in Finland*. Helsinki, THL (INTERLINKS National Report, http://interlinks.euro.centre.org/countries).

Hammar, T., Niemi, M. and H. Finne-Soveri (2011) 'RAI-benchmarking: An instrument for leadership and development', http://interlinks.euro.centre.org/framework, date accessed 10 May 2012.

Hammar, T., Perälä, M-L. and P. Rissanen (2007) 'The effects of integrated home care and discharge practice on functional ability and health-related quality of life: a cluster-randomised trial among home care patients', *International Journal of Integrated Care*, 7 (July), e29.

Hammar, T., Rissanen, P. and M-L. Perälä (2009) 'The cost-effectiveness of integrated home care and discharge practice for home care clients', *Health Policy*, 92, 10–20.

Hartikainen, A. (2009) Vapaaehtoiset vuodeosastolla. Etnografinen tutkimus vanhusten ja vapaaehtoisten koothaamisesta, http://www.espoonvapaaehtoisverkosto.fi, date accessed 30 August 2011.

Hasvold, P.E. and R. Wootton (2011) 'Use of telephone and SMS reminders to improve attendance at hospital appointments: a systematic review', *Journal of Telemedicine and Telecare*, 17, 358–64.

Hasselhorn, H.-M., Müller, B.H., Tackenberg, P., Kümmerling, A. and M. Simon (2005) *Berufsausstieg bei Pflegepersonal*. Dortmund, Berlin, Dresden, Bundesanstalt für Arbeitsschutz und Arbeitsmedizin.

Hauptverband der Sozialversicherungsträger (2006) *Sonderauswertung: Inanspruchnahme von Gesundenuntersuchungen in Österreich im Jahr 2005*. Wien, HSV.

Hearn, J. and I.J. Higginson (1998) 'Do specialist palliative care teams improve outcomes for cancer patients? A systematic literature review', *Palliative Medicine*, 12, 317–32.

Hébert, R., Durand, S., Somme, D. and M. Raîche (2009) 'PRISMA in Québec and France: implementation and impact of a coordination-type integrated service delivery (ISD) system for frail older people', *International Journal of Integrated Care*, 9 (December, Annual Conference Supplement).

Hedman, N.O, Johansson, R. and U. Rosenqvist (2007) 'Clustering and inertia: structural structural integration of home care in Swedish elderly care', *International Journal of Integrated Care*, 7 (September).

Heitmueller, A. and K. Inglis (2007) 'The earnings of informal carers: wage differentials and opportunity costs', *Journal of Health Economics*, 26, 821–41.

HelpAge International (2009) The voices of older people – a contribution to the state of the world's older people 2012, manual for the focus group study. London, HelpAge International (Manuscript).

Hersch, W.R., Hickam, D.H., Severance, S.M., Dana, T.L., Krages, K.P. and M. Helfand (2006) 'Telemedicine for the Medicare Population: Update', *Evidence report/ Technology Assessment*, *131*, 1–41.

Heusinger, J. (2012) '"Wenn ick wat nich will, will ick nich!" Milieuspezifische Ressourcen und Restriktionen für einen selbstbestimmten Alltag trotz Pflegebedarf' in S. Kümpers and J. Heusinger (eds) *Autonomie trotz Armut und Pflegebedarf? Altern unter Bedingungen von Marginalisierung*. Bern, Huber, pp. 77–105.

Hibbard, J.H., Stockard, J. and M. Tusler (2005) Hospital Performance Reports: Impact On Quality, Market Share, And Reputation', *Health Affairs*, 24(4), 1150–60.

Higgs, J. and M. Jones (2000) 'Will evidence-based practice take the reasoning out of practice?' in J. Higgs and M. Jones (eds) *Clinical reasoning in the health professionals*. Oxford, Butterworth Heineman (2nd edn), pp. 307–15.

Hirschman, A. O. (1970) *Exit, voice, and loyalty: Responses to decline in firms, organizations, and states*. Cambridge MA, Harvard University Press.

HM Government (2008) *Carers at the heart of 21st century families and communities*. London, Department of Health (Carers team).

HM Government (2009) *Shaping the future of care together* (Green Paper). London, TSO.

Hodges, R., Wright, M. and K. Keasey (1996) 'Corporate governance in the public services: issues and concepts', *Public Money and Management*, 16(2), 7–13.

Hofmarcher, M.M., Oxley, H. and E. Rusticelli (2007) *Improved Health System Performance through better Care Coordination*. Paris, OECD Publishing (OECD Health Working Papers, No. 30), http://dx.doi.org/10.1787/246446201766, date accessed 23 June 2012.

Holdsworth, L. (2011a) 'Single assessment process', http://interlinks.euro.centre.org/ framework, date accessed 3 March 2012.

Holdsworth, L. (2011b) 'Hospital discharge lounges', http://interlinks.euro.centre. org/framework, date accessed 3 March 2012.

Holdsworth, L. (2011c) 'Direct payments for informal carers', http://interlinks.euro. centre.org/framework, date accessed 3 March 2012.

Holdsworth, L. (2011d) 'Better homes active lives – Extra care housing scheme in Kent', http://interlinks.euro.centre.org/framework, date accessed 22 May 2012.

Holdsworth, L. (2011e) 'Informal carers' rights to an assessment of needs', http:// interlinks.euro.centre.org/framework, date accessed 22 November 2012.

Holdsworth, L. and J. Billings (2009) *United Kingdom national report (England): Quality in long-term care*. Canterbury, University of Kent (INTERLINKS National Report, http://interlinks.euro.centre.org/countries).

Holstein, M. B. and M. Minkler (2003) 'Self, society, and the "new gerontology",' *The Gerontologist*, 43(6), 787–96.

Holt-Lunstad, J., Smith, T.B. and J.B. Layton (2010) 'Social relationships and mortality risk: A meta-analytic review', *PLoS Med*, 7(7), 1–20.

Hotta, S. (2010) 'Professional caregiver issues in Japan' in Social and Economy, Volume 3, *Social Services and Community*. Tokyo, University of Tokyo Press, pp. 149–72.

Houten, D. v. (1999) *De standardmens voorbij. Over care, verzorgingsstaat en burgerschap* [Beyond the average citizen. About care, the welfare state and civil society], Maarssen, Elsevier/De Tijdstroom.

Huber, M., Maucher, M. and B. Sak (2008) *Study on social and health services of general interest in the European Union: final synthesis report.* Brussels, DG Employment, Social Affairs and Equal Opportunities.

Huber, M., Rodrigues, R., Hoffmann, F., Gasior, K. and B. Marin (2009) *Facts and figures on long-term care. Europe and North America.* Vienna, European Centre for Social Welfare and Research.

Huber, M., Stanciole, A., Wahlbeck, K., Tamsma, N., Torres, F., Jelfs, E. and J. Bremner (2008) *Quality in and equality of access to health care services.* Brussels, European Commission.

Hudson, B., Hardy, B., Henwood, M. and G. Wistow (1997) *Inter-agency collaboration: final report.* Leeds, Nuffield Institute for Health.

Huijbers, P. (2011) 'Neighbourhood care: better home care at reduced cost', http://interlinks.euro.centre.org/framework, date accessed 22 May 2012.

Hussein, S. (2011) 'Volunteers in the formal long-term care workforce in England', *Social care workforce periodical, 13,* http://www.kcl.ac.uk, date accessed 31 January 2012.

Hustinx, L. and F. Lammertyn (2004) 'The cultural basis of volunteering: Understanding and predicting attitudinal differences between Flemish Red Cross volunteers', *Nonprofit and Voluntary Sector Quarterly,* 33(4), 548–84.

Hustinx, L. and F. Lammertyn (2003) 'Collective and Reflexive Styles of Volunteeering: A Sociological Modernization Perspective', *Voluntas: International Journal of Voluntary and Nonprofit Organizations,* 14(2), 167–87.

Informal Network of Social Services Providers (INSSP) (2010) *Seminar 'Impact of EU legislation on social services', Brussels 29 September 2009. Updated Report,* Brussels, INSSP (www.feantsa.org).

Institute of older people and social services (2009) *Report 2008: Older people in Spain.* Madrid, Ministry of Education, Social Policy and Sport.

International Labour Organisation/ILO (2011) 'Manual on the measurement of volunteer work', http://www.ilo.org, date accessed 11 February 2012.

Iref-Acli (2007) Il welfare fatto in casa. Indagine nazionale sui collaboratori domestici stranieri che lavorano a sostegno delle famiglie italiane. Rapporto di ricerca. Roma, Iref-Acli.

Italian Government, Equal Opportunities Department (2007) Piano Generale 2007–2013 PON GAS 2007–2013. Asse D Pari opportunità e non discriminazione, http://www.retepariopportunita.it, date accessed 3 March 2012.

Jackson, R. and N. Howe (2003) *The aging vulnerability index. An assessment of the capacity of twelve developed countries to meet the aging challenge.* Washington, Center for Strategic and International Studies and Watson Wyatt Worldwide.

Jané-Llopis, E. and P. Anderson (2005) *Mental health promotion and mental disorder prevention. A policy for Europe.* Nijmegen, Radboud University Nijmegen.

Janlöv, A-C., Hallberg, I.R. and K. Petersson (2005) 'The experience of older people of entering into the phase of asking for public home help – a qualitative study', *International Journal of Social Welfare,* 14(3), 326–36.

Jensen, C. (2008) 'Worlds of welfare services and transfers', *Journal of European Social Policy,* 18(51), 151–62.

Johnson, J., Rolph, S. and R. Smith (2010) 'Uncovering history: private sector care homes for older people in England', *Journal of Social Policy,* 39, 235–53.

Johri, M., Beland, P. and H. Bergman (2003) 'International experiments in integrated care for the elderly: a synthesis of the evidence', *International Journal of Geriatric Psychology,* 18, 222–35.

Jones, K., Forder, J., Caiels, J., Welch, E., Windle, K., Davidson, J., Dolan, P., Glendinning, C., Irvine, A. and D. King (2011) 'The cost of implementing personal health budgets'. Canterbury, University of Kent (PSSRU Discussion Paper).

Julia Johnson, J., Rolph, S. and R.R. Smith (2010) *Residential care transformed: Revisiting the last refuge*. Basingstoke, Palgrave Macmillan.

Jylhä, M. (1985) *Oman terveyden kokeminen eläkeiässä* [Self-perceived health of older people]. Tampere, Tampereen yliopisto.

Kagialaris, G. (2011a) 'Help-at-home', http://interlinks.euro.centre.org/framework, date accessed 21 May 2012.

Kagialaris, G. (2011b) 'The Athens association of Alzheimer's disease and related disorders: AAADRD, http://interlinks.euro.centre.org/framework, date accessed 21 May 2012.

Kapsalis, A. (2009) 'Greece: The occupational promotion of migrant workers', http://www.eurofound.europa.eu/ewco/studies/tn0807038s/gr0807039q.htm, date accessed 27 July 2012.

Karafillidis, A. (2009) 'Entkoppelung und Koppelung. Wie die Netzwerktheorie zur Bestimmung sozialer Grenzen beitragen kann' in R. Häussling (ed.) *Grenzen von Netzwerken*, Wiesbaden, VS Verlag für Sozialwissenschaften, pp. 105–31.

Karlsson, M., L. Mayhew, R. Plumb and B. Rickayzen (2004) *An international comparison of long-term care arrangements*. London, Cass Business School.

Kayser-Jones, J. and B.A. Koenig (1994) 'Ethical issues' in J.F. Gubrium and A. Sanker (eds) *Qualitative methods in aging research*. Thousands Oaks, CA, Sage Publications, pp. 15–32.

Kazepov, Y. (ed.) (2010) *Rescaling social policies: Towards multilevel governance in Europe*. Aldershot, Ashgate.

Keck, W. and C. Saraceno (2009) *Balancing elderly care and employment in Germany*. Berlin, Social Science Research Centre.

Kellehear, A. (1999) 'Health-promoting palliative care: developing a social model for practice', *Mortality*, 4(1), 75–82.

Kelley-Gillespie, N. (2009) 'An integrated conceptual model of quality of life for older adults based on a synthesis of the literature', *Applied Research in Quality of Life*, 4, 259–82.

Kelley-Moore, J. (2010) 'Disability and ageing: the social construction of causality' in D. Dannefer and C. Phillipson (eds) *The SAGE Handbook of Social Gerontology*. London, Sage, pp. 96–110.

Kendall, J. (2000) *The third sector and social care for older people in England: Towards an explanation of its contrasting contributions in residential care, domiciliary care and day care*. London, LSE.

Kettle, M., O'Donnell, J. and S. Newman (2011) 'Getting together and being personal: Building personalisation on a co-production approach', *Journal of Care Services Management*, 5(1), 29–34.

King's Fund (2001) *Consultation meeting on assistive technology*. London, King's Fund.

Kitwood, T. (1997) *Dementia reconsidered: The person comes first*. Buckingham, Open University Press.

Klein, R. (2000) 'From evidence-based medicine to evidence-based policy?' *Journal of Health Services Research and Policy*, 5(2), 65–6.

Klie, T. (2009) 'Sozialpolitische Neuorientierung und Neuakzentuierung rechtlicher Steuerung' in Netzwerk: Soziales neu gestalten (SONG) (Hg.) *Zukunft Quartier – Lebensräume zum Älterwerden*. Gütersloh, Bertelsmann Stiftung, pp. 5–29.

Klie, T., Hoch, H. and T. Pfundstein (2005) *Bürgerschaftliches Engagement für Lebensqualität im Alter*. Schlussbericht zur Heim- und Engagiertenbefragung. http://www. bela-bw.de/bela1-bw/Dokumente/BELA-Erhebung-Schlussberich-aktuell.pdf, date accessed 30 August 2011.

Knapp, M., Hardy, B. and J. Forder (2001) 'Commissioning for quality: ten years of social care markets in England', *Journal of Social Policy*, 30(2), 283–306.

Kneer, G. und A. Nassehi (1993) *Niklas Luhmanns Theorie sozialer Systeme. Eine Einführung*. München, UTB.

Kodner, D. (2009) 'All together now: A conceptual exploration of integrated care', *Healthcare Quarterly*, 13 (Special Issue), 6–15.

Kodner, D.L. (2002) 'The quest for integrated systems of care for frail older persons', *Aging Clinical and Experimental Research*, 14(4), 307–13.

Kodner, D.L. and C. Spreeuwenberg (2002) 'Integrated care: meaning, logic, applications, and implications – a discussion paper', *International Journal of Integrated Care*, 2 (November), 1–6.

Koivukangas, P., Ohinmaa, A. and J. Koivukangas (1995) *Nottingham health profile (NHP), Finnish version*. Saarijärvi, Gummerus Kirjapaino Oy (Stakes raportteja 187).

Konrad, E.L. (1996) 'A multidimensional framework for conceptualizing human services integration intiatiatives' in J.M. Marquart and E.L. Konrad (eds) *Evaluating initiatives to integrate human services*. San Francisco, Jossey-Bass, pp. 127–43.

Kowalska, K. (2007) 'Managed care and a process of integration in health care sector: A case study from Poland', *Health Policy*, 84(2), 308–20.

Kraus, M., Riedel, M., Mot, E., Willemé, P., Röhrling, G. and T. Czypionka (2010) *A typology of long-term care systems in Europe*. Brussels, CEPS (ENEPRI Research Reports, www.ceps.eu).

Kremer, M. (2006) 'Consumers in charge of care: The Dutch Personal Budget and its impact on the market, professionals and the family', *European Societies*, 8(3), 385–401.

Křížová, E. and J. Tošner (2011) Vzorce a hodnoty dobrovolnictví v sociálních a zdravotních službách, *Fórum sociální politiky*, 1, 24–30.

Kuhn, D., Ortigara, A. and R.E. Kasayka (2000) 'Dementia care mapping: an innovative tool to measure person-centered care', *Alzheimer's Care Today*, 1(3), 7–15.

Kümpers, S. (2011a) 'Domiciliary rehabilitation (MoRe)', http://interlinks.euro.centre. org/framework, date accessed 13 June 2012.

Kümpers, S. (2011b) 'Geriatric network and geriatric academy Brandenburg', http:// interlinks.euro.centre.org/framework, date accessed 4 July 2012.

Kümpers, S. and M. Zander (2012) 'Der Autonomiebegriff im Kontext von Hilfe- und Pflegebedürftigkeit und sozialer Benachteiligung' in S. Kümpers and J. Heusinger (eds) *Autonomie trotz Armut und Pflegebedarf? Altern unter Bedingungen von Marginalisierung*. Bern, Huber, pp. 21–38.

Kümpers, S., Allen, K., Campbell, L., Dieterich, A., Glasby, J., Kagialaris, G., Mastroyiannakis, T., Pokrajac, T., Ruppe, G., Turk, E., van der Veen, R. and L. Wagner (2010) *Prevention and rehabilitation within long-term care across Europe. European Overview Report*. Berlin, Odense, Vienna et al., WZB/European Centre et al. (INTERLINKS Report #1, http://interlinks.euro.centre.org/project/reports).

Kümpers, S., I. Mur, Hardy, B., Maarse, H. and A. Van Raak (2006) 'The importance of knowledge transfer between specialist and generic services in improving health care. A cross-national study of dementia care in England and The Netherlands', *The International Journal for Health Planning and Management*, 21, 151–67.

Kushner, C., Baranek, P. and M. Dewar (2008) *Home care: Change we need*, Report on the Ontario Health Coalition's Home Care Hearings, 17 November.

Laberg, T., Aspelund, H. and H. Thygesen (2005) *Smart home technology. Planning and management in municipal services*. Oslo, Social- og helsedirektoratet.

Lafortune, G., Balestat, G. and the Disability Study Expert Group Members (2007) *Trends in severe disability among elderly people: Assessing the evidence in 12 OECD countries and the future implications*. Paris, OECD (OECD Health Working Papers No. 26).

Lamont, M. and V. Molnar (2002) 'The study of boundaries in the social sciences', *Annual Review of Sociology*, 28, 167–95.

Lamura, G., Döhner, H. and C. Kofahl on behalf of the EUROFAMCARE Consortium (eds) (2008) *Family carers of older people in Europe. A six-country comparative study*. Hamburg, LIT Verlag.

Lamura, G., Mnich, E., Bien, B., Krevers, B., McKee, K., Mestheneos, E. and H. Döhner (2007) 'Dimensions of future social service provision in the ageing societies of Europe', *Advances in Gerontology*, 20(3), 13–30.

Lamura, G., Principi, A., Polverini, F., Balducci, C., Melchiorre, M.G. and M.V. Gianelli (2007) *Eurofamcare. Rapporto nazionale*. Ancona, I.N.R.C.A.

Landi, F., Onder, G., Russo, A., Tabaccanti, S. *et al.* (2001) 'A new model of integrated home care for the elderly: impact on hospital use', *Journal of Clinical Epidemiology*, 54, 968–70.

Larsen, T. (2011) 'Clinical continuity by integrated care', http://interlinks.euro.centre. org/framework, date accessed 13 April 2012.

Lawton, M.P. and E.M. Brody (1969) 'Assessment of older people: selfmaintaining and instrumental activities of daily living', *Gerontologist*, 9, 179–86.

Lazenby, J.A.A. (2007) 'Ethics, Identity and Organizational Learning', *World Academy of Science, Engineering and Technology*, 30, 25–30.

Leatherman, S., Berwick, D., Iles, D. *et al.* (2003) 'The Business Case for Quality: Case Studies and an Analysis', *Health Affairs*, 22(2), 17–30.

Le Bihan, B. and C. Martin (eds) (2010) *Working and caring for elderly parents in six European countries. National Reports (France, Germany, Italy, Netherlands, Portugal, Sweden)*. Paris, Drees/Mire France.

Lee, R., Mason, A. and D. Cotlear (2010) *Some economic consequences of global aging. A discussion note for the World Bank*. Washington, World Bank.

Leichsenring, K. (2004a) 'Providing integrated health and social care: a European over-view' in K. Leichsenring and A. Alaszewski (eds) *Providing integrated health and social care for older persons: a European overview of issues at stake*. Aldershot, Ashgate, pp. 9–52.

Leichsenring, K. (2004b) 'Developing integrated health and social care services for older persons in Europe', *International Journal of Integrated Care*, 4 (September), 1–15.

Leichsenring, K., Roth, G., Wolf, M. and A. Sissouras (2005) 'Moments of truth – An overview of pathways to integration and better quality in long-term care' in J. Billings and K. Leichsenring (eds) *Integrating health and social care services for older persons – Evidence from nine European countries*, Aldershot, Ashgate, pp. 13–39.

Leichsenring, K. (2009) *Developing and ensuring quality in LTC in Austria*. Vienna, European Centre for Social Welfare Policy and Research (INTERLINKS National Report, http://interlinks.euro.centre.org/countries).

Leichsenring, K. (2010) 'Achieving quality long-term care in residential facilities'. Discussion Paper for the Peer Review 2010 Germany, http://www.peer-review-social-inclusion.eu, date accessed 3 April 2012.

Leichsenring, K. (2011a) 'Tax-based comprehensive LTC allowance', http://interlinks. euro.centre.org/framework, date accessed 27 May 2012.

Leichsenring, K. (2011b) 'The E-Qalin quality management system', http://interlinks. euro.centre.org/framework, date accessed 9 May 2012.

Leichsenring, K., Barnett, S. and R. Rodrigues (2010) *Contracting for quality*. Brighton, ESN.

Leichsenring, K., Ruppe, G., Rodrigues, R. and M. Huber (2009) *Long-term care and social services in Austria. Paper prepared for the Workshop on Social and Long-term Care at the World Bank office in Vienna*. Vienna, European Centre for Social Welfare Policy and Research.

Leinonen, A. (2006) Vanhusneuvoston funktioita jäljittämässä, Tutkimus maaseutu-maisten kuntien vanhusneuvostoista, https://jyx.jyu.fi/dspace/, date accessed 30 August 2011.

Leitner, S. (2003) 'Varieties of familialism: The caring function of the family in com-parative perspective', *European Societies*, 5(4), 353–75.

Leutz, W. (1999) 'Five laws for integrating medical and social services: lessons from the United States and the United Kingdom', *Milbank Memorial Fund Quarterly*, 77, 77–110.

Leys, M. (2003) 'Health technology assessment: The contribution of qualitative research', *Journal of Technology Assessment in Health Care*, 19, 317–29.

Ligthart, S.A. (2006) *Casemanagement bij dementie*. Utrecht: Kwaliteitsinstituut voor de Gezondheidszorg CBO.

Lilly, M., Laporte, A. and P. Coyle (2007) 'Labor market work and home care's unpaid caregivers: a systematic review of labor force participation rates, predictors of labor market withdrawal, and hours of work', *The Milbank Quarterly*, 85(4), 641–90.

Lines, L. and K.S. Hone (2002) 'Research Methods for Older Adults' in S. Brewster and M. Zajicek (eds) *A new research agenda for older adults*. Proceedings of a Workshop at BCS HCI, London, pp. 36–7.

Ljunggren, G. and T. Emilsson (2009) *Developing and ensuring quality within long-term care – Sweden*. Stockholm, Stockholm County Council (INTERLINKS National Report, http://interlinks.euro.centre.org/countries).

Ljunggren, G. and T. Emilsson (2010) *Governance and finance of long-term care – Sweden*. Stockholm County Council (INTERLINKS National Report, http://interlinks.euro. centre.org/countries).

LOC (2009a) 'The Netherlands and the care sector in 2010–2050 Introduction to a theoretical framework for a long-term vision', http://www.loc.nl, date accessed 6 March 2012.

LOC (2009b) Valuable care. 'The future of healthcare 2010–2050', http://www.loc.nl, date accessed 6 March 2012.

Long, A.F., Grayson, L. and A. Boaz (2006) 'Assessing the quality of knowledge in social care: exploring the potential of a set of generic standards', *British Journal of Social Work*, 36(2), 207–26.

López, D. (2010) 'The securitization of care spaces: lessons from telecare' in M. Schillmeier and M. Domènach (eds) *New Technologies and Emerging Spaces of Care*, Farnham-Burlington, Ashgate, pp. 39–56.

Lüdecke, D. (2009) 'Nachhaltigkeit in der vernetzten Versorgung' in H. Döhner, H. Kaupen-Haasa and O. von dem Knesebeck (eds) (2009) *Medizinsoziologie in Wissenschaft und Praxis. Festschrift für Alf Trojan*. Berlin, Münster, LIT-Verlag, pp. 109–20.

Luherne, M. (2011) 'WeDO: European partnership for the wellbeing and dignity of older people', http://interlinks.euro.centre.org/framework, date accessed 8 February 2012.

Luhmann, N. (1990) 'Der medizinische Code' in N. Luhmann (Hg.) *Soziologische Aufklärung 5: Konstruktivistische Perspektiven.* Frankfurt/M., Suhrkamp, pp. 183–95.

Luhmann, N. (1995) *Social systems.* Stanford, CA, Stanford University Press.

Lundsgaard, J. (2005) *Consumer direction and choice in long-term care for older persons, including payments for informal care: Hcan it help improve care outcomes, employment and fiscal sustainability?* Paris, OECD (OECD Health Working Papers no. 20).

Lutz, H. (ed.) (2008) *Migration and domestic work: A European perspective on a global theme.* Aldershot, Ashgate.

Lyon, D. and M. Glucksmann (2008) Comparative configurations of care work across Europe, *Sociology*, 42(1), 101–18.

Magnusson, L., Hanson, E. and M. Nolan (2005) 'The impact of information and communication technology on family carers of older people and professionals in Sweden', *Ageing and Society*, 25, 693–714.

Mahoney, F.I. and D.W. Barthel (1965) 'Functional Evaluation: The Barthel Index', *Maryland State Medical Journal*, 14(2), 61–5.

Mak, S. (2011a) 'Alzheimer café', http://interlinks.euro.centre.org/framework, date accessed 9 May 2012.

Mak, S. (2011b) 'Case managers for people with dementia and their informal caregivers', http://interlinks.euro.centre.org/framework, date accessed 5 March 2012.

Mak, S. (2011c) 'Meeting centres for people with dementia and their informal caregivers', http://interlinks.euro.centre.org/framework, date accessed 5 March 2012.

Malone, R.E. (2003) 'Vulnerable places: contextualizing health practices', *Social Science & Medicine*, 56, 2243–44.

Martin, C.M. and C. Peterson (2009) 'The social construction of chronicity. A key to understanding chronic care transformation', *Journal of Evaluation in Clinical Practice*, 15 (3), 578–85.

Martín, M., Salvadó, I., Nadal, S., Miji, L.C., Rico, J.M., Lanz, P. and M.I. Taussig (1996) 'Adaptación para nuestro medio de la Escala de Sobrecarga del Cuidador (caregiver burden interview) de Zarit', *Rev Esp Geriatr Gerontol*, 6, 338–46.

Martin, S., Kelly, G., Kernohan, W.G., McCreight, B. and C. Nugent (2009) *Smart home technologies for health and social care support.* London, John Wiley & Sons.

Mastroyiannakis, T. (2011a) 'Archdiocese of Athens social services department – Christian solidarity,' http://interlinks.euro.centre.org/framework, date accessed 12 March 2012.

Mastroyiannakis, T. (2011b) 'E-health unit', http://interlinks.euro.centre.org/framework, date accessed 13 February 2012.

Matrix Insight (2012) *EU level collaboration on forecasting health workforce needs, workforce planning and health workforce trends – a feasibility study.* Brussels, European Commission.

Mauri, L. and A. Pozzi (2005) *Le politiche di long-term care in Italia. I principali nodi del dibattito.* Milano, Synergia.

Mays, N., Pope, C. and J. Popay (2005) 'Systematically reviewing qualitative and quantitative evidence to inform management and policy-making in the health field', *Journal of Health Services Research and Policy*, 10 (suppl. 3), 6–20.

McCoy, D., Godden, S., Pollock, A.M. and C. Bianchessi (2007) 'Carrot and stick? The Community Care Act (2003) and the effect of financial incentives on delays in discharge from hospitals in England', *Journal of Public Health*, 29, 281–287.

McGuire, P. and T. Fulmer (1997) 'Elder abuse' in C.K. Cassel, H.J. Cohen, E.B. Larson *et al.* (eds) *Geriatric medicine.* New York, Springer (3rd edn), 855–64.

Medick, H. (2006) 'Grenzziehungen und die Herstellung des politisch-sozialen Raum. Zur Begriffsgeschichte und politischen Sozialgeschichte der Grenzen in der Frühen Neuzeit' in M. Eigmüller and G. Vobruba (eds) *Grenzsoziologie. Die politische Strukturierung des Raumes.* Wiesbaden, VS Verlag für Sozialwissenschaften, pp. 37–54.

Medizinischer Dienst des Spitzenverbands Bund der Krankenkassen/MDK (2007) *Qualität in der ambulanten und stationären Pflege. 2. Bericht des MDS nach § 118 Abs. 4 SGB XI.* Essen, MDS.

MeSH Database: the US National Library of Medicine's controlled vocabulary used for indexing articles for MEDLINE/PubMed, http://www.ncbi.nlm.nih.gov/mesh, date accessed 11 January 2012.

Mestheneos E. and J. Triantafillou (eds) (2005) *Supporting family carers of older people in Europe – The pan-European background report empirical evidence, policy trends and future perspectives.* Hamburg, Lit Verlag.

Mestheneos, L. (2012) 'New ambient assistive technologies: the user's perspectives' in J.C. Augusto *et al.* (eds) *Handbook of ambient assisted living: Technology for healthcare, rehabilitation and well-being (ambient intelligence and smart environments).* Amsterdam, IOS Press online, pp. 749–62.

Meyer, M. (2007) *Supporting family carers of older people in Europe. The National Background Report for Germany.* Hamburg, Lit Verlag.

Miles, S.H., Koepp, R. and E.P. Weber (1996) 'Advance end-of-life treatment planning: a research review', *Archives of Internal Medicine*, 156(10), 1062–8.

Milligan, C., Mort, M. and C. Roberts (2010) 'Cracks in the door? technology and the shifting topology of care' in M. Schillmeier and M. Domènech (eds) *New technologies and emerging spaces of care.* Farnham-Burlington, Ashgate, pp. 19–38.

Mingot, K. (2011) 'Care leave act', http://interlinks.euro.centre.org/framework, date accessed 10 May 2012.

Ministry of Social Affairs and Health and the Association of Finnish Local and Regional Authorities (2008) *The national framework for high-quality services for older people.* Helsinki, Ministry of Social Affairs and Health publications.

Minkman, M.M.N. (2011) *Developing integrated care. Towards a development model for integrated care.* Deventer, Kluwer.

Minkman, M.M.N., Balsters, H., Mast, J. and M. Kuiper (eds) (2011) *Blijvend zorgen voor beter (Sustainable improvement).* Deventer, Kluwer.

Minkman, M.M.N., Ligthart, S.A. and R. Huijsman (2009) 'Integrated dementia care in the Netherlands: a multiple case study of case management programmes', *Health and Social Care in the Community*, 17(5), 485–94.

MISSOC Secretariat (2009) *MISSOC Analysis 2009 – Long-term care.* Brussels, MISSOC, DG EMployment, Social Affairs and Equal Opportunities.

Mitchell, G.K. (2002) 'How well do general practitioners deliver palliative care? A systematic review', *Palliative Medicine*, 16(6), 457–64.

Mival, O. (2002) 'In search of the cybermuse: Supporting creative activity within product design'. *Proceedings of the Creativity and Cognition Conference*, Loughborough, UK, 4, 20.

Modin, S. and A.K. Furhoff (2004) 'The medical care of patients with primary care home nursing is complex and influenced by non-medical factors: a comprehensive retrospective study from a suburban area in Sweden', *BMC Health Services Research*, 4(22), htpp://biomedcentral.com/1472–6963/4/22, date accessed 27 July 2012.

Møller Andersen, L. (2011) 'A process of using electronic communication between hospital and municipality: SAM:BO', http://interlinks.euro.centre.org/framework, date accessed 10 May 2012.

Mor, V., Leone, T. and A. Maresso (eds) (2012) *The challenges in regulating long-term care quality: An international comparison*. Cambridge, Cambridge University Press.

Moran, N., Arksey, H., Glendinning, C., Jones, K., Netten, A. and P. Rabiee (2011) 'Personalisation and carers: whose rights? Whose benefits?', *British Journal of Social Work*, 41 (June), 1–19.

Moriarty, J., Manthorpe, J., Iliffe, S., Rapaport, J., Clough, R., Bright, L. and M. Cornes (2007) 'Promoting the use of diverse sources of evidence: evaluating progress in the provision of services for people with dementia and their carers', *Evidence and Policy*, 3(3), 385–405.

Morley, J.E., Haren, M.T. and M.J. Rolland (2006) Frailty, *Med Clin N Am*, 90, 837–47.

Mosby's Medical Dictionary (2009) St. Louis, MO, Mosby Elsevier (8th edn).

Motel-Klingebiel, A., Tesch-Roemer, C. and H.-J. von Kondratowitz (2005) 'Welfare states do not crowd out the family: evidence for mixed responsibility from comparative analyses', *Ageing & Society*, 25, 863–82.

MPSVR SR (2011) Správa o sociálnej situácii obyvateľstva SR v roku 2010 (Report on the social situation in Slovakia in 2010). Bratislava, MPSCR SR.

Mullen, K.J., Frank, R.G. and M.B. Rosenthal (2010) 'Can you get what you pay for? Pay-for-performance and the quality of healthcare providers', *The RAND Journal of Economics*, 41(1), 64–91.

Müller, W. (2011) 'The hospital comes to your home – outreach geriatric remobilisation', http://interlinks.euro.centre.org/framework, date accessed 13 April 2012.

Murray, S.J., Holmes, D. and G. Rail (2009) 'On the constitution and status of 'evidence' in the health sciences', *Journal of Research in Nursing*, 13(4), 272–80.

My Home Life (2007) *Quality of life in care homes. A review of the literature*. London, Help the Aged.

Naiditch, M. (2009) *Informal care in the long-term care system: France*. Paris, Irdes (INTERLINKS National Report, http://interlinks.euro.centre.org/countries).

Naiditch, M. (2011a) 'Co-ordinating care for older people (COPA): Team work integrating health and social care professionals in community care', http://interlinks. euro.centre.org/framework, date accessed 10 May 2012.

Naiditch, M. (2011b) 'Equinoxe – A home alarm system linked to volunteering', http://interlinks.euro.centre.org/framework, date accessed 21 May 2012.

Naiditch, M. (2011c) 'Respite care platform: Organizing a range of respite services in the community', http://interlinks.euro.centre.org/framework, date accessed 10 May 2012.

Naiditch, M. (2011d) 'Geographic localisation and safe monitoring of Alzheimer patients – Geoloc', http://interlinks.euro.centre.org, date accessed 10 May 2012.

Naiditch, M. (2011e) 'Governing the building process of care innovation for Alzheimer patients and their families: the MAIA national pilot project', http://interlinks. euro. centre.org, date accessed 8 March 2012.

Naiditch, M. (2012) 'How to preserve an endangered resource? Teachings from a comparison of policies supporting providers of informal care to elderly dependent people in Europe', http://www.irdes.fr/EspaceAnglais/Publications/IrdesPublications/QES176.pdf, date accessed 25 November 2012.

Naiditch, M. and L. Com-Ruelle (2009) *Developing and ensuring quality in LTC in France*. Paris, IRDES (INTERLINKS National Report, http://interlinks.euro.centre. org/countries).

Naue, U. and T. Kroll (2010) 'Bridging policies and practice: challenges and opportunities for the governance of disability and ageing', *International Journal of Integrated Care*, 10 (April), 1–6.

Naughton, M. (2005) 'Evidence-based policy' and the government of the criminal justice system – only if the evidence fits!', *Critical Social Policy*, 25(1), 47–69.

Nelson, H., Hooker, K., Dehart, K., Edwards, J. and K. Lanning (2004) 'Factors important to success in the volunteer long-term care ombudsman role', *The Gerontologist*, 44(1), 116–20.

Network Non Autosufficienza – N.N.A. (ed.) (2009) *L'assistenza agli anziani non autosufficienti. Rapporto 2009.* Santarcangelo, Maggioli Editore.

Newcomer, R., Harrington, C. and A. Friedlob (1990) 'Social health maintenance organisations: Assessing their initial experience', *Health Services Research*, 25(3), 425–54.

Newell, A., Arnott, J., Carmichael, A. and M. Morgan (2007) *Methodologies for involving older adults in the design process.* Bejing, HCII.

Newell, A.F. and P. Gregor (2000) 'User sensitive inclusive design in search of a new paradigm' in J. Scholtz and J. Thomas (eds) *CUU 2000, Proceedings of the First ACM Conference on Universal Usability*, Arlington, pp. 39–44.

Newell, A.F., Carmichael, A. Morgan, M. and A. Dickinson (2006) 'The use of theatre in requirements gathering and usability studies', *Interacting with Computers*, 18, 996–1011.

Newman, A.B., Gottdiener, J.S., McBurnie, M.A. *et al.* (2001) 'Associations of subclinical cardiovascular disease with frailty', *The Journals of Gerontology Series A: Biological Sciences and Medical Sciences*, 56A, M158-M166.

Newman, J. and E. Tonkiens (eds) (2011) *Participation, responsibility and choice: Summoning the active citizen in western European welfare state.* Amsterdam, Amsterdam University Press.

Ngai, L.R. and C.A. Pissarides (2009) *Welfare policy and the distribution of hours of work.* Brussels, CEP (Discussion Paper, No. 962).

Nies, H. (2004a) 'Integrated care: concepts and background' in H. Nies and P. Berman (eds) *Integrating services for older people: A resource book from European experience.* Dublin, European Health Management Association, pp. 17–32.

Nies, H. (2004b) *A European research agenda on integrated care for older people.* Dublin, European Health Management Association.

Nies, H. (2006) 'Managing effective partnerships in older people's services', *Health and Social Care in the Community*, 14(5), 391–9.

Nies, H. (2007) *Integrated care for older people. Is it happening in practice?* Proceedings of the 10th European Health Forum Gastein.

Nies, H. and Berman P.C. (eds) (2004) *Integrating services for older people: A resource book for managers.* Dublin, European Health Management Association (EHMA).

Nies, H., Leichsenring, K., van der Veen, R., Rodrigues, R., Gobet, P., Holdsworth, L., Mak, S., Hirsch Durrett, E., Repetti, M., Naiditch, M., Hammar, T., Mikkola, H., Finne-Soveri, H., Hujanen, T., Carretero, S., Cordero, L., Ferrando, M., Emilsson, T., Ljunggren, G., Di Santo, P., Ceruzzi, F. and Turk, E. (2010) *Quality management and quality assurance in long-term care – european overview paper*, Utrecht/Vienna, Stichting Vilans/European Centre for Social Welfare Policy and Research (INTERLINKS Report #2 – http://interlinks.euro.centre.org/project/reports).

Nies, H., Meerveld, J. and R. Denis (2009) 'Dementia care: linear links and networks', *Healthcare Papers*, 10(1), 34–43.

Nikodemová, K. (2009) *Volunteering in Slovakia. facts and figures.* Brussels, European Volunteer Centre.

OECD (2008) *The looming crisis in the health workforce. How can OECD countries respond?* Paris, OECD.

OECD (2011) *Health expenditure and financing*, Paris, OECD (Statistics).

OECD/European Union (2010) *Health at a glance: Europe 2010*, Paris, OECD Publishing.

Ohinmaa, A. and H. Sintonen (1999) 'Inconsistencies and modelling of the Finnish EuroQol (EQ-5D) preference values' in W. Grainer, V.D. Graft, J. Schulenburg and J. Piercy (eds) *EuroQol plenary meeting, 1–2 October 1998*, Hannover, Health Economics and Health System Research, University of Hannover, 75–4.

Oldman, C. and D. Quilgars (1999) 'The last resort? Revisiting ideas about older people's living arrangements', *Ageing and Society*, 19, 363–84.

Olsson, T.M. (2007) 'Reconstructing evidence-based practice: an investigation of three conceptualisations of EBP', *Evidence and Policy*, 3(2), 271–85.

Orton, J. D. and K.E. Weick (1990) 'Loosely coupled systems. A reconceptualization', *Academy of Management Review*, 15(2), 203–23.

Overmars-Marx, T. (2011) 'Netherlands: "Living comfortably in Menterwolde": Integration of health and social services in the local community for people who need care', http://interlinks.euro.centre.org/framework, date accessed 10 May 2012.

Ovretveit, J. (2011) *Evidence: Does clinical coordination improve quality and save money? Volume 2: A detailed review of the evidence*. London, The Health Foundation.

Pachucki, M. A., Pendergrass, S. and M. Lamont (2007) 'Boundary processes: Recent theoretical developments and new contributions', *Poetics*, 35, 331–51.

Pacione, M. (1997) 'Local exchange trading systems as a response to the globalisation of capitalism', *Urban Studies*, 34(8), 1179-99.

Paré, G., Moqadern, K., Pineau, G. and C. St-Hilaire (2010) 'Clinical effects of Home Telemonitoring in the context of diabetes, asthma, heart failure and hypertension: A systematic review', *Journal of Medical Internet Research*, 12(2), e21.

Parker, S.M., Clayton, J.M., Hancock, K. *et al.* (2007) 'A systematic review of prognostic/end-of-life communication with adults in the advance stages of a life-limiting illness: Patient/caregiver preferences for the content, style, and timing of information', *Journal of Pain and Symptom Management*, 34(1), 81–93.

Parliament of Slovak Republic *(2011) Col. on volunteering*, Act No 406, http://tretisektor.gov.sk/data/files/1523_zakon-406-2011-o-dobrovolnictve.pdf, date accessed 5 May 2012.

Pasquinelli, S. and G. Rusmini (2008) *Badanti: la nuova generazione. Caratteristiche e tendenze del lavoro privato di cura*. Rome, IRS.

Pasquinelli, S. and G. Rusmini (2009) 'I sostegni al lavoro privato di cura' in Network Non Autosufficienza – N.N.A. (ed.) (2009) *L'assistenza agli anziani non autosufficienti. Rapporto 2009*. Santarcangelo, Maggioli Editore, pp. 83–96.

Pasquinelli, S. and G. Rusmini (2010) 'La regolarizzazione delle badanti' in Network Non Autosufficienza – N.N.A. (ed.) *Secondo rapporto sull'assistenza agli anziani non autosufficienti 2010*. Santarcangelo, Maggioli Editore, pp. 77–90.

Payne, S., Kerr, C., Hawker, S., Hardey, M. and J. Powell (2002) 'The communication of information about older people between health and social care practitioners', *Age and Aging*, 31, 107–17.

Pelikan, J. (2007) 'Zur Rekonstruktion und Rehabilitation eines absonderlichen Funktionssystems – Medizin und Krankenbehandlung bei Niklas Luhmann und in der Folgerezeption', *Soziale Systeme*, 13(1+2), 290–303.

Pencheon, D. (2005) 'What's next for evidence-based medicine?', *Evidence-Based Healthcare and Public Health*, 9, 319–21.

Perälä, M.-L. and T. Hammar (2003) *PALKO-malli - Palveluja yhteen sovittava kotiutuminen ja kotihoito organisaatiorajat ylittävänä yhteistyönä* [PALKOmodel - Integrated services in the practice of discharge and home care across organisations]. Helsinki, Stakes (aiheita 29).

Petch, A. (2009) 'The evidence base for integrated care', *Journal of Integrated Care*, 17(3), 23–5.

Petch, A., Cook, A. and E. Miller (2005) 'Focusing on outcomes: their role in partnership and practice', *Journal of Integrated Care*, 13(6), 3–12.

Pew, R.W. and S.V. Van Hemel (2004) (eds) *Technology for adaptive ageing*. Washington, D.C. National Academies Press.

Pfau-Effinger, B. (2005) 'Welfare state policies and the development of care arrangements', *European Societies*, 7(2), 321–47.

Pfau-Effinger, B., Eichler, M. and R. Och (2007) 'Ökonomisierung, Pflegepolitik und Strukturen der Pflege älterer Menschen' in A. Evers and R. Heinze (eds) *Sozialpolitik: Ökonomisierung und Entgrenzung*, Wiesbaden, VS-Verlag für Sozialwissenschaften, pp. 83–98.

Pfeiffer, E. (1975) 'A short portable mental status questionnaire for the assessment of organic brain deficit in elderly patients', *Journal of the American Geriatrics Society*, 23(10), 433–41.

Phillips, J., Bernard, M. and M. Chittenden (2002) *Juggling work and care: the experiences of working carers of older adults*. Bristol, The Policy Press.

Pickard, L. (2008) *Informal care for older people by their adult children: Projections of supply and demand to 2041 in England*, Report to the Strategy Unit (Cabinet Office) and the Department of Health. London, LSE (PSSRU Discussion Paper 2515).

Pijl, M. (1994) 'When private care goes public. an analysis of concepts and principles concerning payments for care' in A. Evers, M. Pijl and C. Ungerson (eds) *Payments for care – A comparative overview*. Aldershot, Avebury, pp. 3–18.

Piperno, F. (2009) *Welfare e immigrazione. Impatto e sostenibilità dei flussi migratori diretti al settore socio-sanitario e della cura*. Roma, CeSPI (Working Papers Nr. 55).

Pleschberger, S. (2007) 'Dignity and the challenge of dying in nursing homes: the residents' view', *Age and Ageing*, 36(2), 197–202.

Pleschberger, S. (2010) *Palliative care, Hospizarbeit und Pflege. Eine Auseinandersetzung mit Forschung und Praxis der Versorgung am Lebensende*. Habilitationsschrift, Wien.

Polisena, J., Tran, K., Cimon, K., Hutton, B., McGill, S., Palmer, K. and R.E. Scott (2010) 'Home telemonitoring for congestive heart failure: a systematic review and meta-analysis', *Journal of Telemedicine and Telecare*, 16, 68–76.

Powell, J.L. and J. Hendricks (2009) 'The sociological construction of ageing: lessons for theorising', *International Journal of Sociology and Social Policy*, 29(1/2), 84–94.

Pressman S. and S. Cohen (2005) 'Does positive affect influence health?', *Psychological Bulletin*, 131(6), pp. 925-71.

Prochazkova, L. and T. Schmid (2009) 'Homecare aid: a challenge for social policy and research' in S. Ramon and D. Zaviršek (eds) *Critical edge issues in social work and social policy. Comparative research perspectives*. Ljubljana, University of Ljubljana (Faculty of Social Work), pp. 139–64.

Prochazkova, L., Rupp, B. and T. Schmid (2008) *Evaluierung der 24-Stunden-Betreuung*. Wien, SFS.

PSIRU/Public Services International Research Unit (2011) 'Private care homes in the UK: bankruptcy, torture, and lack of regulation', PSIRU Brief 13 June 2011, London, University of Greenwich.

Quentin, W., Geissler, A., Scheller-Kreinsen, D. and R. Busse (2010) 'DRG-type hospital payment in Germany: The G-DRG system', *Euro Observer*, 12(3), 4–7.

Radbruch, L. and S. Payne (2009) 'White paper on standards and norms for hospice and palliative care in Europe: part 1', *European Journal of Palliative Care*, 16(6), 278–89.

Ramsay, A. and N. Fulop (2008) *The evidence base for integrated care*. London, Department of Health.

Reilly, S., Hughes, J. and D. Challis (2009) 'Case management for LTC conditions: implementation and processes', *Ageing and Society*, 30, 125–55.

Repetti, M. (2011) 'Neighborhood solidarity', http://interlinks.euro.centre.org/framework, date accessed 9 May 2012.

Repková, K. (2011a) *Who cares? The institutional framework for long-term care benefits, national report Slovakia within the 'Local government and public service reform initiative'*. Budapest, Open Society Institute.

Repková, K. (2011b) 'Social protection of informal carers', http://interlinks.euro.centre.org/framework, date accessed 27 November 2012.

Reynolds, K., Henderson, M., Schulman, A. and L.C. Hanson (2002) 'Needs of the dying in nursing homes', *Journal of Palliative Medicine*, 5(6), 895–901.

Rijksinstituut voor Volksgezondheid en Milieu (RIVM) (2011) *Kosten van Ziekten in Nederland 2007. Trends in de Nederlandse zorguitgaven 1999–2010*. Bilthoven, RIVM.

Robine, J. M., Michel, J. P. and F.R. Herrmann (2007) 'Who will care for oldest people in our ageing society?', *British Medical Journal*, 334, 570–1.

Ródenas, F. (2011) 'Improving the assessment of people with care needs – The RAI system (RAI SPAIN)', http://interlinks.euro.centre.org/framework, date accessed 23 March 2012.

Ródenas, F., Garcés, J., Carretero, S. and M. Megía (2008) 'Case management method applied to older adults in the primary care centres in Burjassot' (Valencian Region, Spain), *European Journal of Ageing*, 5, 57–66.

Rodrigues, R. and A. Schmidt (2010) *Paying for long-term care. Policy brief*. Vienna, European Centre for Social Welfare Policy and Research.

Rodrigues, R., Huber, M., Lamura, G. *et al.* (2012) *Facts and figures on healthy ageing and long-term care*. Vienna and Copenhagen, European Centre for Social Welfare Policy and Research, WHO Europe.

Rosenthal, C. (1997) 'Le soutien des familles canadiennes à leurs membres vieillissants: changements de contexte', *Lien Social et Politiques-RIAC*, 38, 123–31.

Rossi, A. (2004) *Anziani e assistenti immigrate. Strumenti per il welfare locale*. Roma, Ediesse.

Rossi, G. and D. Bramanti (a cura di) (2006) *Anziani non autosufficienti e servizi family friendly*. Milano, Franco Angeli.

Rostgaard, T. (coord.) (2011) *Livindhome: Living independently at Home: reforms in home care in 9 European countries, Research report*. Paris, DREES/Mire.

Rothgang, H. and K. Engelke (2009) 'Long-term care: How to organise affordable, sustainable long-term care given the constraints of collective versus individual arrangements and responsibilities'. Discussion Paper for the Peer Review 2009 in The Netherlands, http://www.peer-review-social-inclusion.eu, date accessed 10 July 2012.

Rubenstein, L.Z., Alessi, C.A., Josephson, K.R., Hoyl, M.T., Harper, J.O. and F.M. Pietrusza (2007) 'A randomized trial of a screening, case finding, and referral system for older veterans in primary care', *The American Geriatrics Society*, 55, 166–74.

Ruffner, A. (2005) A symbolic interactionist approach to boundary maintenance in nurse patient relationships. Paper presented at the annual meeting of the American Sociological Association, Philadelphia, PA, 12 August.

Rumford, C. (2006) 'Introduction. Theorising borders', *European Journal of Social Theory*, 9(2), 155–69.

Ruppe, G. (2006) *'Healthy Ageing' and its future. An exploratory study from Austria*. Master's thesis, Amsterdam, UVA.

Ruppe, G. (2011) 'Mobile palliative teams', http://interlinks.euro.centre.org/framework, date accessed 20 January 2012.

Ruppe, G. and A. Heller (2007) 'Ärztliche Versorgung am Lebensende im Pflegeheim' in A. Heller, K. Heimerl and S. Husebö (eds) *Wenn nichts mehr zu machen ist, ist noch viel zu tun*. Freiburg im B., Lambertus, pp. 259–71.

Ryan, G.W. and H.R. Bernard (2033) 'Techniques to identifie themes', *Field Methods*, 15(1), 85–109.

Rycroft-Malone, J., Seers, K., Titchen, A., Harvey, G., Kitson, A. and B. McCormack (2004) 'What counts as evidence in evidence-based practice?', *Journal of Advanced Nursing*, 47(1), 81–90.

Salamon, L.M. and W. Sokolowski (2001) *Volunteering in cross-national perspective: Evidence from 24 countries*. Baltimore, The Johns Hopkins University.

Saraceno, C. (2010) 'Social inequalities in facing old-age dependency: a bi-generational perspective', *Journal of European Social Policy*, 20(1), 32–44.

Saraceno, C. and W. Keck (2008) 'The institutional framework of intergenerational family obligations in Europe: A conceptual and methodological overview', www.multilinks-project.eu/uploads/papers/0000/0010/Report_Saraceno_Keck_Nov08.pdf, date accessed 10 August 2012.

Sarti, R. (2010) 'Nello spazio aperto della casa: "badanti"' al tempo della crisi', http://www.qualificare.info/home.php?id=480, date accessed 9 January 2012.

Saunders, C. (2006) 'A personal therapeutic journey', *British Medical Journal*, 313, 1599–601.

Scharlach, A.E., Giunta, N.Y. and K. Mills-Dick (2001) *Case management in long-term care integration: an overview of current programs and evaluations*. Berkeley, Center for the Advanced Study of Aging Services, University of California.

Scherer, R.W. (2009) 'Evidence-based health care and the Cochrane collaboration', *Human and Experimental Toxicology*, 28, 109–11.

Schillmeier, M. and M. Domènech (eds) (2010) *New technologies and emerging spaces of care*. Farnham-Burlington, Ashgate.

Schmidt, A., Chiatti, C., Fry, G., Hanson, E., Magnusson, L., Socci, M., Stückler, A., Széman, Z., Barbabella, F., Hoffmann, F. and G. Lamura (2011) *Analysis and mapping of 52 ICT-based initiatives for caregivers*. Vienna, European Centre for Social Welfare Policy and Research, http://is.jrc.ec.europa.eu, date accessed 17 May 2012.

Scholten, C. (2011) 'Zonder cement geen bouwwerk. Frijwilligerswerk in de zorg, nu en in de toekomst'. Utrecht, Vilans, MOVISIE, NOV.

Scourfield, P. (2005) 'Understanding why carers' assessments do not always take place', *Practice*, 17(1), 15–28.

Scourfield, P. (2007) 'Are there reasons to be worried about the "caretelization" of residential care?', *Critical Social Policy*, 27(2), 155–80.

Sedlakova, D. (2011) *Integrated care in a hospital with polyclinic at Revúca*, http://interlinks.euro.centre.org.

Sen, A. (1999) *Development as freedom*. Oxford, Oxford University Press.

Sepúlveda, C., Marlin, A., Yoshida, T. and A. Ullrich (2002) 'Palliative care: the World Health Organization's global perspective', *Journal of Pain and Symptom Management*, 24(2), 91–6.

Sgritta, G.B. (2009) *Badanti e anziani in un welfare senza futuro*. Roma, Edizioni Lavoro.

SHARE (2011) 'Survey on health retirement and employment', http://www.share-project.org, date accessed 3 April 2012.

Simonazzi, A. (2009) 'Care regimes and national employment models', *Cambridge Journal of Economics*, 33, 211–32.

Singer, S., Burgers, J., Friedberg, M., Rosenthal, M.B., Leape, L. and W. Schneider (2011) 'Defining and measuring integrated patient care: promoting the next frontier in health care delivery', *Medical Care Research and Review*, 68, 112–27.

Smith, V.K. and K.A. Terrisca Peterson (2000) *Exemplary practices in primary care case management*. Lawrenceville, Center for Health Care Strategies, The Robert Wood Johnson Foundation.

Spencer, S., Martin, S., Bourgeault, I.L. and E. O'Shea (2010) *The role of migrant care workers in ageing societies: Report on research findings in the United Kingdom, Ireland, Canada and the United States*. Geneva, IOM.

Stame, N. (2004) 'Theory-based evaluation and types of complexity', *Evaluation*, 19(1), 58–76.

Statistical Office of Slovak Republic (2011) *EHIS 2009 Európske zist'ovanie o zdraví 2009 na Slovensku* (European health interview survey 2009 in Slovakia), Bratislava, Statistical Office.

Stein, V.K. and A. Rieder (2009) 'Integrated care at the crossroads – defining the way forward', *International Journal of Integrated Care*, 9 (April), 1–7.

Stevens, K.R. and C.A. Ledbetter (2000) 'Basis of evidence-based practice. Part 1: the nature of the evidence', *Seminars in Perioperative Nursing*, 9(3), 91–7.

Stevens, M., Glendinning, C., Jacobs, S. *et al.* (2011) 'Assessing the role of increasing choice in English social care services', *Journal of Social Policy*, 40(2), 257–74.

Stiehr, K. (2011a) 'Charter of rights for people in need of long-term care and assistance', http://interlinks.euro.centre.org/framework, date accessed 9 May 2012.

Stiehr, K. (2011b) '"We care" – Representative body of informal carers, relatives and friends in Germany', http://interlinks.euro.centre.org/framework, date accessed 21 May 2012.

Stiehr, K. (2011c) 'Network careCompany (Netzerk pflegeBegleitung)', http://interlinks.euro.centre.org/framework, date accessed 21 May 2012.

Stiehr, K. (2011d) 'Forum For Culture-Sensitive Care In Old Age', http://interlinks.euro.centre.org/framework, date accessed 24 November 2012.

Stolt, R., Blomqvist, P. and U. Winblad (2010) 'Privatization of social services: Quality differences in Swedish elderly care', *Social Science & Medicine*, 72(4), 560–7.

Stroetmann, K.A., Artmann, J., Stroetmann, V.N., Protti, D.J., Dumortier, J., Giest, S., Walossek, U. and D. Whitehouse (2011) *European countries on their journey towards national eHealth infrastructures. Final European progress report*. Brussels, European Commission Directorate Information Society and Media Unit (ICT for Health).

Strümpel, C. and J. Billings (2008) *Overview on health promotion for older people*. Vienna, Austrian Red Cross (HealthPROElderly European Summary Report).

Szebehely, M. and G.B. Trydegård (2012) 'Home care for older people in Sweden: a universal model in transition', *Health and Social Care in the Community*, 20(3), 300–9.

Széman, Z. (2012) 'Family Strategies in Hungary: The Role of Undocumented Migrants in the Eldercare', *Journal of Population Ageing*, 5(2), 97–118.

Tang, F. (2006) 'What resources are needed for volunteerism? a life course perspective', *The Journal of Applied Gerontology*, 25(5), 375–90.

Tepponen, M. (2009) *Kotihoidon integrointi ja laatu* [Integration and quality of home care]. Kuopio, Kuopio University Publications.

Tepponen, M. (2011) 'ISISEMD: ICT solutions and new welfare technology facilitating integration', http://interlinks.euro.centre.org/framework, date accessed 12 April 2012.

Tepponen, M. and T. Hammar (2011a) 'Managing client oriented processes in an integrated organisation', http://interlinks.euro.centre.org/framework, date accessed 22 May 2012.

Tepponen, M. and T. Hammar (2011b) 'ICT solutions and new health technology facilitating integration in home care', http://interlinks.euro.centre.org/framework, date accessed 22 May 2012.

Tesch-Römer, C. (2007) 'Freedom of choice and dignity for the elderly'. Discussion Paper presented at the Peer Review and Assessment in Social Inclusion, Stockholm, www.peer-review-social-inclusion.eu, date accessed 12 July 2012.

The Ministry of Social Affairs and Health and the Association of Finnish Local and Regional Authorities (2008) *The national framework for high-quality services for older people*, Helsinki, Ministry of Social Affairs and Health publications.

Theobald, H. (2006) 'Pflegeressourcen, soziale Ausgrenzung und Ungleichheit: Ein europäischer Vergleich', *Zeitschrift für Frauenforschung und Geschlechterstudien*, 24(2+3), 102–16.

THL OSF (2010) *Statistical yearbook on social welfare and health care*. Helsinki, THL.

Thomas, C., Morris, S.M. and D. Clark (2004) 'Place of death: preferences among cancer patients and their carers', *Social Science and Medicine*, 58, 2431–44.

Thomson, S., Foubister, T. and E. Mossialos (2009) *Financing health care in the European Union*. Copenhagen, WHO Europe/European Observatory.

Tilly, C. (2004) 'Social boundary mechanisms', *Philosophy of the Social Sciences*, 34(2), 211–36.

Timonen, V., Convery, J. and S. Cahill (2006) 'Care revolutions in the making? A comparison of cash-for-care in four European countries', *Ageing & Society*, 26, 455–74.

Tinker, A. (1999) *Ageing in place: what can we learn from each other? The Sixth F. Oswald Barnett Oration*, Melbourne, Ecumenical Housing Inc and Copelen.

Tjadens, F. and F. Colombo (2011) 'Long-term care: valuing care providers', *Eurohealth*, 17, 13–15.

Toljamo, M. and M.L. Perälä (2008) *Kotihoidon henkilöstön työn, työtyytyväisyyden ja palvelujen laadun muutokset PALKO -hankkeen aikana. Kysely kotihoidon henkilöstölle vuosina 2001 ja 2003* [The changes in home care personnel's job, job satisfaction and the quality of services during the PALKOproject. A survey of home care personnel in 2001 and 2003]. Helsinki, Stakes (raportteja 7).

Townsend, P. (1962) *The last refuge – A survey of residential institutions and Homes for the aged in England and Wales*. London, Routledge & Paul.

Townsend, P. (2006) 'Policies for the aged in the 21st century: more "structured dependency" or the realisation of human rights?', *Ageing and Society*, 26 (2), 161–79.

Triantafillou, J. (2011) 'Elderly care vocational skill building and certification: ECVC', http://interlinks.euro.centre.org/framework, date accessed 22 May 2012.

Triantafillou, J., Naiditch, M., Repková, K., Stiehr, K., Carretero, S., Emilsson, T., Di Santo, P., Bednárik, R., Brichtova, L., Ceruzzi, F., Cordero, L., Mastroyiannakis, T., Ferrando, M., Mingot, K., Ritter, J. and D. Vlantoni (2010) *Informal care in the long-term care system – European overview paper*, Athens/Vienna, CMT Prooptiki/European Centre for Social Welfare Policy and Research (INTERLINKS Report #3, http://inter-links.euro.centre.org/project/reports).

Triemstra, M., Winters, S., Kool, R.B. and T.A. Wiegers (2010) 'Measuring client experiences in long-term care in the Netherlands: a pilot study with the consumer quality index long-term care', *BMC Health Services Research*, 10(95), 1–11.

Tronto, J. (1993) *Moral boundaries. A political argument for an ethic of care*. New York and London, Routledge.

Trydegård, G.B. and M. Thorslund (2001) 'Inequality in the welfare state? Local variation in care of the elderly – the case of Sweden', *International Journal of Social Welfare*, 10, 174–84.

Trydegård, G.B. and M. Thorslund (2010) 'One uniform welfare state or a multitude of welfare municipalities? The evolution of local variation in Swedish eldercare', *Social Policy & Administration*, 44(4), 495–511.

Turai, T. (2011) 'East Europeans in the global care market', Paper presented at the Conference 'Challenges of Ageing Societies in the Visegrad Countries', Budapest, 17–18 November 2011.

Turk, E. (2009) *Developing and ensuring quality in LTC in Slovenia*. Ljubljana, Institute of Public Health (INTERLINKS National Report, http://interlinks.euro.centre.org/countries).

Twigg, J. and K. Atkin (1994) *Carers perceived: policy and practice in informal care*. Buckingham, Open University Press.

Ufficio Studi Confartigianato (2011) Elaboration based on Eurostat data, www.confartigianato.an.it, date accessed 15 July 2012.

Ungerson, C. (2004) 'Whose empowerment and independence? A cross-national perspective on "cash for care" schemes', *Ageing & Society*, 24(2), 189–212.

Ungerson, C. and S. Yeandle (eds) (2007) *Cash-for-care in developed welfare states*. Basingstoke, Palgrave Macmillan.

United Nations (2001) *United Nations volunteers report*, prepared for the UN General Assembly Special Session on Social Development, Geneva, UN.

University of Wales Lampeter (2008) 'Volunteering and health: what impact does it really have?', http://www.volunteering.org.uk, date accessed 22 August 2011.

Upshur, R.E.G. (2001) 'The status of qualitative research as evidence' in J. Morse, J.M. Swanson and A.J. Kuzel (eds) *The nature of qualitative evidence*. Thousand Oaks, Sage Publications, pp. 5–26.

Vaarama, M., Pieper, R. and A. Sixsmith (eds) (2008) *Care-related quality of life in old age: Concepts, models and empirical findings*. Berlin, Springer.

Valencian Government (2004) *Plan para la mejora de la atención domiciliaria en la Comunidad Valenciana (IMAD) 2004-07* [Plan to improve home care in the Valencia Region]. Valencia, Valencian Government, Health Department.

Van den Bergh, B., Ferrer, I. and A. Carbonell (2007) 'Monetary valuation of informal care: the well-being valuation method', *Health Economics*, 16, 1227–44.

van der Veen, R. and S. Mak (2009) *Developing and ensuring quality within long-term care: Netherlands*. Utrecht, Vilans (INTERLINKS National Report, http://interlinks.euro.centre.org/countries).

van Hooren, F. (2010) 'When families need immigrants: the exceptional position of migrant domestic workers and care assistants in Italian immigration policy', *Bulletin of Italian Politics*, 2(2), 21–38.

van Raak, A., Mur-Veeman, I., Hardy, B., Steenbergen, M. and A. Paulus (eds) (2003) *Integrated care in Europe: Description and comparison of integrated care in six EU countries*. Maarssen, Elsevier.

Vávrová, S. and R. Polepilová (2011) 'Dobrovolnictví v hospicích', *Sociální práce/Sociálna práca*, 11(4), 79–91.

Viitanen, T. (2007) *Informal and formal care in Europe*. Sheffield, University of Sheffield and IZA (Discussion paper No. 2648).

Viitanen, T. (2005) *Informal elderly care and female labour force participation across Europe*. The Hague, CPB (ENEPRI Research Report No. 13).

Vincent, J. and L. Fortunati (eds) (2009) *Electronic emotion. The mediation of emotion via information and communication technologies*. Oxford, Peter Lang.

Vincent, J.A. (2011) 'Anti-ageing and scientific avoidance of death' in M. Gott and C. Ingleton (eds) *Living with ageing and dying: Palliative and end of life care for older people*. New York, Oxford University Press, pp. 29–41.

Vondeling, H. (2004) 'Economic evaluation of integrated care: an introduction', *Inernational Journal of Integrated Care*, 4(March), 1–10.

Vuorenkoski, L., Mladovsky, P. and E. Mossialos (2008) *Finland: Health system review. Health Systems in Transition*, 10(4), 1–168, http://www.euro.who.int/data/assets/pdf_file/0007/80692/E91937.pdf

Wagner, E.H., Austin, B.T., Davis, C., Hindmarsh, M., Schaefer, J. and A. Bonomi (2001) 'Improving chronic illness care: translating evidence into action', *Health Affairs*, 20(6), 64–77.

Wagner, L. (1997) 'Long term care in the Danish health care system' in H. Jolt and M.M. Leibovici (eds) *Long term care: Concept or reality*. Philadelphia, PA, Hanley & Belfus, pp. 149–56.

Wagner, L. (2006) 'Two decades of integrated health care in Denmark', *Journal of Nursing Research* (Tidsskrift for Sygeplejeforskning), 2, 13–20.

Wagner, L. (2011a) 'Bridging the gap between nursing home and community care (the Skævinge Project)', http://interlinks.euro.centre.org/framework, date accessed 22 May 2012.

Wagner, L. (2011b) 'Follow-up home visits after discharge from hospital', http://interlinks.euro.centre.org/framework, date accessed 22 May 2012.

Walker, A. and C. Walker (1998) 'Normalisation and "normal" ageing: the social construction of dependency among older people with learning difficulties', *Disability & Society*, 13(1), 125–42.

Walsh, E.G. and W.D. Clark (2002) 'Managed care and dually eligible beneficiaries: challenges in coordination', *Health Care Financ Rev*, 24(1), 63–82.

Weicht, B. (2010) 'Embodying the ideal carer: the Austrian discourse on migrant carers', *International Journal of Ageing and Later Life*, 5(2), 17–52.

Weigl, B. (2011) 'Social work with older people – Frequent home visits for socially isolated older people with lower levels of care needs', http://interlinks.euro.centre.org/framework, date accessed 21 May 2012.

Weigl, B. (2012) *Freiwilligenarbeit in ambulanten Sorge- und Pflegearrangements – Ein Vergleich von Betreuungs- und Teilhabekonzepten aus England und Deutschland*. Hamburg, Verlag Dr. Kovač.

Weiss, E., Anderson, R. and R. Lasker (2002) 'Making the most of collaboration: Exploring the relationship between partnership synergy and partnership functioning', *Health Education & Behavior*, 29(6), 683–98.

Weritz-Hanf, P. (2011) 'Outcome indicators for rating the quality of care provided by care homes for older people', http://interlinks.euro.centre.org/framework, date accessed 25 July 2012.

WHO – World Health Organisation (1990) *Cancer pain relief and palliative care. Report of a WHO Expert Committee*, Geneva, World Health Organization.

WHO – World Health Organisation (1998) *A health telematics policy in support of WHO's health-for-all strategy for global health development: report of the WHO group consultation on health telematics*. Geneva, WHO, pp. 11–16.

WHO – World Health Organisation (2000) *Towards an international consensus on policy for long-term care for the ageing*. Geneva, WHO and Millbank Memorial Fund.

WHO – World Health Organisation (2001) *International classification of functioning, disability and health (ICF)*. Geneva, World Health Organisation.

WHO – World Health Organisation (2002) *Lessons for long-term care policy. The cross-cluster initiative on long-term care*. Geneva, World Health Organization.

WHO – World Health Organisation (2004a) 'A glossary of terms for community health care and services for older persons', http://whqlibdoc.who.int/wkc/2004/WHO_WKC_Tech.Ser._04.2.pdf, date accessed 13 June 2012.

WHO – World Health Organisation (2004b) *Better palliative care for older people*, Copenhagen, WHO Regional Office for Europe.

WHO – World Health Organisation (2010a) *Telemedicine. Opportunities and developments in member states. Report on the second global survey on eHealth, Global Observatory for eHealth Series – Volume 2*. Geneva, WHO.

WHO – World Health Organisation (2010b) *WHO Global Code of Practice on the International Recruitment of Health Personnel*, http://www.who.int/hrh/migration/code/WHO_global_code_of_practice_EN.pdf, date accessed 10 July 2012.

WHO – World Health Organisation (2011) *WHO Definition of Palliative Care*, http://www.who.int/cancer/palliative/definition/en/, date accessed 4 May 2012.

Wiener Heimkommission (2009) Bericht der bei der Wiener Pflege-, Patientinnen- und Patientenanwaltschaft eingerichtetenWiener Heimkommission 2008, http://www.wien.gv.at/gesundheit/wppa/pdf/heimkommission-2008.pdf.

Wilberforce, M., Glendinning, C., Challis, D. *et al.* (2011) 'Implementing consumer choice in long-term care: the impact of individual budgets on social care providers in England', *Social Policy & Administration*, 45(5), 593–612.

Williams, P. and H. Sullivan (2009) 'Faces of integration', *International Journal of Integrated Care*, 9 (December), 1–13.

Windt, W. (van der), Smeets, R.C.K.H. and E.J. Arnold (2009) *Arbeidsmarkt van verpleegkundigen, verzorgenden en sociaalagogen 2009-2013*. Utrecht, Prismant.

Wismar, M., Maier, C.B., Glinos, I.A., Dussault, G. and J. Figueras (eds) (2011) *Health professional mobility and health systems. Evidence from 17 European countries*. Copenhagen, WHO/European Observatory on Health Systems and Policies.

Witherington, E.M.A., Pirzada, O.M. and A.J. Avery (2008) 'Communication gaps and readmissions to hospital for patients aged 75 years and older: observational study', *Quality and Safety in Health Care*, 17, 71–5.

Wittchen, H.U., Jacobi, F., Rehm, J., Gustavsson, A. *et al.* (2011) ECNP/EBC REPORT 2011. The size and burden of mental disorders and other disorders of the brain in Europe 2010, *European Neuropsychopharmacology*, 21, 655–79.

Wittenberg, R., Sandhu, B. and M. Knapp (2002) 'Funding long-term care: the public and private options' in E. Mossialos, A. Dixon, J. Figueras and J. Kutzin (eds) *Funding Health Care: Options for Europe*. Buckingham, Open University Press, pp. 226–49.

Wootton, R., Bahaadinbeigy, K. and D. Hailey (2011) 'Estimating travel reduction associated with the use of telemedicine by patients and health care professionals: proposal for quantitative synthesis in a systematic review', *BMC Health Services Research*, 11, 185.

Yeandle, S. and G. Fry (2010) *The potential of ICT in supporting domiciliary care in England*. Luxembourg, Publications Office of the European Union (European Commission Joint Research Centre Scientific and Technical Reports).

Yeates, N. (2009) *Globalizing care economies and migrant workers. Explorations in global care chains*. Basingstoke, Palgrave Macmillan.

Yuen, H., Huang, P., Burik, J. and T. Smith (2008) 'Impact of participating in volunteer activities for residents living in long-term care facilities', *The American Journal of Occupational Therapy*, 62(1), 71–6.

Zaidi, A. (2010) *Poverty risks for older people in EU countries – An update*. Vienna, European Centre for Social Welfare Policy and Research (Policy Brief).

Zarit, S.H. (1998) *Dementia: caregivers and stress*. Victoria, Centre on Aging, University of Victoria (Community paper series nr. 8).

Zarit, S.H. (2002) 'Caregiver's Burden' in S. Andrieu and J.P. Aquino (eds) *Family and professional carers: findings lead to action*. Paris, Serdi Edition and Fondation Médéric Alzheimer.

Zarit, S.H., Reever, K.E. and J. Bach-Peterson (1980) 'Relatives of the impaired elderly: correlates of feelings of burden', *Gerontologist*, 20, 649–54.

Zimmermann, C. (2007) 'Death denial: obstacle or instrument for palliative care? An analysis of clinical literature', *Sociology of Health & Illness*, 29(2), 297–314.

Index